THE
MATHEMATICAL THEORY
OF
RELATIVITY

THE MATHEMATICAL THEORY OF RELATIVITY

BY

A. S. EDDINGTON

CAMBRIDGE
AT THE UNIVERSITY PRESS
1963

CAMBRIDGE UNIVERSITY PRESS
Cambridge, New York, Melbourne, Madrid, Cape Town, Singapore, São Paulo

Cambridge University Press
The Edinburgh Building, Cambridge CB2 8RU, UK

Published in the United States of America by Cambridge University Press, New York

www.cambridge.org
Information on this title: www.cambridge.org/9780521091657

First published 1923
Second Edition 1924
Reprinted 1963
Re-issued in this digitally printed version 2007

A catalogue record for this publication is available from the British Library

ISBN 978-0-521-09165-7 paperback

PREFACE

A FIRST draft of this book was published in 1921 as a mathematical supplement to the French Edition of *Space, Time and Gravitation*. During the ensuing eighteen months I have pursued my intention of developing it into a more systematic and comprehensive treatise on the mathematical theory of Relativity. The matter has been rewritten, the sequence of the argument rearranged in many places, and numerous additions made throughout; so that the work is now expanded to three times its former size. It is hoped that, as now enlarged, it may meet the needs of those who wish to enter fully into these problems of reconstruction of theoretical physics.

The reader is expected to have a general acquaintance with the less technical discussion of the theory given in *Space, Time and Gravitation*, although there is not often occasion to make direct reference to it. But it is eminently desirable to have a general grasp of the revolution of thought associated with the theory of Relativity before approaching it along the narrow lines of strict mathematical deduction. In the former work we explained how the older conceptions of physics had become untenable, and traced the gradual ascent to the ideas which must supplant them. Here our task is to formulate mathematically this new conception of the world and to follow out the consequences to the fullest extent.

The present widespread interest in the theory arose from the verification of certain minute deviations from Newtonian laws. To those who are still hesitating and reluctant to leave the old faith, these deviations will remain the chief centre of interest; but for those who have caught the spirit of the new ideas the observational predictions form only a minor part of the subject. It is claimed for the theory that it leads to an understanding of the world of physics clearer and more penetrating than that previously attained, and it has been my aim to develop the theory in a form which throws most light on the origin and significance of the great laws of physics.

It is hoped that difficulties which are merely analytical have been minimised by giving rather fully the intermediate steps in all the proofs with abundant cross-references to the auxiliary formulae used.

For those who do not read the book consecutively attention may be called to the following points in the notation. The summation convention (p. 50) is used. German letters always denote the product of the corresponding English letter by $\sqrt{-g}$ (p. 111). $\mathfrak{h}$ is the symbol for "Hamiltonian differentiation" introduced on p. 139. An asterisk is prefixed to symbols generalised so as to be independent of or covariant with the gauge (p. 203).

A selected list of original papers on the subject is given in the Bibliography at the end, and many of these are sources (either directly or at second-hand) of the developments here set forth. To fit these into a continuous chain of deduction has involved considerable modifications from their original form, so that it has not generally been found practicable to indicate the sources of the separate sections. A frequent cause of deviation in treatment is the fact that in the view of most contemporary writers the Principle of Stationary Action is the final governing law of the world; for reasons explained in the text I am unwilling to accord it so exalted a position. After the original papers of Einstein, and those of de Sitter from which I first acquired an interest in the theory, I am most indebted to Weyl's *Raum, Zeit, Materie.* Weyl's influence will be especially traced in §§ 49, 58, 59, 61, 63, as well as in the sections referring to his own theory.

I am under great obligations to the officers and staff of the University Press for their help and care in the intricate printing.

A. S. E.

10 *August* 1922.

CONTENTS

CHAPTER I

ELEMENTARY PRINCIPLES

CHAPTER II

THE TENSOR CALCULUS

CHAPTER III

THE LAW OF GRAVITATION

CHAPTER IV

RELATIVITY MECHANICS

CHAPTER V

CURVATURE OF SPACE AND TIME

CHAPTER VI

ELECTRICITY

CHAPTER VII

WORLD GEOMETRY

PART I. WEYL'S THEORY

PART II. GENERALISED THEORY

INTRODUCTION

THE subject of this mathematical treatise is not pure mathematics but physics. The vocabulary of the physicist comprises a number of words such as length, angle, velocity, force, work, potential, current, etc., which we shall call briefly "physical quantities." Some of these terms occur in pure mathematics also; in that subject they may have a generalised meaning which does not concern us here. The pure mathematician deals with ideal quantities defined as having the properties which he deliberately assigns to them. But in an experimental science we have to discover properties not to assign them; and physical quantities are defined primarily according to the way in which we recognise them when confronted by them in our observation of the world around us.

Consider, for example, a length or distance between two points. It is a numerical quantity associated with the two points; and we all know the procedure followed in practice in assigning this numerical quantity to two points in nature. A definition of distance will be obtained by stating the exact procedure; that clearly must be the primary definition if we are to make sure of using the word in the sense familiar to everybody. The pure mathematician proceeds differently; he defines distance as an attribute of the two points which obeys certain laws—the axioms of the geometry which he happens to have chosen—and he is not concerned with the question how this "distance" would exhibit itself in practical observation. So far as his own investigations are concerned, he takes care to use the word self-consistently; but it does not necessarily denote the thing which the rest of mankind are accustomed to recognise as the distance of the two points.

To find out any physical quantity we perform certain practical operations followed by calculations; the operations are called experiments or observations according as the conditions are more or less closely under our control. The physical quantity so discovered is primarily the result of the operations and calculations; it is, so to speak, *a manufactured article*—manufactured by our operations. But the physicist is not generally content to believe that the quantity he arrives at is something whose nature is inseparable from the kind of operations which led to it; he has an idea that if he could become a god contemplating the external world, he would see his manufactured physical quantity forming a distinct feature of the picture. By finding that he can lay x unit measuring-rods in a line between two points, he has manufactured the quantity x which he calls the distance between the points; but he believes that that distance x is something already existing in the picture of the world —a gulf which would be apprehended by a superior intelligence as existing in itself without reference to the notion of operations with measuring-rods.

Yet he makes curious and apparently illogical discriminations. The parallax of a star is found by a well-known series of operations and calculations; the distance across the room is found by operations with a tape-measure. Both parallax and distance are quantities manufactured by our operations; but for some reason we do not expect parallax to appear as a distinct element in the true picture of nature in the same way that distance does. Or again, instead of cutting short the astronomical calculations when we reach the parallax, we might go on to take the cube of the result, and so obtain another manufactured quantity, a "cubic parallax." For some obscure reason we expect to see distance appearing plainly as a gulf in the true world-picture; parallax does not appear directly, though it can be exhibited as an angle by a comparatively simple construction; and cubic parallax is not in the picture at all. The physicist would say that he *finds* a length, and *manufactures* a cubic parallax; but it is only because he has inherited a preconceived theory of the world that he makes the distinction. We shall venture to challenge this distinction.

Distance, parallax and cubic parallax have the same kind of potential existence even when the operations of measurement are not actually made—*if* you will move sideways you will be able to determine the angular shift, *if* you will lay measuring-rods in a line to the object you will be able to count their number. Any one of the three is an indication to us of some existent condition or relation in the world outside us—a condition not created by our operations. But there seems no reason to conclude that this world-condition *resembles* distance any more closely than it resembles parallax or cubic parallax. Indeed any notion of "resemblance" between physical quantities and the world-conditions underlying them seems to be inappropriate. If the length AB is double the length CD, the parallax of B from A is half the parallax of D from C; there is undoubtedly some world-relation which is different for AB and CD, but there is no reason to regard the world-relation of AB as being better represented by double than by half the world-relation of CD.

The connection of manufactured physical quantities with the existent world-condition can be expressed by saying that the physical quantities are *measure-numbers* of the world-condition. Measure-numbers may be assigned according to any code, the only requirement being that the same measure-number always indicates the same world-condition and that different world-conditions receive different measure-numbers. Two or more physical quantities may thus be measure-numbers of the same world-condition, *but in different codes*, e.g. parallax and distance; mass and energy; stellar magnitude and luminosity. The constant formulae connecting these pairs of physical quantities give the relation between the respective codes. But in admitting that physical quantities can be used as measure-numbers of world-conditions existing independently of our operations, we do not alter their status as manufactured quantities. The same series of operations will naturally manufacture the same result when world-conditions are the same, and different results when they are different. (Differences of world-conditions which do not influence the results of experiment and observation are *ipso facto* excluded from the domain of physical knowledge.) The size to which a crystal grows may be a measure-number of the temperature of the mother-liquor; but it is none the less a manufactured size, and we do not conclude that the true nature of size is caloric.

The study of physical quantities, although they are the results of our own operations (actual or potential), gives us some kind of knowledge of the world-conditions, since the same operations will give different results in different world-conditions. It seems that this indirect knowledge is all that we can ever attain, and that it is only through its influences on such operations that we can represent to ourselves a "condition of the world." Any attempt to describe a condition of the world otherwise is either mathematical symbolism or meaningless jargon. To grasp a condition of the world as completely as it is in our power to grasp it, we must have in our minds a symbol which comprehends at the same time its influence on the results of all possible kinds of operations. Or, what comes to the same thing, we must contemplate its measures according to all possible measure-codes—of course, without confusing the different codes. It might well seem impossible to realise so comprehensive an outlook; but we shall find that the mathematical calculus of tensors does represent and deal with world-conditions precisely in this way. A tensor expresses simultaneously the whole group of measure-numbers associated with any world-condition; and machinery is provided for keeping the various codes distinct. For this reason the somewhat difficult tensor calculus is not to be regarded as an evil necessity in this subject, which ought if possible to be replaced by simpler analytical devices; our knowledge of conditions in the external world, as it comes to us through observation and experiment, is precisely of the kind which can be expressed by a tensor and not otherwise. And, just as in arithmetic we can deal freely with a billion objects without trying to visualise the enormous collection; so the tensor calculus enables us to deal with the world-condition in the totality of its aspects without attempting to picture it.

Having regard to this distinction between physical quantities and world-conditions, we shall not define a physical quantity as though it were a feature in the world-picture which had to be sought out. *A physical quantity is defined by the series of operations and calculations of which it is the result.* The tendency to this kind of definition had progressed far even in pre-relativity physics. Force had become "mass × acceleration," and was no longer an invisible agent in the world-picture, at least so far as its definition was concerned. Mass is defined by experiments on inertial properties, no longer as "quantity of matter." But for some terms the older kind of definition (or lack of definition) has been obstinately adhered to; and for these the relativity theory must find new definitions. In most cases there is no great difficulty in framing them. We do not need to ask the physicist what conception he attaches to "length"; we watch him measuring length, and frame our definition according to the operations he performs. There may sometimes be cases in which theory outruns experiment and requires us to decide between two definitions, either of which would be consistent with present experimental practice; but usually we can foresee which of them corresponds to the ideal which the experimentalist has set before himself. For example, until recently the practical man was never confronted with problems of non-Euclidean space, and it might be suggested that he would be uncertain how to construct a straight line when so confronted; but as a matter of fact he showed no hesitation, and the eclipse observers measured without ambiguity the bending of light from the "straight line." The appropriate practical definition was so obvious that there was never any danger of different people meaning different loci by this term. Our guiding rule will be that a physical quantity must be defined by prescribing operations and calculations which will lead to an unambiguous result, and that due heed must be paid to existing practice; the last clause should secure that everyone uses the term to denote the same *quantity*, however much disagreement there may be as to the *conception* attached to it.

When defined in this way, there can be no question as to whether the operations give us the real physical quantity or whether some theoretical correction (not mentioned in the definition) is needed. The physical quantity is the measure-number of a world-condition in some code; we cannot assert that a code is right or wrong, or that a measure-number is real or unreal; what we require is that the code should be the accepted code, and the measure-number the number in current use. For example, what is the real difference of time between two events at distant places? The operation of determining time has been entrusted to astronomers, who (perhaps for mistaken reasons) have elaborated a regular procedure. If the times of the two events are found in accordance with this procedure, the difference must be the real difference of time; the phrase has no other meaning. But there is a certain generalisation to be noticed. In cataloguing the operations of the astronomers, so as to obtain a definition of time, we remark that one condition is adhered to in practice evidently from necessity and not from design—the observer and his apparatus are placed on the earth and move with the earth. This condition is so accidental and parochial that we are reluctant to insist on it in our definition of time; yet it so happens that the motion of the apparatus makes an important difference in the measurement, and without this restriction the operations lead to no definite result and cannot define anything. We adopt what seems to be the commonsense solution of the difficulty. We decide that time is *relative to an observer*; that is to say, we admit that an observer on another star, who carries out all the rest of the operations and calculations as specified in our definition, is also measuring time—not our time, but a time relative to himself. The same relativity affects the great majority of elementary physical quantities*; the description of the operations is insufficient to lead to a unique answer unless we arbitrarily prescribe a particular motion of the observer and his apparatus.

In this example we have had a typical illustration of "relativity," the recognition of which has had far-reaching results revolutionising the outlook of physics. Any operation of measurement involves a comparison between a measuring-appliance and the thing measured. Both play an equal part in the comparison and are theoretically, and indeed often practically, interchangeable; for example, the result of an observation with the meridian circle gives the right ascension of the star or the error of the clock indifferently, and we can regard either the clock or the star as the instrument or the object of measurement. Remembering that physical quantities are results of comparisons of this kind, it is clear that they cannot be considered to belong solely to one partner in the comparison. It is true that we standardise the measuring appliance as far as possible (the method of standardisation being explained or implied in the definition of the physical quantity) so that in general the variability of the measurement can only indicate a variability of the object measured. To that extent there is no great practical harm in regarding the measurement as belonging solely to the second partner in the relation. But even so we have often puzzled ourselves needlessly over paradoxes, which disappear when we realise that the physical quantities are not properties of certain external objects but are relations between these objects and something else. Moreover, we have seen that the standardisation of the measuring-appliance is usually left incomplete, as regards the specification of its motion; and rather than complete it in a way which would be arbitrary and pernicious, we prefer to recognise explicitly that our physical quantities belong not solely to the objects measured but have reference also to the particular frame of motion that we choose.

The principle of relativity goes still further. Even if the measuring-appliances were standardised completely, the physical quantities would still involve the properties of the constant standard. We have seen that the world-condition or object which is surveyed can only be apprehended in our knowledge as the sum total of all the measurements in which it can be concerned; any *intrinsic* property of the object must appear as a uniformity or law in these measures. When one partner in the comparison is fixed and the other partner varied widely, whatever is common to all the measurements may be ascribed exclusively to the first partner and regarded as an intrinsic property of it. Let us apply this to the converse comparison; that is to say, keep the measuring-appliance constant or standardised, and vary as widely as possible the objects measured—or, in simpler terms, make a particular

* The most important exceptions are number (of discrete entities), action, and entropy.

kind of measurement in all parts of the field. Intrinsic properties of the measuring-appliance should appear as uniformities or laws in these measures. We are familiar with several such uniformities; but we have not generally recognised them as properties of the measuring-appliance. We have called them *laws of nature*!

The development of physics is progressive, and as the theories of the external world become crystallised, we often tend to replace the elementary physical quantities defined through operations of measurement by theoretical quantities believed to have a more fundamental significance in the external world. Thus the *vis viva* mv^2, which is immediately determinable by experiment, becomes replaced by a generalised energy, virtually defined by having the property of conservation; and our problem becomes inverted—we have not to discover the properties of a thing which we have recognised in nature, but to discover how to recognise in nature a thing whose properties we have assigned. This development seems to be inevitable; but it has grave drawbacks especially when theories have to be reconstructed. Fuller knowledge may show that there is nothing in nature having precisely the properties assigned; or it may turn out that the thing having these properties has entirely lost its importance when the new theoretical standpoint is adopted*. When we decide to throw the older theories into the melting-pot and make a clean start, it is best to relegate to the background terminology associated with special hypotheses of physics. Physical quantities defined by operations of measurement are independent of theory, and form the proper starting-point for any new theoretical development.

Now that we have explained how physical quantities are to be defined, the reader may be surprised that we do not proceed to give the definitions of the leading physical quantities. But to catalogue all the precautions and provisos in the operation of determining even so simple a thing as length, is a task which we shirk. We might take refuge in the statement that the task though laborious is straightforward, and that the practical physicist knows the whole procedure without our writing it down for him. But it is better to be more cautious. I should be puzzled to say off-hand what is the series of operations and calculations involved in measuring a length of 10^{-15} cm.; nevertheless I shall refer to such a length when necessary as though it were a quantity of which the definition is obvious. We cannot be forever examining our foundations; we look particularly to those places where it is reported to us that they are insecure. I may be laying myself open to the charge that I am doing the very thing I criticise in the older physics—using terms that

* We shall see in § 59 that this has happened in the case of energy. The dead-hand of a superseded theory continues to embarrass us, because in this case the recognised terminology still has implicit reference to it. This, however, is only a slight drawback to set off against the many advantages obtained from the classical generalisation of energy as a step towards the more complete theory.

have no definite observational meaning, and mingling with my physical quantities things which are not the results of any conceivable experimental operation. I would reply—

By all means explore this criticism if you regard it as a promising field of inquiry. I here assume that you will probably find me a justification for my 10^{-15} cm.; but you may find that there is an insurmountable ambiguity in defining it. In the latter event you may be on the track of something which will give a new insight into the fundamental nature of the world. Indeed it has been suspected that the perplexities of quantum phenomena may arise from the tacit assumption that the notions of length and duration, acquired primarily from experiences in which the average effects of large numbers of quanta are involved, are applicable in the study of individual quanta. There may need to be much more excavation before we have brought to light all that is of value in this critical consideration of experimental knowledge. Meanwhile I want to set before you the treasure which has already been unearthed in this field.

CHAPTER I

ELEMENTARY PRINCIPLES

1. Indeterminateness of the space-time frame.

It has been explained in the early chapters of *Space, Time and Gravitation* that observers with different motions use different reckonings of space and time, and that no one of these reckonings is more fundamental than another. Our problem is to construct a method of description of the world in which this indeterminateness of the space-time frame of reference is formally recognised.

Prior to Einstein's researches no doubt was entertained that there existed a "true even-flowing time" which was unique and universal. The moving observer, who adopts a time-reckoning different from the unique true time, must have been deluded into accepting a fictitious time with a fictitious space-reckoning modified to correspond. The compensating behaviour of electromagnetic forces and of matter is so perfect that, so far as present knowledge extends, there is no test which will distinguish the true time from the fictitious. But since there are many fictitious times and, according to this view, only one true time, some kind of distinction is implied although its nature is not indicated.

Those who still insist on the existence of a unique "true time" generally rely on the possibility that the resources of experiment are not yet exhausted and that some day a discriminating test may be found. But the off-chance that a future generation may discover a significance in our utterances is scarcely an excuse for making meaningless noises.

Thus in the phrase *true time*, "true" is an epithet whose meaning has yet to be discovered. It is a blank label. We do not know what is to be written on the label, nor to which of the apparently indistinguishable time-reckonings it ought to be attached. There is no way of progress here. We return to firmer ground, and note that in the mass of experimental knowledge which has accumulated, the words *time* and *space* refer to one of the "fictitious" times and spaces—primarily that adopted by an observer travelling with the earth, or with the sun—and our theory will deal directly with these space-time frames of reference, which are admittedly fictitious or, in the more usual phrase, *relative to an observer with particular motion.*

The observers are studying the same external events, notwithstanding their different space-time frames. The space-time frame is therefore something overlaid by the observer on the external world; the partitions representing his space and time reckonings are imaginary surfaces drawn in the world like the lines of latitude and longitude drawn on the earth. They do not follow the natural lines of structure of the world, any more than the meridians follow the lines of geological structure of the earth. Such a mesh-system is of great utility and convenience in describing phenomena, and we shall continue to employ it; but we must endeavour not to lose sight of its fictitious and arbitrary nature.

It is evident from experience that a four-fold mesh-system must be used; and accordingly an event is located by four coordinates, generally taken as x, y, z, t. To understand the significance of this location, we first consider the simple case of two dimensions. If we describe the points of a plane figure by their rectangular coordinates x, y, the description of the figure is complete and would enable anyone to construct it; but it is also more than complete, because it specifies an arbitrary element, the orientation, which is irrelevant to the intrinsic properties of the figure and ought to be cast aside from a description of those properties. Alternatively we can describe the figure by stating the distances between the various pairs of points in it; this description is also complete, and it has the merit that it does not prescribe the orientation or contain anything else irrelevant to the intrinsic properties of the figure. The drawback is that it is usually too cumbersome to use in practice for any but the simplest figures.

Similarly our four coordinates x, y, z, t may be expected to contain an arbitrary element, analogous to an orientation, which has nothing to do with the properties of the configuration of events. A different set of values of x, y, z, t may be chosen in which this arbitrary element of the description is altered, but the configuration of events remains unchanged. It is this arbitrariness in coordinate specification which appears as the indeterminateness of the space-time frame. The other method of description, by giving the distances between every pair of events (or rather certain relations between pairs of events which are analogous to distance), contains all that is relevant to the configuration of events and nothing that is irrelevant. By adopting this latter method we can strip away the arbitrary part of the description, leaving only that which has an exact counterpart in the configuration of the external world.

To put the contrast in another form, in our common outlook the idea of position or *location* seems to be fundamental. From it we derive distance or *extension* as a subsidiary notion, which covers part but not all of the conceptions which we associate with location. Position is looked upon as the physical fact—a coincidence with what is vaguely conceived of as an identifiable point of space—whereas distance is looked upon as an abstraction or a computational result calculable when the positions are known. The view which we are going to adopt reverses this. Extension (distance, interval) is now fundamental; and the location of an object is a computational result summarising the physical fact that it is at certain intervals from the other objects in the world. Any idea contained in the concept location which is not expressible by reference to distances from other objects, must be dismissed from our minds. Our ultimate analysis of space leads us not to a "here" and a "there," but to an extension such as that which relates "here" and "there." To put the conclusion rather crudely—space is not a lot of points close together; it is a lot of distances interlocked.

Accordingly our fundamental hypothesis is that—

Everything connected with location which enters into observational knowledge—everything we can know about the configuration of events—is contained in a relation of extension between pairs of events.

This relation is called the *interval*, and its measure is denoted by ds.

If we have a system S consisting of events $A, B, C, D, \ldots$, and a system S' consisting of events $A', B', C', D', \ldots$, then the fundamental hypothesis implies that the two systems will be exactly alike observationally if, and only if, all pairs of corresponding intervals in the two systems are equal, $AB = A'B'$, $AC = A'C', \ldots$. In that case if S and S' are material systems they will appear to us as precisely similar bodies or mechanisms; or if S and S' correspond to the same material body at different times, it will appear that the body has not undergone any change detectable by observation. But the position, motion, or orientation of the body may be different; that is a change detectable by observation, not of the system S, but of a wider system comprising S and surrounding bodies.

Again let the systems S and S' be abstract coordinate-frames of reference, the events being the corners of the meshes; if all corresponding intervals in the two systems are equal, we shall recognise that the coordinate-frames are of precisely the same kind—rectangular, polar, unaccelerated, rotating, etc.

2. The fundamental quadratic form.

We have to keep side by side the two methods of describing the configurations of events by coordinates and by the mutual intervals, respectively—the first for its conciseness, and the second for its immediate absolute significance. It is therefore necessary to connect the two modes of description by a formula which will enable us to pass readily from one to the other. The particular formula will depend on the coordinates chosen as well as on the absolute properties of the region of the world considered; but it appears that in all cases the formula is included in the following general form—

The interval ds between two neighbouring events with coordinates (x_1, x_2, x_3, x_4) and $(x_1 + dx_1, x_2 + dx_2, x_3 + dx_3, x_4 + dx_4)$ in any coordinate-system is given by

$$ds^2 = g_{11} dx_1^2 + g_{22} dx_2^2 + g_{33} dx_3^2 + g_{44} dx_4^2 + 2g_{12} dx_1 dx_2 + 2g_{13} dx_1 dx_3 + 2g_{14} dx_1 dx_4 + 2g_{23} dx_2 dx_3 + 2g_{24} dx_2 dx_4 + 2g_{34} dx_3 dx_4 \quad \ldots\ldots(2 \cdot 1),$$

where the coefficients g_{11}, etc. are functions of x_1, x_2, x_3, x_4. That is to say, ds^2 is some quadratic function of the differences of coordinates.

This is, of course, not the most general case conceivable; for example, we might have a world in which the interval depended on a general quartic function of the dx's. But, as we shall presently see, the quadratic form (2·1) is definitely indicated by observation as applying to the actual world. Moreover near the end of our task (§ 97) we shall find in the general theory of relation-structure a precise reason why a quadratic function of the coordinate-differences should have this paramount importance.

Whilst the form of the right-hand side of (2·1) is that required by observation, the insertion of ds^2 on the left, rather than some other function of ds, is merely a convention. The quantity ds is a measure of the interval. It is necessary to consider carefully how measure-numbers are to be affixed to the different intervals occurring in nature. We have seen in the last section that equality of intervals can be tested observationally; but so far as we have yet gone, intervals are merely either equal or unequal, and their differences have not been further particularised. Just as wind-strength may be measured by velocity, or by pressure, or by a number on the Beaufort scale, so the relation of extension between two events could be expressed numerically according to many different plans. To conform to (2·1) a particular code of measure-numbers must be adopted; the nature and advantages of this code will be explained in the next section.

The pure geometry associated with the general formula (2·1) was studied by Riemann, and is generally called Riemannian geometry. It includes Euclidean geometry as a special case.

3. Measurement of intervals.

Consider the operation of proving by measurement that a distance AB is equal to a distance CD. We take a configuration of events $LMNOP$..., viz. a measuring-scale, and lay it over AB, and observe that A and B coincide with two particular events P, Q (scale-divisions) of the configuration. We find that the same configuration* can also be arranged so that C and D coincide with P and Q respectively. Further we apply all possible tests to the measuring-scale to see if it has "changed" between the two measurements; and we are only satisfied that the measures are correct if no observable difference can be detected. According to our fundamental axiom, the absence of any observable difference between the two configurations (the structure of the measuring-scale in its two positions) signifies that the intervals are unchanged; in particular the interval between P and Q is unchanged. It follows that the interval A to B is equal to the interval C to D. We consider that the experiment proves equality of distance; but it is primarily a test of equality of interval.

* The logical point may be noticed that the measuring-scale in two positions (necessarily at different times) represents the same *configuration* of events, not the same events.

In this experiment time is not involved; and we conclude that in space considered apart from time the test of equality of distance is equality of interval. There is thus a one-to-one correspondence of distances and intervals. We may therefore adopt the same measure-number for the interval as is in general use for the distance, thus settling our plan of affixing measure-numbers to intervals. It follows that, when time is not involved, the interval reduces to the distance.

It is for this reason that the quadratic form (2·1) is needed in order to agree with observation, for it is well known that in three dimensions the square of the distance between two neighbouring points is a quadratic function of their infinitesimal coordinate-differences—a result depending ultimately on the experimental law expressed by Euclid I, 47.

When time is involved other appliances are used for measuring intervals. If we have a mechanism capable of cyclic motion, its cycles will measure equal intervals provided the mechanism, its laws of behaviour, and all relevant surrounding circumstances, remain precisely similar. For the phrase "precisely similar" means that no observable differences can be detected in the mechanism or its behaviour; and that, as we have seen, requires that all corresponding intervals should be equal. In particular the interval between the events marking the beginning and end of the cycle is unaltered. Thus a clock primarily measures equal intervals; it is only under more restricted conditions that it also measures the time-coordinate t.

In general any repetition of an operation under similar conditions, but for a different time, place, orientation and velocity (attendant circumstances which have a relative but not an absolute significance*), tests equality of interval.

It is obvious from common experience that intervals which can be measured with a clock cannot be measured with a scale, and *vice versa*. We have thus two varieties of intervals, which are provided for in the formula (2·1), since ds^2 may be positive or negative and the measure of the interval will accordingly be expressed by a real or an imaginary number. The abbreviated phrase "imaginary interval" must not be allowed to mislead; there is nothing imaginary in the corresponding relation; it is merely that in our arbitrary code an imaginary number is assigned as its measure-number. We might have adopted a different code, and have taken, for example, the antilogarithm of ds^2 as the measure of the interval; in that case space-intervals would have received code-numbers from 1 to ∞, and time-intervals numbers from 0 to 1. When we encounter $\sqrt{-1}$ in our investigations, we must remember that it has been introduced by our choice of measure-code, and must not think of it as occurring with some mystical significance in the external world.

* They express relations to events which are not concerned in the test, e.g. to the sun and stars.

4. Rectangular coordinates and time.

Suppose that we have a small region of the world throughout which the g's can be treated as constants*. In that case the right-hand side of (2·1) can be broken up into the sum of four squares, admitting imaginary coefficients if necessary. Thus writing

$$y_1 = a_1 x_1 + a_2 x_2 + a_3 x_3 + a_4 x_4,$$
$$y_2 = b_1 x_1 + b_2 x_2 + b_3 x_3 + b_4 x_4;\ \text{etc.},$$

so that $$dy_1 = a_1 dx_1 + a_2 dx_2 + a_3 dx_3 + a_4 dx_4;\ \text{etc.},$$

we can choose the constants a_1, b_1, ... so that (2·1) becomes

$$ds^2 = dy_1^2 + dy_2^2 + dy_3^2 + dy_4^2 \quad \text{......................(4·1).}$$

For, substituting for the dy's and comparing coefficients with (2·1), we have only 10 equations to be satisfied by the 16 constants. There are thus many ways of making the reduction. Note, however, that the reduction to the sum of four squares of complete differentials is not in general possible for a *large* region, where the g's have to be treated as functions, not constants.

Consider all the events for which y_4 has some specified value. These will form a three-dimensional world. Since dy_4 is zero for every pair of these events, their mutual intervals are given by

$$ds^2 = dy_1^2 + dy_2^2 + dy_3^2 \quad \text{......................(4·2).}$$

But this is exactly like familiar space in which the interval (which we have shown to be the same as the distance for space without time) is given by

$$ds^2 = dx^2 + dy^2 + dz^2 \quad \text{..........................(4·3),}$$

where x, y, z are rectangular coordinates.

Hence a section of the world by $y_4 = \text{const.}$ will appear to us as space, and y_1, y_2, y_3 will appear to us as rectangular coordinates. The coordinate-frames y_1, y_2, y_3, and x, y, z, are examples of the systems S and S' of § 1, for which the intervals between corresponding pairs of mesh-corners are equal. The two systems are therefore exactly alike observationally; and if one appears to us to be a rectangular frame in space, so also must the other. One proviso must be noted; the coordinates y_1, y_2, y_3 for real events must be real, as in familiar space, otherwise the resemblance would be only formal.

Granting this proviso, we have reduced the general expression to

$$ds^2 = dx^2 + dy^2 + dz^2 + dy_4^2 \quad \text{....................(4·4),}$$

where x, y, z will be recognised by us as rectangular coordinates in space. Clearly y_4 must involve the time, otherwise our location of events by the four coordinates would be incomplete; but we must not too hastily identify it with the time t.

* It will be shown in § 36 that it is always possible to transform the coordinates so that the first derivatives of the g's vanish at a selected point. We shall suppose that this preliminary transformation has already been made, in order that the constancy of the g's may be a valid approximation through as large a region as possible round the selected point.

I suppose that the following would be generally accepted as a satisfactory (pre-relativity) definition of equal time-intervals:—if we have a mechanism capable of cyclic motion, its cycles will measure equal durations of time *anywhere* and *anywhen*, provided the mechanism, its laws of behaviour, and all outside influences remain precisely similar. To this the relativist would add the condition that the mechanism (as a whole) must be at rest in the space-time frame considered, because it is now known that a clock in motion goes slow in comparison with a fixed clock. The non-relativist does not disagree in fact, though he takes a slightly different view; he regards the proviso that the mechanism must be at rest as already included in his enunciation, because for him motion involves progress through the aether, which (he considers) directly affects the behaviour of the clock, and is one of those "outside influences" which have to be kept "precisely similar."

Since then it is agreed that the mechanism as a whole is to be at rest, and the moving parts return to the same positions after a complete cycle, we shall have for the two events marking the beginning and end of the cycle

$$dx,\ dy,\ dz = 0.$$

Accordingly (4·4) gives for this case

$$ds^2 = dy_4^2.$$

We have seen in § 3 that the cycles of the mechanism in all cases correspond to equal intervals ds; hence they correspond to equal values of dy_4. But by the above definition of time they also correspond to equal lapses of time dt; hence we must have dy_4 proportional to dt, and we express this proportionality by writing

$$dy_4 = ic\,dt \quad \text{.................................(4·5)},$$

where $i = \sqrt{-1}$, and c is a constant. It is, of course, possible that c may be an imaginary number, but provisionally we shall suppose it real. Then (4·4) becomes

$$ds^2 = dx^2 + dy^2 + dz^2 - c^2dt^2 \quad \text{.....................(4·6)}.$$

A further discussion is necessary before it is permissible to conclude that (4·6) is the most general possible form for ds^2 in terms of ordinary space and time coordinates. If we had reduced (2·1) to the rather more general form

$$ds^2 = dx^2 + dy^2 + dz^2 - c^2dt^2 - 2c\alpha\,dx\,dt - 2c\beta\,dy\,dt - 2c\gamma\,dz\,dt \quad \text{...(4·7)},$$

this would have agreed with (4·6) in the only two cases yet discussed, viz. (1) when $dt = 0$, and (2) when $dx, dy, dz = 0$. To show that this more general form is inadmissible we must examine pairs of events which differ both in time and place.

In the preceding pre-relativity definition of t our clocks had to remain stationary and were therefore of no use for comparing time at different places. What did the pre-relativity physicist mean by the difference of time dt between two events at different places? I do not think that we can attach any meaning to his hazy conception of what dt signified; but we know one or two ways in which he was accustomed to determine it. One method which he used was that of transport of chronometers. Let us examine then what happens when we move a clock from $(x_1, 0, 0)$ at the time t_1 to another place $(x_2, 0, 0)$ at the time t_2.

We have seen that the clock, whether at rest or in motion, provided it remains a precisely similar mechanism, records equal *intervals*; hence the difference of the clock-readings at the beginning and end of the journey will be proportional to the integrated interval

$$\int_1^2 ds \quad \text{.............................. (4·81).}$$

If the transport is made in the direct line $(dy=0,\ dz=0)$, we shall have according to (4·7)

$$-ds^2 = c^2dt^2 + 2c\alpha\, dx\, dt - dx^2$$

$$= c^2dt^2\left\{1 + \frac{2\alpha}{c}\frac{dx}{dt} - \frac{1}{c^2}\left(\frac{dx}{dt}\right)^2\right\}.$$

Hence the difference of the clock-readings (4·81) is proportional to

$$\int_{t_1}^{t_2} dt\left(1 + \frac{2\alpha u}{c} - \frac{u^2}{c^2}\right)^{\frac{1}{2}} \quad \text{...................... (4·82),}$$

where $u = dx/dt$, i.e the velocity of the clock. The integral will not in general reduce to $t_2 - t_1$; so that the difference of time at the two places is not given correctly by the reading of the clock. Even when $\alpha = 0$, the moving clock does not record correct time.

Now introduce the condition that the velocity u is very small, remembering that $t_2 - t_1$ will then become very large. Neglecting u^2/c^2, (4·82) becomes

$$\int_{t_1}^{t_2} dt\left(1 + \frac{\alpha}{c}\frac{dx}{dt}\right) \quad \text{approximately}$$

$$= (t_2 - t_1) + \frac{\alpha}{c}(x_2 - x_1).$$

The clock, if moved sufficiently slowly, will record the correct time-difference if, and only if, $\alpha = 0$. Moving it in other directions, we must have, similarly, $\beta = 0$, $\gamma = 0$. Thus (4·6) is the most general formula for the interval, when the time at different places is compared by slow transport of clocks from one place to another.

I do not know how far the reader will be prepared to accept the condition that it must be possible to correlate the times at different places by moving a clock from one to the other with infinitesimal velocity. The method employed in accurate work is to send an electromagnetic signal from one to the other, and we shall see in § 11 that this leads to the same formulae. We can scarcely consider that either of these methods of comparing time at different places is an essential part of our primitive notion of time in the same way that measurement at one place by a cyclic mechanism is; therefore they are best regarded as conventional. Let it be understood, however, that although the relativity theory has formulated the convention explicitly, the usage of the word *time-difference* for the quantity fixed by this convention is in accordance with the long established practice in experimental physics and astronomy.

Setting $\alpha = 0$ in (4·82), we see that the accurate formula for the clock-reading will be

$$\int_{t_1}^{t_2} dt\,(1 - u^2/c^2)^{\frac{1}{2}}$$

$$= (1 - u^2/c^2)^{\frac{1}{2}}\,(t_2 - t_1) \qquad \text{.........................(4·9)}$$

for a uniform velocity u. Thus a clock travelling with finite velocity gives too small a reading—the clock goes slow compared with the time-reckoning conventionally adopted.

To sum up the results of this section, if we choose coordinates such that the general quadratic form reduces to

$$ds^2 = dy_1^2 + dy_2^2 + dy_3^2 + dy_4^2 \qquad \text{.................(4·95)},$$

then y_1, y_2, y_3 and $y_4\sqrt{-1}$ will represent ordinary rectangular coordinates and time. If we choose coordinates for which

$$ds^2 = dy_1^2 + dy_2^2 + dy_3^2 + dy_4^2 + 2\alpha\,dy_1dy_4 + 2\beta\,dy_2dy_4 + 2\gamma\,dy_3dy_4 \;\text{...(4·96)},$$

these coordinates also will agree with rectangular coordinates and time so far as the more primitive notions of time are concerned; but the reckoning by this formula of differences of time at different places will not agree with the reckoning adopted in physics and astronomy according to long established practice. For this reason it would only introduce confusion to admit these coordinates as a permissible space and time system.

We who regard all coordinate-frames as equally fictitious structures have no special interest in ruling out the more general form (4·96). It is not a question of ascribing greater significance to one frame than to another, but of discovering which frame corresponds to the space and time reckoning generally accepted and used in standard works such as the Nautical Almanac.

As far as § 14 our work will be subject to the condition that we are dealing with a region of the world in which the g's are constant, or approximately constant. A region having this property is called *flat*. The theory of this case is called the "special" theory of relativity; it was discussed by Einstein in 1905—some ten years before the general theory. But it becomes much simpler when regarded as a special case of the general theory, because it is no longer necessary to defend the conditions for its validity as being essential properties of space-time. For a given region these conditions may hold, or they may not. The special theory applies only if they hold; other cases must be referred to the general theory.

5. The Lorentz transformation.

Make the following transformation of coordinates

$$x = \beta(x' - ut'),\quad y = y',\quad z = z',\quad t = \beta(t' - ux'/c^2) \quad ......(5\cdot1),$$

$$\beta = (1 - u^2/c^2)^{-\frac{1}{2}},$$

where u is any real constant not greater than c.

We have by (5·1)

$$dx^2 - c^2dt^2 = \beta^2\{(dx' - u\,dt')^2 - c^2(dt' - u\,dx'/c^2)^2\}$$

$$= \beta^2\left\{\left(1 - \frac{u^2}{c^2}\right)dx'^2 - (c^2 - u^2)\,dt'^2\right\}$$

$$= dx'^2 - c^2dt'^2.$$

Hence from (4·6)

$$ds^2 = dx^2 + dy^2 + dz^2 - c^2dt^2 = dx'^2 + dy'^2 + dz'^2 - c^2dt'^2 \quad ......(5\cdot2).$$

The accented and unaccented coordinates give the same formula for the interval, so that the intervals between corresponding pairs of mesh-corners will be equal, and therefore in all observable respects they will be alike. We shall recognise x', y', z' as rectangular coordinates in space, and t' as the associated time. We have thus arrived at another possible way of reckoning space and time—another fictitious space-time frame, equivalent in all its properties to the original one. For convenience we say that the first reckoning is that of an observer S and the second that of an observer S', both observers being at rest in their respective spaces*.

The constant u is easily interpreted. Since S is at rest in his own space, his location is given by $x = $ const. By (5·1) this becomes, in S''s coordinates, $x' - ut' = $ const.; that is to say, S is travelling in the x'-direction with velocity u. Accordingly the constant u is interpreted as the velocity of S relative to S'.

It does not follow immediately that the velocity of S' relative to S is $-u$; but this can be proved by algebraical solution of the equations (5·1) to determine x', y', z', t'. We find

$$x' = \beta(x + ut),\quad y' = y,\quad z' = z,\quad t' = \beta(t + ux/c^2) \quad ......(5\cdot3),$$

showing that an interchange of S and S' merely reverses the sign of u.

The essential property of the foregoing transformation is that it leaves the formula for ds^2 unaltered (5·2), so that the coordinate-systems which it connects are alike in their properties. Looking at the matter more generally, we have already noted that the reduction to the sum of four squares can be made in many ways, so that we can have

$$ds^2 = dy_1^2 + dy_2^2 + dy_3^2 + dy_4^2 = dy_1'^2 + dy_2'^2 + dy_3'^2 + dy_4'^2 \quad ......(5\cdot4).$$

* This is partly a matter of nomenclature. A sentient observer can force himself to "recollect that he is moving" and so adopt a space in which he is not at rest; but he does not so readily adopt the time which properly corresponds; unless he uses the space-time frame in which he is at rest, he is likely to adopt a hybrid space-time which leads to inconsistencies. There is no ambiguity if the "observer" is regarded as merely an involuntary measuring apparatus, which by the principles of § 4 naturally partitions a space and time with respect to which it is at rest.

The determination of the necessary connection between any two sets of coordinates satisfying this equation is a problem of pure mathematics; we can use freely the conceptions of four-dimensional geometry and imaginary rotations to find this connection, whether the conceptions have any physical significance or not. We see from (5·4) that ds is the distance between two points in four-dimensional Euclidean space, the coordinates (y_1, y_2, y_3, y_4) and (y_1', y_2', y_3', y_4') being rectangular systems (real or imaginary) in that space. Accordingly these coordinates are related by the general transformations from one set of rectangular axes to another in four dimensions, viz. translations and rotations. Translation, or change of origin, need not detain us; nor need a rotation of the space-axes (y_1, y_2, y_3) leaving time unaffected. The interesting case is a rotation in which y_4 is involved, typified by

$$y_1 = y_1' \cos\theta - y_4' \sin\theta, \quad y_4 = y_1' \sin\theta + y_4' \cos\theta.$$

Writing $u = ic \tan\theta$, so that $\beta = \cos\theta$, this leads to the Lorentz transformation (5·1).

Thus, apart from obvious trivial changes of axes, the Lorentz transformations are the only ones which leave the form (4·6) unaltered.

Historically this transformation was first obtained for the particular case of electromagnetic equations. Its more general character was pointed out by Einstein in 1905.

6. The velocity of light.

Consider a point moving along the x-axis whose velocity measured by S' is v', so that

$$v' = \frac{dx'}{dt'} \qquad \text{(6·1)}.$$

Then by (5·1) its velocity measured by S is

$$v = \frac{dx}{dt} = \frac{\beta(dx' - u\,dt')}{\beta(dt' - u\,dx'/c^2)}$$

$$= \frac{v' - u}{1 - uv'/c^2} \text{ by (6·1)} \qquad \text{(6·2)}.$$

In non-relativity kinematics we should have taken it as axiomatic that $v = v' - u$.

If two points move relatively to S' with equal velocities in opposite directions $+v'$ and $-v'$, their velocities relative to S are

$$\frac{v' - u}{1 - uv'/c^2} \text{ and } -\frac{v' + u}{1 + uv'/c^2}.$$

As we should expect, these speeds are usually unequal; but there is an exceptional case when $v' = c$. The speeds relative to S are then also equal, both in fact being equal to c.

Again it follows from (5·2) that when

$$\left(\frac{dx'}{dt'}\right)^2 + \left(\frac{dy'}{dt'}\right)^2 + \left(\frac{dz'}{dt'}\right)^2 = c^2,$$

$ds = 0$, and hence

$$\left(\frac{dx}{dt}\right)^2 + \left(\frac{dy}{dt}\right)^2 + \left(\frac{dz}{dt}\right)^2 = c^2.$$

Thus when the resultant velocity relative to S' is c, the velocity relative to S is also c, whatever the direction. We see that the velocity c has a unique and very remarkable property.

According to the older views of absolute time this result appears incredible. Moreover we have not yet shown that the formulae have practical significance, since c might be imaginary. But experiment has revealed a real velocity with this remarkable property, viz. 299,860 km. per sec. We shall call this the *fundamental velocity*.

By good fortune there is an entity—light—which travels with the fundamental velocity. It would be a mistake to suppose that the existence of such an entity is responsible for the prominence accorded to the fundamental velocity c in our scheme; but it is helpful in rendering it more directly accessible to experiment. The Michelson-Morley experiment detected no difference in the velocity of light in two directions at right angles. Six months later the earth's orbital motion had altered the observer's velocity by 60 km. per sec., corresponding to the change from S' to S, and there was still no difference. Hence the velocity of light has the distinctive property of the fundamental velocity.

Strictly speaking the Michelson-Morley experiment did not prove directly that the velocity of light was constant in all directions, but that the average to-and-fro velocity was constant in all directions. The experiment compared the times of a journey "there-and-back." If $v(\theta)$ is the velocity of light in the direction θ, the experimental result is

$$\left.\begin{aligned} \frac{1}{v(\theta)} + \frac{1}{v(\theta+\pi)} &= \text{const.} = C \\ \frac{1}{v'(\theta)} + \frac{1}{v'(\theta+\pi)} &= \text{const.} = C' \end{aligned}\right\} \quad \text{......(6·3)}$$

for all values of θ. The constancy has been established to about 1 part in 10^{10}.

It is exceedingly unlikely that the first equation could hold unless

$$v(\theta) = v(\theta + \pi) = \text{const.};$$

and it is fairly obvious that the existence of the second equation excludes the possibility altogether. However, on account of the great importance of the identification of the fundamental velocity with the velocity of light, we give a formal proof.

Let a ray travelling with velocity v traverse a distance R in a direction θ, so that

$$dt = R/v, \quad dx = R\cos\theta, \quad dy = R\sin\theta.$$

Let the relative velocity of S and S' be small so that u^2/c^2 is neglected. Then by (5·3)

$$dt' = dt + u\,dx/c^2, \quad dx' = dx + u\,dt, \quad dy' = dy.$$

Writing δR, $\delta\theta$, δv for the change in R, θ, v when a transformation is made to S''s system, we obtain

$$\delta\,(R/v) = dt' - dt = uR\cos\theta/c^2,$$
$$\delta\,(R\cos\theta) = dx' - dx = uR/v,$$
$$\delta\,(R\sin\theta) = dy' - dy = 0.$$

Whence the values of δR, $\delta\theta$, $\delta\,(1/v)$ are found as follows:

$$\delta R = uR\cos\theta/v,$$
$$\delta\theta = -u\sin\theta/v,$$
$$\delta\left(\frac{1}{v}\right) = u\cos\theta\left(\frac{1}{c^2} - \frac{1}{v^2}\right).$$

Here $\delta\,(1/v)$ refers to a comparison of velocities in the directions θ in S's system and θ' in S''s system. Writing $\Delta\,(1/v)$ for a comparison when the direction is θ in both systems

$$\Delta\left(\frac{1}{v}\right) = \delta\left(\frac{1}{v}\right) - \frac{\partial}{\partial\theta}\left(\frac{1}{v}\right).\delta\theta$$
$$= \frac{u}{c^2}\cos\theta - \frac{u}{v^2}\cos\theta + \frac{u\sin\theta}{v}\frac{\partial}{\partial\theta}\left(\frac{1}{v}\right)$$
$$= \frac{u}{c^2}\cos\theta + \tfrac{1}{2}u\sin^3\theta\frac{\partial}{\partial\theta}\left(\frac{1}{v^2\sin^2\theta}\right).$$

Hence

$$\Delta\left(\frac{1}{v(\theta)} + \frac{1}{v(\theta+\pi)}\right) = \tfrac{1}{2}u\sin^3\theta\frac{\partial}{\partial\theta}\left\{\frac{1}{\sin^2\theta}\left(\frac{1}{v^2(\theta)} - \frac{1}{v^2(\theta+\pi)}\right)\right\}.$$

By (6·3) the left-hand side is independent of θ, and equal to the constant $C' - C$. We obtain on integration

$$\frac{1}{v^2(\theta)} - \frac{1}{v^2(\theta+\pi)} = \frac{C'-C}{u}(\sin^2\theta\,.\log\tan\tfrac{1}{2}\theta - \cos\theta),$$

or

$$\frac{1}{v(\theta)} - \frac{1}{v(\theta+\pi)} = \frac{C'-C}{C}\cdot\frac{1}{u}(\sin^2\theta\,.\log\tan\tfrac{1}{2}\theta - \cos\theta).$$

It is clearly impossible that the difference of $1/v$ in opposite directions should be a function of θ of this form; because the origin of θ is merely the direction of relative motion of S and S', which may be changed at will in different experiments, and has nothing to do with the propagation of light relative to S. Hence $C' - C = 0$, and $v(\theta) = v(\theta+\pi)$. Accordingly by (6·3) $v(\theta)$ is independent of θ; and similarly $v'(\theta)$ is independent of θ. Thus the velocity of light is uniform in all directions for both observers and is therefore to be identified with the fundamental velocity.

When this proof is compared with the statement commonly (and correctly) made that the equality of the forward and backward velocity of light cannot be deduced from experiment, regard must be paid to the context. The use of the Michelson-Morley experiment to fill a particular gap in a generally deductive argument must not be confused with its use (e.g. in *Space, Time and Gravitation*) as the basis of a pure induction from experiment. Here we have not even used the fact that it is a second-order experiment. We have deduced the Lorentz transformation from the fundamental hypothesis of § 1, and have already introduced a conventional system of time-reckoning explained in § 4. The present argument shows that the convention that time is defined by the slow transport of chronometers is equivalent to the convention that the forward velocity of light is equal to the backward velocity. The proof of this equivalence is mainly deductive except for one hiatus—the connection of the propagation of light and the fundamental velocity—and for that step appeal is made to the Michelson-Morley experiment.

The law of composition of velocities (6·2) is well illustrated by Fizeau's experiment on the propagation of light along a moving stream of water. Let the observer S' travel with the stream of water, and let S be a fixed observer. The water is at rest relatively to S' and the velocity of the light relative to him will thus be the ordinary velocity of propagation in still water, viz. $v' = c/\mu$, where μ is the refractive index. The velocity of the stream being w, $-w$ is the velocity of S relative to S'; hence by (6·2) the velocity v of the light relative to S is

$$v = \frac{v' + w}{1 + wv'/c^2} = \frac{c/\mu + w}{1 + w/\mu c}$$
$$= c/\mu + w(1 - 1/\mu^2) \text{ approximately,}$$

neglecting the square of w/c.

Accordingly the velocity of the light is not increased by the full velocity of the stream in which it is propagated, but by the fraction $(1 - 1/\mu^2)\,w$. For water this is about $0·44\,w$. The effect can be measured by dividing a beam of light into two parts which are sent in opposite directions round a circulating stream of water. The factor $(1 - 1/\mu^2)$ is known as Fresnel's convection-coefficient; it was confirmed experimentally by Fizeau in 1851.

If the velocity of light *in vacuo* were a constant c' differing from the fundamental velocity c, the foregoing calculation would give for Fresnel's convection-coefficient

$$1 - \frac{c'^2}{c^2} \cdot \frac{1}{\mu^2}.$$

Thus Fizeau's experiment provides independent evidence that the fundamental velocity is at least approximately the same as the velocity of light. In the most recent repetitions of this experiment made by Zeeman* the agreement between theory and observation is such that c' cannot differ from c by more than 1 part in 500.

* *Amsterdam Proceedings*, vol. XVIII, pp. 398 and 1240.

7. Timelike and spacelike intervals.

We make a slight change of notation, the quantity hitherto denoted by ds^2 being in all subsequent formulae replaced by $-ds^2$, so that (4·6) becomes

$$ds^2 = c^2dt^2 - dx^2 - dy^2 - dz^2 \quad\text{.....................(7·1).}$$

There is no particular advantage in this change of sign; it is made in order to conform to the customary notation.

The formula may give either positive or negative values of ds^2, so that the interval between real events may be a real or an imaginary number. We call real intervals timelike, and imaginary intervals spacelike.

From (7·1)
$$\left(\frac{ds}{dt}\right)^2 = c^2 - \left(\frac{dx}{dt}\right)^2 - \left(\frac{dy}{dt}\right)^2 - \left(\frac{dz}{dt}\right)^2$$
$$= c^2 - v^2 \quad\text{...(7·2),}$$

where v is the velocity of a point describing the track along which the interval lies. The interval is thus real or imaginary according as v is less than or greater than c. Assuming that a material particle cannot travel faster than light, the intervals along its track must be timelike. We ourselves are limited by material bodies and therefore can only have direct experience of timelike intervals. We are immediately aware of the passage of time without the use of our external senses; but we have to infer from our sense perceptions the existence of spacelike intervals outside us.

From any event x, y, z, t, intervals radiate in all directions to other events; and the real and imaginary intervals are separated by the cone

$$0 = c^2dt^2 - dx^2 - dy^2 - dz^2,$$

which is called the *null-cone*. Since light travels with velocity c, the track of any light-pulse proceeding from the event lies on the null-cone. When the g's are not constants and the fundamental quadratic form is not reducible to (7·1), there is still a null-surface, given by $ds = 0$ in (2·1), which separates the timelike and spacelike intervals. There can be little doubt that in this case also the light-tracks lie on the null-surface, but the property is perhaps scarcely self-evident, and we shall have to justify it in more detail later.

The formula (6·2) for the composition of velocities in the same straight line may be written

$$\tanh^{-1} v/c = \tanh^{-1} v'/c - \tanh^{-1} u/c \quad\text{..............(7·3).}$$

The quantity $\tanh^{-1} v/c$ has been called by Robb the *rapidity* corresponding to the velocity v. Thus (7·3) shows that relative rapidities in the same direction compound according to the simple addition-law. Since $\tanh^{-1} 1 = \infty$, the velocity of light corresponds to infinite rapidity. We cannot reach infinite rapidity by adding any finite number of finite rapidities; therefore we cannot reach the velocity of light by compounding any finite number of relative velocities less than that of light.

There is an essential discontinuity between speeds greater than and less than that of light which is illustrated by the following example. If two points move in the same direction with velocities

$$v_1 = c + \epsilon, \quad v_2 = c - \epsilon$$

respectively, their relative velocity is by (6·2)

$$\frac{v_1 - v_2}{1 - v_1 v_2/c^2} = \frac{2\epsilon}{1 - (c^2 - \epsilon^2)/c^2} = \frac{2c^2}{\epsilon},$$

which tends to infinity as ϵ is made infinitely small! If the fundamental velocity is exactly 300,000 km. per sec., and two points move in the same direction with speeds of 300,001 and 299,999 km. per sec., the speed of one relative to the other is 180,000,000,000 km. per sec. The barrier at 300,000 km. per sec. is not to be crossed by approaching it. A particle which is aiming to reach a speed of 300,001 km. per sec. might naturally hope to attain its object by continually increasing its speed; but when it has reached 299,999 km. per sec., and takes stock of the position, it sees its goal very much farther off than when it started.

A particle of matter is a structure whose linear extension is timelike. We might perhaps imagine an analogous structure ranged along a spacelike track. That would be an attempt to picture a particle travelling with a velocity greater than that of light; but since the structure would differ fundamentally from matter as known to us, there seems no reason to think that it would be recognised by us as a particle of matter, even if its existence were possible. For a suitably chosen observer a spacelike track can lie wholly in an instantaneous space. The structure would exist along a line in space at one moment; at preceding and succeeding moments it would be non-existent. Such instantaneous intrusions must profoundly modify the continuity of evolution from past to future. In default of any evidence of the existence of these spacelike particles we shall assume that they are impossible structures.

8. Immediate consciousness of time.

Our minds are immediately aware of a "flight of time" without the intervention of external senses. Presumably there are more or less cyclic processes occurring in the brain, which play the part of a material clock, whose indications the mind can read. The rough measures of duration made by the internal time-sense are of little use for scientific purposes, and physics is accustomed to base time-reckoning on more precise external mechanisms. It is, however, desirable to examine the relation of this more primitive notion of time to the scheme developed in physics.

Much confusion has arisen from a failure to realise that time as currently used in physics and astronomy deviates widely from the time recognised by the primitive time-sense. In fact the time of which we are immediately conscious is not in general physical time, but the more fundamental quantity which we have called interval (confined, however, to timelike intervals).

Our time-sense is not concerned with events outside our brains; it relates only to the linear chain of events along our own track through the world. We may learn from another observer similar information as to the time-succession of events along his track. Further we have inanimate observers—clocks—from which we may obtain similar information as to their local time-successions. The combination of these linear successions along different tracks into a complete ordering of the events in relation to one another is a problem that requires careful analysis, and is not correctly solved by the haphazard intuitions of pre-relativity physics. Recognising that both clocks and time-sense measure ds between pairs of events along their respective tracks, we see that the problem reduces to that which we have already been studying, viz. to pass from a description in terms of intervals between pairs of events to a description in terms of coordinates.

The external events which we see appear to fall into our own local time-succession; but in reality it is not the events themselves, but the sense-impressions to which they indirectly give rise, which take place in the time-succession of our consciousness. The popular outlook does not trouble to discriminate between the external events themselves and the events constituted by their light-impressions on our brains; and hence events throughout the universe are crudely located in our private time-sequence. Through this confusion the idea has arisen that the instants of which we are conscious extend so as to include external events, and are world-wide; and the enduring universe is supposed to consist of a succession of instantaneous states. This crude view was disproved in 1675 by Römer's celebrated discussion of the eclipses of Jupiter's satellites; and we are no longer permitted to locate external events in the instant of our visual perception of them. The whole foundation of the idea of world-wide instants was destroyed 250 years ago, and it seems strange that it should still survive in current physics. But, as so often happens, the theory was patched up although its original *raison d'être* had vanished. Obsessed with the idea that the external events had to be put somehow into the instants of our private consciousness, the physicist succeeded in removing the pressing difficulties by placing them not in the instant of visual perception but in a suitable preceding instant. Physics borrowed the idea of world-wide instants from the rejected theory, and constructed mathematical continuations of the instants in the consciousness of the observer, making in this way time-partitions throughout the four-dimensional world. We need have no quarrel with this very useful construction which gives physical time. We only insist that its artificial nature should be recognised, and that the original demand for a *world-wide* time arose through a mistake. We should probably have had to invent universal time-partitions in any case in order to obtain a complete mesh-system; but it might have saved confusion if we had arrived at it as a deliberate invention instead of an inherited misconception. If it is found that physical time has properties which would ordinarily be regarded as contrary to common sense, no surprise need be felt; this highly technical construct of physics is not to be confounded with the time of common sense. It is important for us to discover the exact properties of physical time; but those properties were put into it by the astronomers who invented it.

9. The "3 + 1 dimensional" world.

The constant c^2 in (7·1) is positive according to experiments made in regions of the world accessible to us. The 3 minus signs with 1 plus sign particularise the world in a way which we could scarcely have predicted from first principles. H. Weyl expresses this specialisation by saying that the world is 3 + 1 dimensional. Some entertainment may be derived by considering the properties of a 2 + 2 or a 4 + 0 dimensional world. A more serious question is, Can the world change its type? Is it possible that in making the reduction of (2·1) to the sum or difference of squares for some region remote in space or time, we might have 4 minus signs? I think not; because if the region exists it must be separated from our 3 + 1 dimensional region by some boundary. On one side of the boundary we have

$$ds^2 = -dx^2 - dy^2 - dz^2 + c_1^2 dt^2,$$

and on the other side

$$ds^2 = -dx^2 - dy^2 - dz^2 - c_2^2 dt^2.$$

The transition can only occur through a boundary where

$$ds^2 = -dx^2 - dy^2 - dz^2 + 0\,dt^2,$$

so that the fundamental velocity is zero. Nothing can move at the boundary, and no influence can pass from one side to another. The supposed region beyond is thus not in any spatio-temporal relation to our own universe—which is a somewhat pedantic way of saying that it does not exist.

This barrier is more formidable than that which stops the passage of light round the world in de Sitter's spherical space-time (*Space, Time and Gravitation*, p. 160). The latter stoppage was relative to the space and time of a distant observer; but everything went on normally with respect to the space and time of an observer at the region itself. But here we are contemplating a barrier which does not recede as it is approached.

The passage to a 2 + 2 dimensional world would occur through a transition region where

$$ds^2 = -dx^2 - dy^2 + 0\,dz^2 + c^2 dt^2.$$

Space here reduces to two dimensions, but there does not appear to be any barrier. The conditions on the far side, where time becomes two-dimensional, defy imagination.

10. The FitzGerald contraction.

We shall now consider some of the consequences deducible from the Lorentz transformation.

The first equation of (5·3) may be written

$$x'/\beta = x + ut,$$

which shows that S, besides making the allowance ut for the motion of his origin, divides by β all lengths in the x-direction measured by S'. On the other hand the equation $y' = y$ shows that S accepts S''s measures in directions transverse to their relative motion. Let S' take his standard metre (at rest relative to him, and therefore moving relative to S) and point it first in the transverse direction y' and then in the longitudinal direction x'. For S' its length is 1 metre in each position, since it is his standard; for S the length is 1 metre in the transverse position and $1/\beta$ metres in the longitudinal position. Thus S finds that a moving rod contracts when turned from the transverse to the longitudinal position.

The question remains, How does the length of this moving rod compare with the length of a similarly constituted rod at rest relative to S? The answer is that the transverse dimensions are the same whilst the longitudinal dimensions are contracted. We can prove this by a *reductio ad absurdum*. For suppose that a rod moving transversely were longer than a similar rod at rest. Take two similar transverse rods A and A' at rest relatively to S and S' respectively. Then S must regard A' as the longer, since it is moving relatively to him; and S' must regard A as the longer, since it is moving relatively to him. But this is impossible since, according to the equation $y = y'$, S and S' agree as to transverse measures.

We see that the Lorentz transformation (5·1) requires that (x, y, z, t) and (x', y', z', t') should be measured with standards of identical material constitution, but moving respectively with S and S'. This was really implicit in our deduction of the transformation, because the property of the two systems is that they give the same formula (5·2) for the interval; and the test of complete similarity of the standards is equality of all corresponding intervals occurring in them.

The fourth equation of (5·1) is

$$t = \beta (t' - ux'/c^2).$$

Consider a clock recording the time t', which accordingly is at rest in S''s system (x' = const.). Then for any time-lapse by this clock, we have

$$\delta t = \beta \delta t',$$

since $\delta x' = 0$. That is to say, S does not accept the time as recorded by this moving clock, but multiplies its readings by β, as though the clock were going slow. This agrees with the result already found in (4·9).

It may seem strange that we should be able to deduce the contraction of a material rod and the retardation of a material clock from the general geometry of space and time. But it must be remembered that the contraction and retardation do not imply any absolute change in the rod and clock. The "configuration of events" constituting the four-dimensional structure which we call a rod is unaltered; all that happens is that the observer's space and time partitions cross it in a different direction.

Further we make no prediction as to what would happen to the rod set in motion in an actual experiment. There may or may not be an absolute change of the configuration according to the circumstances by which it is set in motion. Our results apply to the case in which the rod after being set in motion is (according to all experimental tests) found to be similar to the rod in its original state of rest*.

When a number of phenomena are connected together it becomes somewhat arbitrary to decide which is to be regarded as the explanation of the others. To many it will seem easier to regard the strange property of the fundamental velocity as *explained* by these differences of behaviour of the observers' clocks and scales. They would say that the observers arrive at the same value of the velocity of light because they omit the corrections which would allow for the different behaviour of their measuring-appliances. That is the relative point of view, in which the relative quantities, length, time, etc., are taken as fundamental. From the absolute point of view, which has regard to intervals only, the standards of the two observers are equal and behave similarly; the so-called *explanations* of the invariance of the velocity of light only lead us away from the root of the matter.

Moreover the recognition of the FitzGerald contraction does not enable us to avoid paradox. From (5·3) we found that S''s longitudinal measuring-rods were contracted relatively to those of S. From (5·1) we can show similarly that S's rods are contracted relatively to those of S'. There is complete reciprocity between S and S'. This paradox is discussed more fully in *Space, Time and Gravitation*, p. 55.

11. Simultaneity at different places.

It will be seen from the fourth equation of (5·1), viz.

$$t = \beta (t' - ux'/c^2),$$

that events at different places which are simultaneous for S' are not in general simultaneous for S. In fact, if $dt' = 0$,

$$dt = -\beta u dx'/c^2 \quad \text{.........................(11·1)}.$$

It is of some interest to examine in detail how this difference of reckoning of simultaneity arises. It has been explained in § 4 that by convention the time at two places is compared by transporting a clock from one to the other with infinitesimal velocity. Our formulae are based on this convention; and, of course, (11·1) will only be true if the convention is adhered to. The fact that infinitesimal velocity relative to S' is not the same as infinitesimal velocity relative to S, leaves room for the discrepancy of reckoning of simultaneity to creep in. Consider two points A and B at rest relative to S', and distant x' apart. Take a clock at A and move it gently to B by giving it an

* It may be impossible to change the motion of a rod without causing a rise of temperature. Our conclusions will then not apply until the temperature has fallen again, i.e. until the temperature-test shows that the rod is precisely similar to the rod before the change of motion.

infinitesimal velocity du' for a time x'/du'. Owing to the motion, the clock will by (4·9) be retarded in the ratio $(1-du'^2/c^2)^{-\frac{1}{2}}$; this continues for a time x'/du' and the total loss is thus

$$\{1-(1-du'^2/c^2)^{\frac{1}{2}}\}\,x'/du',$$

which tends to zero when du' is infinitely small. S' may accordingly accept the result of the comparison without applying any correction for the motion of the clock.

Now consider S's view of this experiment. For him the clock had already a velocity u, and accordingly the time indicated by the clock is only $(1-u^2/c^2)^{\frac{1}{2}}$ of the true time for S. By differentiation, an additional velocity du* causes a supplementary loss

$$(1-u^2/c^2)^{-\frac{1}{2}}\,u\,du/c^2 \text{ clock seconds} \quad \text{..............(11·2)}$$

per true second. Owing to the FitzGerald contraction of the length AB, the distance to be travelled is x'/β, and the journey will occupy a time

$$x'/\beta\,du \text{ true seconds} \quad \text{......................(11·3).}$$

Multiplying (11·2) and (11·3), the total loss due to the journey is

$$ux'/c^2 \text{ clock seconds,}$$

or

$$\beta ux'/c^2 \text{ true seconds for } S \quad \text{....................(11·4).}$$

Thus, whilst S' accepts the uncorrected result of the comparison, S has to apply a correction $\beta ux'/c^2$ for the disturbance of the chronometer through transport. This is precisely the difference of their reckonings of simultaneity given by (11·1).

In practice an accurate comparison of time at different places is made, not by transporting chronometers, but by electromagnetic signals—usually wireless time-signals for places on the earth, and light-signals for places in the solar system or stellar universe. Take two clocks at A and B, respectively. Let a signal leave A at clock-time t_1, reach B at time t_B by the clock at B, and be reflected to reach A again at time t_2. The observer S', who is at rest relatively to the clocks, will conclude that the instant t_B at B was simultaneous with the instant $\frac{1}{2}(t_1+t_2)$ at A, because he assumes that the forward velocity of light is equal to the backward velocity. But for S the two clocks are moving with velocity u; therefore he calculates that the outward journey will occupy a time $x/(c-u)$ and the homeward journey a time $x/(c+u)$. Now

$$\frac{x}{c-u}=\frac{x(c+u)}{c^2-u^2}=\frac{\beta^2 x}{c^2}(c+u),$$

$$\frac{x}{c+u}=\frac{x(c-u)}{c^2-u^2}=\frac{\beta^2 x}{c^2}(c-u).$$

Thus the instant t_B of arrival at B must be taken as $\beta^2 xu/c^2$ later than the half-way instant $\frac{1}{2}(t_1+t_2)$. This correction applied by S, but not by S', agrees with (11·4) when we remember that owing to the FitzGerald contraction $x=x'/\beta$.

* Note that du will not be equal to du'.

Thus the same difference in the reckoning of simultaneity by S and S' appears whether we use the method of transport of clocks or of light-signals. In either case a convention is introduced as to the reckoning of time-differences at different places; this convention takes in the two methods the alternative forms—

(1) A clock moved with infinitesimal velocity from one place to another continues to read the correct time at its new station, *or*

(2) The forward velocity of light along any line is equal to the backward velocity*.

Neither statement is by itself a statement of observable fact, nor does it refer to any intrinsic property of clocks or of light; it is simply an announcement of the rule by which we propose to extend fictitious time-partitions through the world. But the mutual agreement of the two statements is a fact which could be tested by observation, though owing to the obvious practical difficulties it has not been possible to verify it directly. We have here given a theoretical proof of the agreement, depending on the truth of the fundamental axiom of § 1.

The two alternative forms of the convention are closely connected. In general, in any system of time-reckoning, a change du in the velocity of a clock involves a change of rate proportional to du, but there is a certain turning-point for which the change of rate is proportional to du^2. In adopting a time-reckoning such that this stationary point corresponds to his own motion, the observer is imposing a symmetry on space and time with respect to himself, which may be compared with the symmetry imposed in assuming a constant velocity of light in all directions. Analytically we imposed the same general symmetry by adopting (4·6) instead of (4·7) as the form for ds^2, making our space-time reckoning symmetrical with respect to the interval and therefore with respect to all observational criteria.

12. Momentum and mass.

Besides possessing extension in space and time, matter possesses inertia. We shall show in due course that *inertia, like extension, is expressible in terms of the interval relation*; but that is a development belonging to a later stage of our theory. Meanwhile we give an elementary treatment based on the empirical laws of conservation of momentum and energy rather than on any deep-seated theory of the nature of inertia.

For the discussion of space and time we have made use of certain ideal apparatus which can only be imperfectly realised in practice—rigid scales and

* The chief case in which we require for practical purposes an accurate convention as to the reckoning of time at places distant from the earth, is in calculating the elements and mean places of planets and comets. In these computations the velocity of light in any direction is taken to be 300,000 km. per sec., an assumption which rests on the convention (2). All experimental methods of measuring the velocity of light determine only an average to-and-fro velocity.

perfect cyclic mechanisms or clocks, which always remain similar configurations from the absolute point of view. Similarly for the discussion of inertia we require some ideal material object, say a perfectly elastic billiard ball, whose condition as regards inertial properties remains constant from an absolute point of view. The difficulty that actual billiard balls are not perfectly elastic must be surmounted in the same way as the difficulty that actual scales are not rigid. To the ideal billiard ball we can affix a constant number, called the *invariant mass**, which will denote its absolute inertial properties; and this number is supposed to remain unaltered throughout the vicissitudes of its history, or, if temporarily disturbed during a collision, is restored at the times when we have to examine the state of the body.

With the customary definition of momentum, the components

$$M\frac{dx}{dt},\quad M\frac{dy}{dt},\quad M\frac{dz}{dt} \qquad (12\cdot1)$$

cannot satisfy a general law of conservation of momentum unless the mass M is allowed to vary with the velocity. But with the slightly modified definition

$$m\frac{dx}{ds},\quad m\frac{dy}{ds},\quad m\frac{dz}{ds} \qquad (12\cdot2)$$

the law of conservation can be satisfied simultaneously in all space-time systems, m being an invariant number. This was shown in *Space, Time and Gravitation*, p. 142.

Comparing (12·1) and (12·2), we have

$$M = m\frac{dt}{ds} \qquad (12\cdot3).$$

We call m the *invariant mass*, and M the *relative mass*, or simply the *mass*.

The term "invariant" signifies unchanged for any transformation of coordinates, and, in particular, the same for all observers; constancy during the life-history of the body is an additional property of m attributed to our ideal billiard balls, but not assumed to be true for matter in general.

Choosing units of length and time so that the velocity of light is unity, we have by (7·2)

$$\frac{ds}{dt} = (1 - v^2)^{\frac{1}{2}}.$$

Hence by (12·3)

$$M = m(1 - v^2)^{-\frac{1}{2}} \qquad (12\cdot4).$$

The mass increases with the velocity by the same factor as that which gives the FitzGerald contraction; and when $v = 0$, $M = m$. The invariant mass is thus equal to the mass at rest.

It is natural to extend (12·2) by adding a fourth component, thus

$$m\frac{dx}{ds},\quad m\frac{dy}{ds},\quad m\frac{dz}{ds},\quad m\frac{dt}{ds} \qquad (12\cdot5).$$

* Or *proper-mass*.

By (12·3) the fourth component is equal to M. Thus the momenta and mass (relative mass) form together a symmetrical expression, the momenta being space-components, and the mass the time-component. We shall see later that the expression (12·5) constitutes a vector, and the laws of conservation of momentum and mass assert the conservation of this vector.

The following is an analytical proof of the law of variation of mass with velocity directly from the principle of conservation of mass and momentum. Let M_1, M_1' be the mass of a body as measured by S and S' respectively, v_1, v_1' being its velocity in the x-direction. Writing

$$\beta_1 = (1 - v_1^2/c^2)^{-\frac{1}{2}}, \quad \beta_1' = (1 - v_1'^2/c^2)^{-\frac{1}{2}}, \quad \beta = (1 - u^2/c^2)^{-\frac{1}{2}},$$

we can easily verify from (6·2) that

$$\beta_1 v_1 = \beta\beta_1' (v_1' - u) \quad \text{.........................(12·6).}$$

Let a number of such particles be moving in a straight line subject to the conservation of mass and momentum as measured by S', viz.

$$\Sigma M_1' \text{ and } \Sigma M_1' v_1' \text{ are conserved}$$

Since β and u are constants it follows that

$$\Sigma M_1' \beta (v_1' - u) \text{ is conserved.}$$

Therefore by (12·6)

$$\Sigma M_1' \beta_1 v_1 / \beta_1' \text{ is conserved} \quad \text{..................(12·71).}$$

But since momentum must also be conserved for the observer S

$$\Sigma M_1 v_1 \text{ is conserved} \quad \text{.....................(12·72).}$$

The results (12·71) and (12·72) will agree if

$$M_1/\beta_1 = M_1'/\beta_1',$$

and it is easy to see that there can be no other general solution. Hence for different values of v_1, M_1 is proportional to β_1, or

$$M = m(1 - v^2/c^2)^{-\frac{1}{2}},$$

where m is a constant for the body.

It requires a greater impulse to produce a given change of velocity δv in the original direction of motion than to produce an equal change δw at right angles to it. For the momenta in the two directions are initially

$$mv(1 - v^2/c^2)^{-\frac{1}{2}}, \quad 0,$$

and after a change δv, δw, they become

$$m(v + \delta v)[1 - \{(v + \delta v)^2 + (\delta w)^2\}/c^2]^{-\frac{1}{2}}, \quad m\delta w[1 - \{(v + \delta v)^2 + (\delta w)^2\}/c^2]^{-\frac{1}{2}}.$$

Hence to the first order in δv, δw the changes of momentum are

$$m(1 - v^2/c^2)^{-\frac{3}{2}} \delta v, \quad m(1 - v^2/c^2)^{-\frac{1}{2}} \delta w,$$

or

$$M\beta^2 \delta v, \quad M\delta w,$$

where β is the FitzGerald factor for velocity v. The coefficient $M\beta^2$ was formerly called the *longitudinal mass*, M being the *transverse mass*; but the longitudinal mass is of no particular importance in the general theory, and the term is dropping out of use.

13. Energy.

When the units are such that $c = 1$, we have

$$M = m(1 - v^2)^{-\frac{1}{2}}$$
$$= m + \tfrac{1}{2}mv^2 \text{ approximately} \quad \text{.................}(13\cdot1),$$

if the speed is small compared with the velocity of light. The second term is the kinetic energy, so that the change of mass is the same as the change of energy, when the velocity alters. This suggests the identification of mass with energy. It may be recalled that in mechanics the total energy of a system is left vague to the extent of an arbitrary additive constant, since only changes of energy are defined. In identifying energy with mass we fix the additive constant m for each body, and m may be regarded as the internal energy of constitution of the body.

The approximation used in (13·1) does not invalidate the argument. Consider two ideal billiard balls colliding. The conservation of mass (relative mass) states that

$$\Sigma m(1 - v^2)^{-\frac{1}{2}} \text{ is unaltered.}$$

The conservation of energy states that

$$\Sigma m(1 + \tfrac{1}{2}v^2) \text{ is unaltered.}$$

But if both statements were exactly true we should have two equations determining unique values of the speeds of the two balls; so that these speeds could not be altered by the collision. The two laws are not independent, but one is an approximation to the other. The first is the accurate law since it is independent of the space-time frame of reference. Accordingly the expression $\frac{1}{2}mv^2$ for the kinetic energy in elementary mechanics is only an approximation in which terms in v^4, etc. are neglected.

When the units of length and time are not restricted by the condition $c = 1$, the relation between the mass M and the energy E is

$$M = E/c^2 \quad \text{.............................}(13\cdot2).$$

Thus the energy corresponding to a gram is $9 . 10^{20}$ ergs. This does not affect the identity of mass and energy—that both are measures of the same world-condition. A world-condition can be examined by different kinds of experimental tests, and the units gram and erg are associated with different tests of the mass-energy condition. But when once the measure has been made it is of no consequence to us which of the experimental methods was chosen, and grams or ergs can be used indiscriminately as the unit of mass. In fact, measures made by energy-tests and by mass-tests are convertible like measures made with a yard-rule and a metre-rule.

The principle of conservation of mass has thus become merged in the principle of conservation of energy. But there is another independent phenomenon which perhaps corresponds more nearly to the original idea of Lavoisier when he enunciated the law of conservation of matter. I refer to the permanence of *invariant mass* attributed to our ideal billiard balls but not supposed to be a general property of matter. The conservation of m is an accidental property like rigidity; the conservation of M is an invariable law of nature.

When radiant heat falls on a billiard ball so that its temperature rises, the increased energy of motion of the molecules causes an increase of mass M. The invariant mass m also increases since it is equal to M for a body at rest. There is no violation of the conservation of M, because the radiant heat has mass M which it transfers to the ball; but we shall show later that the electromagnetic waves have no invariant mass, and the addition to m is created out of nothing. Thus invariant mass is not conserved in general.

To some extent we can avoid this failure by taking the microscopic point of view. The billiard ball can be analysed into a very large number of constituents—electrons and protons—each of which is believed to preserve the same invariant mass for life. But the invariant mass of the billiard ball is not exactly equal to the sum of the invariant masses of its constituents*. The permanence and permanent similarity of all electrons seems to be the modern equivalent of Lavoisier's "conservation of matter." It is still uncertain whether it expresses a universal law of nature; and we are willing to contemplate the possibility that occasionally a positive and negative electron may coalesce and annul one another. In that case the mass M would pass into the electromagnetic waves generated by the catastrophe, whereas the invariant mass m would disappear altogether. Again if ever we are able to synthesise helium out of hydrogen, 0·8 per cent. of the invariant mass will be annihilated, whilst the corresponding proportion of relative mass will be liberated as radiant energy.

It will thus be seen that although in the special problems considered the quantity m is usually supposed to be permanent, its conservation belongs to an altogether different order of ideas from the universal conservation of M.

14. Density and temperature.

Consider a volume of space delimited in some invariant way, e.g. the content of a material box. The counting of a number of discrete particles continually within (i.e. moving with) the box is an absolute operation; let the absolute number be N. The volume V of the box will depend on the space-reckoning, being decreased in the ratio β for an observer moving relatively to the box and particles, owing to the FitzGerald contraction of one of the dimensions of the box. Accordingly the particle-density $\sigma = N/V$ satisfies

$$\sigma' = \sigma\beta \qquad (14{\cdot}1),$$

* This is because the invariant mass of each electron is its relative mass referred to axes moving with it; the invariant mass of the billiard ball is the relative mass referred to axes at rest in the billiard ball as a whole.

where σ' is the particle-density for an observer in relative motion, and σ the particle-density for an observer at rest relative to the particles.

It follows that the mass-density ρ obeys the equation

$$\rho' = \rho\beta^2 \qquad (14\cdot2),$$

since the mass of each particle is increased for the moving observer in the ratio β.

Quantities referred to the space-time system of an observer moving with the body considered are often distinguished by the prefix *proper-* (German, *Eigen-*), e.g. proper-length, proper-volume, proper-density, proper-mass = invariant mass.

The transformation of temperature for a moving observer does not often concern us. In general the word obviously means proper-temperature, and the motion of the observer does not enter into consideration. In thermometry and in the theory of gases it is essential to take a standard with respect to which the matter is at rest on the average, since the indication of a thermometer moving rapidly through a fluid is of no practical interest. But thermodynamical temperature is defined by

$$dS = dM/T \qquad (14\cdot3),$$

where dS is the change of entropy for a change of energy dM. The temperature T defined by this equation will depend on the observer's frame of reference. Entropy is clearly meant to be an invariant, since it depends on the probability of the statistical state of the system compared with other states which might exist. Hence T must be altered by motion in the same way as dM, that is to say

$$T' = \beta T \qquad (14\cdot4).$$

But it would be useless to apply such a transformation to the adiabatic gas-equation

$$T = k\rho^{\gamma-1},$$

for, in that case, T is evidently intended to signify the proper-temperature and ρ the proper-density.

In general it is unprofitable to apply the Lorentz transformation to the *constitutive equations* of a material medium and to coefficients occurring in them (permeability, specific inductive capacity, elasticity, velocity of sound). Such equations naturally take a simpler and more significant form for axes moving with the matter. The transformation to moving axes introduces great complications without any evident advantages, and is of little interest except as an analytical exercise.

15. General transformations of coordinates.

We obtain a transformation of coordinates by taking new coordinates x_1', x_2', x_3', x_4' which are any four functions of the old coordinates x_1, x_2, x_3, x_4. Conversely, x_1, x_2, x_3, x_4 are functions of x_1', x_2', x_3', x_4'. It is assumed that multiple values are excluded, at least in the region considered, so that values of (x_1, x_2, x_3, x_4) and (x_1', x_2', x_3', x_4') correspond one to one.

If $\quad x_1 = f_1(x_1', x_2', x_3', x_4')$; $x_2 = f_2(x_1', x_2', x_3', x_4')$; etc.,

$$dx_1 = \frac{\partial f_1}{\partial x_1'} dx_1' + \frac{\partial f_1}{\partial x_2'} dx_2' + \frac{\partial f_1}{\partial x_3'} dx_3' + \frac{\partial f_1}{\partial x_4'} dx_4'; \text{ etc.} \quad \ldots\ldots(15 \cdot 1),$$

or it may be written simply,

$$dx_1 = \frac{\partial x_1}{\partial x_1'} dx_1' + \frac{\partial x_1}{\partial x_2'} dx_2' + \frac{\partial x_1}{\partial x_3'} dx_3' + \frac{\partial x_1}{\partial x_4'} dx_4'; \text{ etc.} \quad \ldots\ldots(15 \cdot 2).$$

Substituting from (15·2) in (2·1) we see that ds^2 will be a homogeneous quadratic function of the differentials of the new coordinates; and the new coefficients g_{11}', g_{22}', etc. could be written down in terms of the old, if desired.

For an example consider the usual transformation to axes revolving with constant angular velocity ω, viz.

$$\left.\begin{aligned} x &= x_1' \cos \omega x_4' - x_2' \sin \omega x_4' \\ y &= x_1' \sin \omega x_4' + x_2' \cos \omega x_4' \\ z &= x_3' \\ t &= x_4' \end{aligned}\right\} \quad \ldots\ldots\ldots\ldots\ldots\ldots(15 \cdot 3).$$

Hence

$$\begin{aligned} dx &= dx_1' \cos \omega x_4' - dx_2' \sin \omega x_4' + \omega\,(-x_1' \sin \omega x_4' - x_2' \cos \omega x_4')\, dx_4', \\ dy &= dx_1' \sin \omega x_4' + dx_2' \cos \omega x_4' + \omega\,(x_1' \cos \omega x_4' - x_2' \sin \omega x_4')\, dx_4', \\ dz &= dx_3', \\ dt &= dx_4'. \end{aligned}$$

Taking units of space and time so that $c = 1$, we have for our original fixed coordinates by (7·1)

$$ds^2 = -dx^2 - dy^2 - dz^2 + dt^2.$$

Hence, substituting the values found above,

$$\begin{aligned} ds^2 = -dx_1'^2 - dx_2'^2 - dx_3'^2 &+ \{1 - \omega^2 (x_1'^2 + x_2'^2)\}\, dx_4'^2 \\ &+ 2\omega x_2' dx_1' dx_4' - 2\omega x_1' dx_2' dx_4' \quad \ldots\ldots(15 \cdot 4). \end{aligned}$$

Remembering that all observational differences of coordinate-systems must arise *via* the interval, this formula must comprise everything which distinguishes the rotating system from a fixed system of coordinates.

In the transformation (15·3) we have paid no attention to any contraction of the standards of length or retardation of clocks due to motion with the rotating axes. The formulae of transformation are those of elementary kinematics, so that x_1', x_2', x_3', x_4' are quite strictly the coordinates used in the ordinary theory of rotating axes. But it may be suggested that elementary kinematics is now seen to be rather crude, and that it would be worth while to touch up the formulae (15·3) so as to take account of these small changes of the standards. A little consideration shows that the suggestion is impracticable. It was shown in § 4 that if x_1', x_2', x_3', x_4' represent rectangular coordinates and time as partitioned by direct readings of scales and clocks, then

$$ds^2 = -dx_1'^2 - dx_2'^2 - dx_3'^2 + c^2dx_4'^2 \quad \text{..............(15·45)},$$

so that coordinates which give any other formula for the interval cannot represent the immediate indications of scales and clocks. As shown at the end of § 5, the only transformations which give (15·45) are Lorentz transformations. If we wish to make a transformation of a more general kind, such as that of (15·3), we must necessarily abandon the association of the coordinate-system with uncorrected scale and clock readings. It is useless to try to "improve" the transformation to rotating axes, because the supposed improvement could only lead us back to a coordinate-system similar to the fixed axes with which we started.

The inappropriateness of rotating axes to scale and clock measurements can be regarded from a physical point of view. We cannot keep a scale or clock at rest in the rotating system unless we constrain it, i.e. subject it to molecular bombardment—an "outside influence" whose effect on the measurements must not be ignored.

In the x, y, z, t system of coordinates the scale and clock are the natural equipment for exploration. In other systems they will, if unconstrained, continue to measure ds; but the reading of ds is no longer related in a simple way to the differences of coordinates which we wish to determine; it depends on the more complicated calculations involved in (2·1). The scale and clock to some extent lose their pre-eminence, and since they are rather elaborate appliances it may be better to refer to some simpler means of exploration. We consider then two simpler test-objects—the moving particle and the light-pulse.

In ordinary rectangular coordinates and time x, y, z, t an undisturbed particle moves with uniform velocity, so that its track is given by the equations

$$x = a + bt, \qquad y = c + dt, \qquad z = e + ft \quad \text{...........(15·5)},$$

i.e. the equations of a straight line in four dimensions. By substituting from (15·3) we could find the equations of the track in rotating coordinates; or by substituting from (15·2) we could obtain the differential equations for any desired coordinates. But there is another way of proceeding. The differential equations of the track may be written

$$\frac{d^2x}{ds^2}, \frac{d^2y}{ds^2}, \frac{d^2z}{ds^2}, \frac{d^2t}{ds^2} = 0 \quad \text{....................(15·6)},$$

which on integration, having regard to the condition (7·1), give equations (15·5).

The equations (15·6) are comprised in the single statement

$$\int ds \text{ is stationary} \quad \text{..........................(15·7)}$$

for all arbitrary small variations of the track which vanish at the initial and final limits—a well-known property of the straight line.

In arriving at (15·7) we use freely the geometry of the x, y, z, t system given by (7·1); but the final result does not allude to coordinates at all, and must be unaltered whatever system of coordinates we are using. To obtain explicit equations for the track in any desired system of coordinates, we substitute in (15·7) the appropriate expression (2·1) for ds and apply the calculus of variations. The actual analysis will be given in § 28.

The track of a light-pulse, being a straight line in four dimensions, will also satisfy (15·7); but the light-pulse has the special velocity c which gives the additional condition obtained in § 7, viz.

$$ds = 0 \qquad\qquad (15\cdot8).$$

Here again there is no reference to any coordinates in the final result.

We have thus obtained equations (15·7) and (15·8) for the behaviour of the moving particle and light-pulse which must hold good whatever the coordinate-system chosen. The indications of our two new test-bodies are connected with the interval, just as in § 3 the indications of the scale and clock were connected with the interval. It should be noticed however that whereas the use of the older test-bodies depends only on the truth of the fundamental axiom, the use of the new test-bodies depends on the truth of the empirical laws of motion and of light-propagation. In a deductive theory this appeal to empirical laws is a blemish which we must seek to remove later.

16. Fields of force.

Suppose that an observer has chosen a definite system of space-coordinates and of time-reckoning (x_1, x_2, x_3, x_4) and that the geometry of these is given by

$$ds^2 = g_{11}dx_1^2 + g_{22}dx_2^2 + \ldots + 2g_{12}dx_1dx_2 + \ldots \qquad (16\cdot1).$$

Let him be under the mistaken impression that the geometry is

$$ds_0^2 = -dx_1^2 - dx_2^2 - dx_3^2 + dx_4^2 \qquad (16\cdot2),$$

that being the geometry with which he is most familiar in pure mathematics. We use ds_0 to distinguish his mistaken value of the interval. Since intervals can be compared by experimental methods, he ought soon to discover that his ds_0 cannot be reconciled with observational results, and so realise his mistake. But the mind does not so readily get rid of an obsession. It is more likely that our observer will continue in his opinion, and attribute the discrepancy of the observations to some influence which is present and affects the behaviour of his test-bodies. He will, so to speak, introduce a supernatural agency which he can blame for the consequences of his mistake. Let us examine what name he would apply to this agency.

Of the four test-bodies considered the moving particle is in general the most sensitive to small changes of geometry, and it would be by this test that the observer would first discover discrepancies. The path laid down for it by our observer is

$$\int ds_0 \text{ is stationary,}$$

i.e. a straight line in the coordinates (x_1, x_2, x_3, x_4). The particle, of course, pays no heed to this, and moves in the different track

$$\int ds \text{ is stationary.}$$

Although apparently undisturbed it deviates from "uniform motion in a straight line." The name given to any agency which causes deviation from uniform motion in a straight line is *force* according to the Newtonian definition of force. Hence the agency invoked through our observer's mistake is described as a "field of force."

The field of force is not always introduced by inadvertence as in the foregoing illustration. It is sometimes introduced deliberately by the mathematician, e.g. when he introduces the centrifugal force. There would be little advantage and many disadvantages in banishing the phrase "field of force" from our vocabulary. We shall therefore regularise the procedure which our observer has adopted. We call (16·2) the *abstract geometry* of the system of coordinates (x_1, x_2, x_3, x_4); it may be chosen arbitrarily by the observer. The *natural geometry* is (16·1).

A field of force represents the discrepancy between the natural geometry of a coordinate-system and the abstract geometry arbitrarily ascribed to it.

A field of force thus arises from an attitude of mind. If we do not take our coordinate-system to be something different from that which it really is, there is no field of force. If we do not regard our rotating axes as though they were non-rotating, there is no centrifugal force.

Coordinates for which the natural geometry is

$$ds^2 = -dx_1^2 - dx_2^2 - dx_3^2 + dx_4^2$$

are called Galilean coordinates. They are the same as those we have hitherto called ordinary rectangular coordinates and time (the velocity of light being unity). Since this geometry is familiar to us, and enters largely into current conceptions of space, time and mechanics, we usually choose Galilean geometry when we have to ascribe an abstract geometry. Or we may use slight modifications of it, e.g. substitute polar for rectangular coordinates.

It has been shown in § 4 that when the g's are constants, coordinates can be chosen so that Galilean geometry is actually the natural geometry. There is then no need to introduce a field of force in order to enjoy our accustomed outlook; and if we deliberately choose non-Galilean coordinates and attribute to them abstract Galilean geometry, we recognise the artificial character of the field of force introduced to compensate the discrepancy. But in the more general case it is not possible to make the reduction of § 4 accurately throughout the region explored by our experiments; and no Galilean coordinates exist. In that case it has been usual to select some system (preferably an approximation to a Galilean system) and ascribe to it the abstract geometry of the Galilean system. The field of force so introduced is called "Gravitation."

It should be noticed that the rectangular coordinates and time in current use can scarcely be regarded as a close approximation to the Galilean system, since the powerful force of terrestrial gravitation is needed to compensate the error.

The naming of coordinates (e.g. time) usually follows the *abstract geometry* attributed to the system. In general the natural geometry is of some complicated kind for which no detailed nomenclature is recognised. Thus when we call a coordinate t the "time," we may either mean that it fulfils the observational conditions discussed in § 4, or we may mean that any departure from those conditions will be ascribed to the interference of a field of force. In the latter case "time" is an arbitrary name, useful because it fixes a consequential nomenclature of velocity, acceleration, etc.

To take a special example, an observer at a station on the earth has found a particular set of coordinates x_1, x_2, x_3, x_4 best suited to his needs. He calls them x, y, z, t in the belief that they are actually rectangular coordinates and time, and his terminology—straight line, circle, density, uniform velocity, etc.—follows from this identification. But, as shown in § 4, this nomenclature can only agree with the measures made by clocks and scales provided (16·2) is satisfied; and if (16·2) is satisfied, the tracks of undisturbed particles must be straight lines. Experiment immediately shows that this is not the case; the tracks of undisturbed particles are parabolas. But instead of accepting the verdict of experiment and admitting that x_1, x_2, x_3, x_4 are not what he supposed they were, our observer introduces a field of force to explain why his test is not fulfilled. A certain part of this field of force might have been avoided if he had taken originally a different set of coordinates (not rotating with the earth); and in so far as the field of force arises on this account it is generally recognised that it is a mathematical fiction—the centrifugal force. But there is a residuum which cannot be got rid of by any choice of coordinates; there exists no extensive coordinate-system having the simple properties which were ascribed to x, y, z, t. The intrinsic nature of space-time near the earth is not of the kind which admits coordinates with Galilean geometry. This irreducible field of force constitutes the field of terrestrial gravitation. The statement that space-time round the earth is "curved"—that is to say, that it is not of the kind which admits Galilean coordinates—is not an hypothesis; it is an equivalent expression of the observed fact that an irreducible field of force is present, having regard to the Newtonian definition of force. It is this fact of observation which demands the introduction of non-Galilean space-time and non-Euclidean space into the theory.

17. The Principle of Equivalence.

In § 15 we have stated the laws of motion of undisturbed material particles and of light-pulses in a form independent of the coordinates chosen. Since a great deal will depend upon the truth of these laws it is desirable to consider what justification there is for believing them to be both accurate and universal. Three courses are open:

(*a*) It will be shown in Chapters IV and VI that these laws follow rigorously from a more fundamental discussion of the nature of matter and of electromagnetic fields; that is to say, the hypotheses underlying them may be pushed a stage further back.

(*b*) The track of a moving particle or light-pulse under specified initial conditions is unique, and it does not seem to be possible to specify any unique tracks in terms of intervals only other than those given by equations (15·7) and (15·8).

(*c*) We may arrive at these laws by induction from experiment.

If we rely solely on experimental evidence we cannot claim exactness for the laws. It goes without saying that there always remains a possibility of small amendments of the laws too slight to affect any observational tests yet tried. Belief in the perfect accuracy of (15·7) and (15·8) can only be justified on the theoretical grounds (*a*) or (*b*). But the more important consideration is the universality, rather than the accuracy, of the experimental laws; we have to guard against a spurious generalisation extended to conditions intrinsically dissimilar from those for which the laws have been established observationally.

We derived (15·7) from the equations (15·5) which describe the observed behaviour of a particle moving under no field of force. We assume that the result holds in all circumstances. The risky point in the generalisation is not in introducing a field of force, because that may be due to an attitude of mind of which the particle has no cognizance. The risk is in passing from regions of the world where Galilean coordinates (x, y, z, t) are possible to intrinsically dissimilar regions where no such coordinates exist—from flat space-time to space-time which is not flat.

The *Principle of Equivalence* asserts the legitimacy of this generalisation. It is essentially an hypothesis to be tested by experiment as opportunity offers. Moreover it is to be regarded as a suggestion, rather than a dogma admitting of no exceptions. It is likely that some of the phenomena will be determined by comparatively simple equations in which the components of curvature of the world do not appear; such equations will be the same for a curved region as for a flat region. It is to these that the Principle of Equivalence applies. It is a plausible suggestion that the undisturbed motion of a particle and the propagation of light are governed by laws of this specially simple type; and accordingly (15·7) and (15·8) will apply in all circumstances. But there are more complex phenomena governed by equations in which the curvatures of the world are involved; terms containing these curvatures will vanish in the equations summarising experiments made in a flat region, and would have to be reinstated in passing to the general equations. Clearly there must be some phenomena of this kind which discriminate between a flat world and a curved world; otherwise we could have no knowledge of world-curvature. For these the Principle of Equivalence breaks down.

The Principle of Equivalence thus asserts that *some* of the chief differential equations of physics are the same for a curved region of the world as for an osculating flat region*. There can be no infallible rule for generalising experimental laws; but the Principle of Equivalence offers a suggestion for trial, which may be expected to succeed sometimes, and fail sometimes.

The Principle of Equivalence has played a great part as a guide in the original building up of the generalised relativity theory; but now that we have reached the new view of the nature of the world it has become less necessary. Our present exposition is in the main deductive. We start with a general theory of world-structure and work down to the experimental consequences, so that our progress is from the general to the special laws, instead of *vice versa*.

18. Retrospect.

The investigation of the external world in physics is a quest for *structure* rather than *substance*. A structure can best be represented as a complex of relations and relata; and in conformity with this we endeavour to reduce the phenomena to their expressions in terms of the relations which we call intervals and the relata which we call events.

If two bodies are of identical structure as regards the complex of interval relations, they will be exactly similar as regards observational properties†, if our fundamental hypothesis is true. By this we show that experimental measurements of lengths and duration are equivalent to measurements of the interval relation.

To the events we assign four identification-numbers or coordinates according to a plan which is arbitrary within wide limits. The connection between our physical measurements of interval and the system of identification-numbers is expressed by the general quadratic form (2·1). In the particular case when these identification-numbers can be so assigned that the product terms in the quadratic form disappear leaving only the four squares, the coordinates have the metrical properties belonging to rectangular coordinates and time, and are accordingly so identified. If any such system exists an infinite number of others exist connected with it by the Lorentz transformation, so that there is no unique space-time frame. The relations of these different space-time reckonings have been considered in detail. It is

* The correct equations for a curved world will necessarily include as a special case those already obtained for a flat world. The practical point on which we seek the guidance of the Principle of Equivalence is whether the equations already obtained for a flat world will serve as they stand or will require generalisation.

† At present this is limited to extensional properties (in both space and time). It will be shown later that all mechanical properties are also included. Electromagnetic properties require separate consideration.

shown that there must be a particular speed which has the remarkable property that its value is the same for all these systems; and by appeal to the Michelson-Morley experiment or to Fizeau's experiment it is found that this is a distinctive property of the speed of light.

But it is not possible throughout the world to choose coordinates fulfilling the current definitions of rectangular coordinates and time. In such cases we usually relax the definitions, and attribute the failure of fulfilment to a field of force pervading the region. We have now no definite guide in selecting what coordinates to take as rectangular coordinates and time; for whatever the discrepancy, it can always be ascribed to a suitable field of force. The field of force will vary according to the system of coordinates selected; but in the general case it is not possible to get rid of it altogether (in a large region) by any choice of coordinates. This irreducible field of force is ascribed to gravitation. It should be noticed that the gravitational influence of a massive body is not properly expressed by a definite field of force, but by the property of irreducibility of the field of force. We shall find later that the irreducibility of the field of force is equivalent to what in geometrical nomenclature is called a curvature of the continuum of space-time.

For the fuller study of these problems we require a special mathematical calculus which will now be developed *ab initio*.

CHAPTER II

THE TENSOR CALCULUS

19. Contravariant and covariant vectors.

We consider the transformation from one system of coordinates x_1, x_2, x_3, x_4 to another system x_1', x_2', x_3', x_4'.

The differentials (dx_1, dx_2, dx_3, dx_4) are transformed according to the equations (15·2), viz.

$$dx_1' = \frac{\partial x_1'}{\partial x_1} dx_1 + \frac{\partial x_1'}{\partial x_2} dx_2 + \frac{\partial x_1'}{\partial x_3} dx_3 + \frac{\partial x_1'}{\partial x_4} dx_4; \text{ etc.}$$

which may be written shortly

$$dx_\mu' = \sum_{\alpha=1}^{4} \frac{\partial x_\mu'}{\partial x_\alpha} dx_\alpha,$$

four equations being obtained by taking $\mu = 1, 2, 3, 4$, successively.

Any set of four quantities transformed according to this law is called a *contravariant vector*. Thus if (A^1, A^2, A^3, A^4) becomes (A'^1, A'^2, A'^3, A'^4) in the new coordinate-system, where

$$A'^\mu = \sum_{\alpha=1}^{4} \frac{\partial x_\mu'}{\partial x_\alpha} A^\alpha \quad \text{.........................(19·1)},$$

then (A^1, A^2, A^3, A^4), denoted briefly as A^μ, is a contravariant vector. The upper position of the suffix (which is, of course, not an exponent) is reserved to indicate contravariant vectors.

If ϕ is an invariant function of position, i.e. if it has a fixed value at each point independent of the coordinate-system employed, the four quantities

$$\left(\frac{\partial \phi}{\partial x_1}, \frac{\partial \phi}{\partial x_2}, \frac{\partial \phi}{\partial x_3}, \frac{\partial \phi}{\partial x_4}\right)$$

are transformed according to the equations

$$\frac{\partial \phi}{\partial x_1'} = \frac{\partial x_1}{\partial x_1'}\frac{\partial \phi}{\partial x_1} + \frac{\partial x_2}{\partial x_1'}\frac{\partial \phi}{\partial x_2} + \frac{\partial x_3}{\partial x_1'}\frac{\partial \phi}{\partial x_3} + \frac{\partial x_4}{\partial x_1'}\frac{\partial \phi}{\partial x_4}; \text{ etc.}$$

which may be written shortly

$$\frac{\partial \phi}{\partial x_\mu'} = \sum_{\alpha=1}^{4} \frac{\partial x_\alpha}{\partial x_\mu'}\frac{\partial \phi}{\partial x_\alpha}.$$

Any set of four quantities transformed according to this law is called a *covariant vector*. Thus if A_μ is a covariant vector, its transformation law is

$$A_\mu' = \sum_{\alpha=1}^{4} \frac{\partial x_\alpha}{\partial x_\mu'} A_\alpha \quad \text{.......................(19·2)}.$$

We have thus two varieties of vectors which we distinguish by the upper or lower position of the suffix. The first illustration of a contravariant vector, dx_μ, forms rather an awkward exception to the rule that a lower suffix indicates covariance and an upper suffix contravariance. There is no other exception likely to mislead the reader, so that it is not difficult to keep in mind this peculiarity of dx_μ; but we shall sometimes find it convenient to indicate its contravariance explicitly by writing

$$dx_\mu \equiv (dx)^\mu \quad \text{.................................(19·3).}$$

A vector may either be a single set of four quantities associated with a special point in space-time, or it may be a set of four functions varying continuously with position. Thus we can have an "isolated vector" or a "vector-field."

For an illustration of a covariant vector we considered the gradient of an invariant, $\partial\phi/\partial x_\mu$; but a covariant vector is not necessarily the gradient of an invariant.

The reader will probably be already familiar with the term vector, but the distinction of covariant and contravariant vectors will be new to him. This is because in the elementary analysis only rectangular coordinates are contemplated, and for transformations from one rectangular system to another the laws (19·1) and (19·2) are equivalent to one another. From the geometrical point of view, the *contravariant vector* is the vector with which everyone is familiar; this is because a displacement, or directed distance between two points, is regarded as representing (dx_1, dx_2, dx_3)* which, as we have seen, is contravariant. The covariant vector is a new conception which does not so easily lend itself to graphical illustration.

20. The mathematical notion of a vector.

The formal definitions in the preceding section do not help much to an understanding of what the notion of a vector really is. We shall try to explain this more fully, taking first the mathematical notion of a vector (with which we are most directly concerned) and leaving the more difficult physical notion to follow.

We have a set of four numbers (A_1, A_2, A_3, A_4) which we associate with some point (x_1, x_2, x_3, x_4) and with a certain system of coordinates. We make a change of the coordinate-system, and we ask, What will these numbers become in the new coordinates? The question is meaningless; they do not automatically "become" anything. Unless we interfere with them they stay as they were. But the mathematician may say "When I am using the coordinates x_1, x_2, x_3, x_4, I want to talk about the numbers A_1, A_2, A_3, A_4; and when I am using x_1', x_2', x_3', x_4' I find that at the corresponding stage of my work I shall want to talk about four different numbers A_1', A_2', A_3', A_4'.

* The customary resolution of a displacement into components in oblique directions assumes this.

So for brevity I propose to call both sets of numbers by the same symbol $\mathfrak{A}$." We reply "That will be all right, provided that you tell us just what numbers will be denoted by $\mathfrak{A}$ for *each* of the coordinate-systems you intend to use. Unless you do this we shall not know what you are talking about."

Accordingly the mathematician begins by giving us a list of the numbers that $\mathfrak{A}$ will signify in the different coordinate-systems. We here denote these numbers by letters. $\mathfrak{A}$ will mean*

X, Y, Z for certain rectangular coordinates x, y, z,

R, Θ, Φ for certain polar coordinates r, θ, ϕ,

Λ, M, N for certain ellipsoidal coordinates λ, μ, ν.

"But," says the mathematician, "I shall never finish at this rate. There are an infinite number of coordinate-systems which I want to use. I see that I must alter my plan. I will give you a general rule to find the new values of $\mathfrak{A}$ when you pass from one coordinate-system to another; so that it is only necessary for me to give you one set of values and you can find all the others for yourselves."

In mentioning a *rule* the mathematician gives up his arbitrary power of making $\mathfrak{A}$ mean anything according to his fancy at the moment. He binds himself down to some kind of regularity. Indeed we might have suspected that our orderly-minded friend would have some principle in his assignment of different meanings to $\mathfrak{A}$. But even so, can we make any guess at the rule he is likely to adopt unless we have some idea of the problem he is working at in which $\mathfrak{A}$ occurs? I think we can; it is not necessary to know anything about the nature of his problem, whether it relates to the world of physics or to something purely conceptual; it is sufficient that we know a little about the nature of a mathematician.

What kind of rule could he adopt? Let us examine the quantities which can enter into it. There are first the two sets of numbers to be connected, say, X, Y, Z and R, Θ, Φ. Nothing has been said as to these being analytical functions of any kind; so far as we know they are isolated numbers. Therefore there can be no question of introducing their derivatives. They are regarded as located at some point of space (x, y, z) and (r, θ, ϕ), otherwise the question of coordinates could scarcely arise. They are changed because the coordinate-system has changed *at this point*, and that change is defined by quantities like $\frac{\partial r}{\partial x}, \frac{\partial^2 \theta}{\partial x \partial y}$, and so on. The integral coordinates themselves, x, y, z, r, θ, ϕ, cannot be involved; because they express relations to a distant origin, whereas we are concerned only with changes at the spot where (X, Y, Z) is located. Thus the rule must involve only the numbers X, Y, Z, R, Θ, Φ combined with the mutual derivatives of x, y, z, r, θ, ϕ.

* For convenience I take a three-dimensional illustration.

One such rule would be

$$\left.\begin{aligned} R &= \frac{\partial r}{\partial x} X + \frac{\partial r}{\partial y} Y + \frac{\partial r}{\partial z} Z \\ \Theta &= \frac{\partial \theta}{\partial x} X + \frac{\partial \theta}{\partial y} Y + \frac{\partial \theta}{\partial z} Z \\ \Phi &= \frac{\partial \phi}{\partial x} X + \frac{\partial \phi}{\partial y} Y + \frac{\partial \phi}{\partial z} Z \end{aligned}\right\} \quad \text{......................(20·1)}.$$

Applying the same rule to the transformation from (r, θ, ϕ) to (λ, μ, ν) we have

$$\Lambda = \frac{\partial \lambda}{\partial r} R + \frac{\partial \lambda}{\partial \theta} \Theta + \frac{\partial \lambda}{\partial \phi} \Phi \quad \text{......................(20·2)},$$

whence, substituting for R, Θ, Φ from (20·1) and collecting terms,

$$\begin{aligned} \Lambda = &\left(\frac{\partial \lambda}{\partial r}\frac{\partial r}{\partial x} + \frac{\partial \lambda}{\partial \theta}\frac{\partial \theta}{\partial x} + \frac{\partial \lambda}{\partial \phi}\frac{\partial \phi}{\partial x}\right) X + \left(\frac{\partial \lambda}{\partial r}\frac{\partial r}{\partial y} + \frac{\partial \lambda}{\partial \theta}\frac{\partial \theta}{\partial y} + \frac{\partial \lambda}{\partial \phi}\frac{\partial \phi}{\partial y}\right) Y \\ &+ \left(\frac{\partial \lambda}{\partial r}\frac{\partial r}{\partial z} + \frac{\partial \lambda}{\partial \theta}\frac{\partial \theta}{\partial z} + \frac{\partial \lambda}{\partial \phi}\frac{\partial \phi}{\partial z}\right) Z \\ = &\frac{\partial \lambda}{\partial x} X + \frac{\partial \lambda}{\partial y} Y + \frac{\partial \lambda}{\partial z} Z \quad \text{..(20·3)}, \end{aligned}$$

which is the same formula as we should have obtained by applying the rule to the direct transformation from (x, y, z) to (λ, μ, ν). The rule is thus self-consistent. But this is a happy accident, pertaining to this particular rule, and depending on the formula

$$\frac{\partial \lambda}{\partial x} = \frac{\partial \lambda}{\partial r}\frac{\partial r}{\partial x} + \frac{\partial \lambda}{\partial \theta}\frac{\partial \theta}{\partial x} + \frac{\partial \lambda}{\partial \phi}\frac{\partial \phi}{\partial x},$$

and amid the apparently infinite choice of formulae it will not be easy to find others which have this self-consistency.

The above rule is that already given for the contravariant vector (19·1). The rule for the covariant vector is also self-consistent. There do not appear to be any other self-consistent rules for the transformation of a set of three numbers (or four numbers for four coordinates) *.

We see then that unless the mathematician disregards the need for self-consistency in his rule, he must inevitably make his quantity $\mathfrak{A}$ either a contravariant or a covariant vector. The choice between these is entirely at his discretion. He might obtain a wider choice by disregarding the property of self-consistency—by selecting a particular coordinate-system, x, y, z, and insisting that values in other coordinate-systems must always be obtained by

* Except that we may in addition multiply by any power of the Jacobian of the transformation. This is self-consistent because

$$\frac{\partial (x, y, z)}{\partial (r, \theta, \phi)} \cdot \frac{\partial (r, \theta, \phi)}{\partial (\lambda, \mu, \nu)} = \frac{\partial (x, y, z)}{\partial (\lambda, \mu, \nu)}.$$

Sets of numbers transformed with this additional multiplication are degenerate cases of tensors of higher rank considered later. See §§ 48, 49.

applying the rule immediately to X, Y, Z, and not permitting intermediate transformations. In practice he does not do this, perhaps because he can never make up his mind that any particular coordinates are deserving of this special distinction.

We see now that a mathematical vector is a common name for an infinite number of sets of quantities, each set being associated with one of an infinite number of systems of coordinates. The arbitrariness in the association is removed by postulating that some method is followed, and that no one system of coordinates is singled out for special distinction. In technical language the transformations must form a *Group. The quantity* (R, Θ, Φ) *is in no sense the same quantity as* (X, Y, Z); they have a common name and a certain analytical connection, but the idea of anything like identity is entirely excluded from the mathematical notion of a vector.

21. The physical notion of a vector.

The components of a force (X, Y, Z), (X', Y', Z'), etc. in different systems of Cartesian coordinates, rectangular or oblique, form a contravariant vector. This is evident because in elementary mechanics a force is resolved into components according to the parallelogram law just as a displacement dx_μ is resolved, and we have seen that dx_μ is a contravariant vector. So far as the mathematical notion of the vector is concerned, the quantities (X, Y, Z) and (X', Y', Z') are not to be regarded as in any way identical; but in physics we conceive that both quantities express some kind of condition or relation of the world, and this condition is the same whether expressed by (X, Y, Z) or by (X', Y', Z'). The physical vector is this vaguely conceived entity, which is independent of the coordinate-system, and is at the back of our measurements of force.

A world-condition cannot appear directly in a mathematical equation; only the *measure* of the world-condition can appear. Any number or set of numbers which can serve to specify uniquely a condition of the world may be called a measure of it. In using the phrase "condition of the world" I intend to be as non-committal as possible; whatever in the external world determines the values of the physical quantities which we observe, will be included in the phrase.

The simplest case is when the condition of the world under consideration can be indicated by a single measure-number. Take two such conditions underlying respectively the wave-length λ and period T of a light-wave. We have the equation

$$\lambda = 3 \,.\, 10^{10} T \quad \text{.............................(21·1)}.$$

This equation holds only for one particular plan of assigning measure-numbers (the C.G.S. system). But it may be written in the more general form

$$\lambda = cT \quad \text{.................................(21·2)},$$

where c is a velocity having the value $3 \,.\, 10^{10}$ in the C.G.S. system. This comprises any number of particular equations of the form (21·1). For each measure-plan, or system of units, c has a different numerical value. The method of determining the necessary change of c when a new measure-plan is adopted, is well known; we assign to it the *dimensions* length ÷ time, and by a simple rule we know how it must be changed when the units of λ and T are changed. For any general equation the total dimensions of every term ought to be the same.

The tensor calculus extends this *principle of dimensions* to changes of measure-code much more general than mere changes of units. There are conditions of the world which cannot be specified by a single measure-number; some require 4, some 16, some 64, etc., measure-numbers. Their variety is such that they cannot be arranged in a single serial order. Consider then an equation between the measure-numbers of two conditions of the world which require 4 measure-numbers. The equation, if it is of the necessary general type, must hold for every possible measure-code; this will be the case if, when we transform the measure-code, both sides of the equation are transformed in the same way, i.e. if we have to perform the same series of mathematical operations on both sides.

We can here make use of the mathematical vector of § 20. Let our equation in some measure-code be

$$A_1, A_2, A_3, A_4 = B_1, B_2, B_3, B_4 \qquad (21\cdot3).$$

Now let us change the code so that the left-hand side becomes *any* four numbers A_1', A_2', A_3', A_4'. We identify this with the transformation of a covariant vector by associating with the change of measure-code the corresponding transformation of coordinates from x_μ to x_μ' as in (19·2). But since (21·3) is to hold in all measure-codes, the transformation of the right-hand side must involve the same set of operations; and the change from B_1, B_2, B_3, B_4 to B_1', B_2', B_3', B_4' will also be the transformation of a covariant vector associated with the *same* transformation of coordinates from x_μ to x_μ'.

We thus arrive at the result that in an equation which is independent of the measure-plan both sides must be covariant or both contravariant vectors. We shall extend this later to conditions expressed by 16, 64, ..., measure-numbers; the general rule is that both sides of the equation must have the same elements of covariance and contravariance. Covariance and contravariance are a kind of generalised dimension, showing how the measure of one condition of the world is changed when the measure of another condition is changed. The ordinary theory of change of units is merely an elementary case of this.

Coordinates are the identification-numbers of the points of space-time. There is no fundamental distinction between measure-numbers and identification-numbers, so that we may regard the change of coordinates as part of the general change applied to all measure-numbers. The change of coordinates no longer leads the way, as it did in § 20; it is placed on the same level with the other changes of measure.

When we applied a change of measure-code to (21·3) we associated with it a change of coordinates; but it is to be noted that the change of coordinates was then ambiguous, since the two sides of the equation might have been taken as both contravariant instead of both covariant; and further the change did not refer explicitly to coordinates *in* the world—it was a mere entry in the mathematician's note-book in order that he might have the satisfaction of calling A_μ and B_μ vectors consistently with his definition. Now if the measure-plan of a condition A_μ is changed the measures of other conditions and relations associated with it will be changed. Among these is a certain relation of two events which we may call the *aspect** of one from the other; and this relation requires four measure-numbers to specify it. Somewhat arbitrarily we decide to make the aspect a contravariant vector, and the measure-numbers assigned to it are denoted by $(dx)^\mu$. That settles the ambiguity once for all. For obscure psychological reasons the mind has singled out this transcendental relation of aspect for graphical representation, so that it is conceived by us as a *displacement* or difference of location in a frame of space-time. Its measure-numbers $(dx)^\mu$ are represented graphically as coordinate-differences dx_μ, and so for each measure-code of aspect we get a corresponding coordinate-frame of location. This "real" coordinate-frame can now replace the abstract frame in the mathematician's note-book, because as we have seen in (19·1) the actual transformation of coordinates *resulting* in a change of dx_μ is the same as the transformation *associated* with the change of dx_μ according to the law of a contravariant vector.

I do not think it is too extravagant to claim that the method of the tensor calculus, which presents all physical equations in a form independent of the choice of measure-code, is the only possible means of studying the conditions of the world which are at the basis of physical phenomena. The physicist is accustomed to insist (sometimes quite unnecessarily) that all equations should be stated in a form independent of the units employed. Whether this is desirable depends on the purpose of the formulae. But whatever additional insight into underlying causes is gained by stating equations in a form independent of units, must be gained to a far greater degree by stating them in a form altogether independent of the measure-code. An equation of this general form is called a *tensor equation.*

When the physicist is attacking the everyday problems of his subject, he may use any form of the equations—any specialised measure-plan—which will shorten the labour of calculation; for in these problems he is concerned with the outward significance rather than the inward significance of his

* The relation of *aspect* (or in its graphical conception *displacement*) with four measure-numbers seems to be derived from the relation of *interval* with one measure-number, by taking account not only of the mutual interval between the two events but also of their intervals from all surrounding events.

formulae. But once in a while he turns to consider their inward significance —to consider that relation of things in the world-structure which is the origin of his formulae. The only intelligible idea we can form of such a structural relation is that it exists between the world-conditions themselves and not between the measure-numbers of a particular code. A law of nature resolves itself into a constant relation, or even an identity, of the two world-conditions to which the different classes of observed quantities forming the two sides of the equation are traceable. Such a constant relation independent of measure-code is only to be expressed by a tensor equation.

It may be remarked that if we take a force (X, Y, Z) and transform it to polar coordinates, whether as a covariant or a contravariant vector, in neither case do we obtain the quantities called polar components in elementary mechanics. The latter are not in our view the true polar components; they are merely rectangular components in three new directions, viz. radial and transverse. In general the elementary definitions of physical quantities do not contemplate other than rectangular components, and they may need to be supplemented before we can decide whether the physical vector is covariant or contravariant. Thus if we define force as "mass × acceleration," the force turns out to be contravariant; but if we define it by "work = force × displacement," the force is covariant. With the latter definition, however, we have to abandon the method of resolution into *oblique* components adopted in elementary mechanics.

In what follows it is generally sufficient to confine attention to the mathematical notion of a vector. Some idea of the physical notion will probably give greater insight, but is not necessary for the formal proofs.

22. The summation convention.

We shall adopt the convention that whenever a literal suffix appears twice in a term that term is to be summed for values of the suffix 1, 2, 3, 4. For example, (2·1) will be written

$$ds^2 = g_{\mu\nu} dx_\mu dx_\nu \quad (g_{\nu\mu} = g_{\mu\nu}) \quad \text{.................(22·1).}$$

Here, since μ and ν each appear twice, the summation

$$\sum_{\mu=1}^{4} \sum_{\nu=1}^{4}$$

is indicated; and the result written out in full gives (2·1).

Again, in the equation

$$A_\mu' = \frac{\partial x_\alpha}{\partial x_\mu'} A_\alpha,$$

the summation on the right is with respect to α only (μ appearing only once). The equation is equivalent to (19·2).

The convention is not merely an abbreviation but an immense aid to the analysis, giving it an impetus which is nearly always in a profitable direction. Summations occur in our investigations without waiting for our tardy approval.

A useful rule may be noted—

Any literal suffix appearing twice in a term is a dummy suffix, which may be changed freely to any other letter not already appropriated in that term. Two or more dummy suffixes can be interchanged*. For example

$$g_{\alpha\beta}\frac{\partial^2 x_\alpha}{\partial x_\mu'\partial x_\nu'}\frac{\partial x_\beta}{\partial x_\lambda'}=g_{\alpha\beta}\frac{\partial^2 x_\beta}{\partial x_\mu'\partial x_\nu'}\frac{\partial x_\alpha}{\partial x_\lambda'}\quad\text{.................(22·2)}$$

by interchanging the dummy suffixes α and β, remembering that $g_{\beta\alpha}=g_{\alpha\beta}$.

For a further illustration we shall prove that

$$\left.\begin{aligned}\frac{\partial x_\mu}{\partial x_\alpha'}\frac{\partial x_\alpha'}{\partial x_\nu}=\frac{dx_\mu}{dx_\nu}&=0,\quad\text{if }\mu\neq\nu\\&=1,\quad\text{if }\mu=\nu\end{aligned}\right\}\quad\text{.................(22·3).}$$

The left-hand side written in full is

$$\frac{\partial x_\mu}{\partial x_1'}\frac{\partial x_1'}{\partial x_\nu}+\frac{\partial x_\mu}{\partial x_2'}\frac{\partial x_2'}{\partial x_\nu}+\frac{\partial x_\mu}{\partial x_3'}\frac{\partial x_3'}{\partial x_\nu}+\frac{\partial x_\mu}{\partial x_4'}\frac{\partial x_4'}{\partial x_\nu},$$

which by the usual theory gives the change dx_μ consequent on a change dx_ν. But x_μ and x_ν are coordinates of the same system, so that their variations are independent; hence dx_μ is zero unless x_μ and x_ν are the same coordinate, in which case, of course, $dx_\mu=dx_\nu$. Thus the theorem is proved.

The multiplier $\dfrac{\partial x_\mu}{\partial x_\alpha'}\dfrac{\partial x_\alpha'}{\partial x_\nu}$ acts as a *substitution-operator*. That is to say if $A(\mu)$ is any expression involving the suffix μ

$$\frac{\partial x_\mu}{\partial x_\alpha'}\frac{\partial x_\alpha'}{\partial x_\nu}A(\mu)=A(\nu)\quad\text{.......................(22·4).}$$

For on the left the summation with respect to μ gives four terms corresponding to the values 1, 2, 3, 4 of μ. One of these values will agree with ν. Denote the other three values by σ, τ, ρ. Then by (22·3) the result is

$$\begin{aligned}&1\,.\,A(\nu)+0\,.\,A(\sigma)+0\,.\,A(\tau)+0\,.\,A(\rho)\\=\;&A(\nu).\end{aligned}$$

The multiplier accordingly has the effect of substituting ν for μ in the multiplicand.

23. Tensors.

The two laws of transformation given in § 19 are now written—

Contravariant vectors $$A'^\mu=\frac{\partial x_\mu'}{\partial x_\alpha}A^\alpha\quad\text{..........................(23·11).}$$

Covariant vectors $$A_\mu'=\frac{\partial x_\alpha}{\partial x_\mu'}A_\alpha\quad\text{.........................(23·12).}$$

We can denote by $A_{\mu\nu}$ a quantity with 16 components obtained by giving μ and ν the values from 1 to 4 independently. Similarly $A_{\mu\nu\sigma}$ has 64 com-

* At first we shall call attention to such changes when we employ them; but the reader will be expected gradually to become familiar with the device as a common process of manipulation.

ponents. By a generalisation of the foregoing transformation laws we classify quantities of this kind as follows—

Contravariant tensors $$A'^{\mu\nu} = \frac{\partial x_\mu'}{\partial x_\alpha} \frac{\partial x_\nu'}{\partial x_\beta} A^{\alpha\beta} \quad\quad (23{\cdot}21).$$

Covariant tensors $$A'_{\mu\nu} = \frac{\partial x_\alpha}{\partial x_\mu'} \frac{\partial x_\beta}{\partial x_\nu'} A_{\alpha\beta} \quad\quad (23{\cdot}22).$$

Mixed tensors $$A'^{\nu}_{\mu} = \frac{\partial x_\alpha}{\partial x_\mu'} \frac{\partial x_\nu'}{\partial x_\beta} A^{\beta}_{\alpha} \quad\quad (23{\cdot}23).$$

The above are called tensors of the second rank. We have similar laws for tensors of higher ranks. E.g.

$$A'^{\tau}_{\mu\nu\sigma} = \frac{\partial x_\alpha}{\partial x_\mu'} \frac{\partial x_\beta}{\partial x_\nu'} \frac{\partial x_\gamma}{\partial x_\sigma'} \frac{\partial x_\tau'}{\partial x_\delta} A^{\delta}_{\alpha\beta\gamma} \quad\quad (23{\cdot}3).$$

It may be worth while to remind the reader that (23·3) typifies 256 distinct equations each with a sum of 256 terms on the right-hand side.

It is easily shown that these transformation laws fulfil the condition of self-consistency explained in § 20, and it is for this reason that quantities governed by them are selected for special nomenclature.

If a tensor vanishes, i.e. if all its components vanish, in one system of coordinates, it will continue to vanish when any other system of coordinates is substituted. This is clear from the linearity of the above transformation laws.

Evidently the sum of two tensors of the same covariant or contravariant character is a tensor. Hence a law expressed by the vanishing of the sum of a number of tensors, or by the equality of two tensors of the same kind, will be independent of the coordinate-system used.

The product of two tensors such as $A_{\mu\nu}$ and B^{τ}_{σ} is a tensor of the character indicated by $A^{\tau}_{\mu\nu\sigma}$. This is proved by showing that the transformation law of the product is the same as (23·3).

The general term *tensor* includes vectors (tensors of the first rank) and invariants or scalars* (tensors of zero rank).

A tensor of the second or higher rank need not be expressible as a product of two tensors of lower rank.

A simple example of an expression of the second rank is afforded by the stresses in a solid or viscous fluid. The component of stress denoted by p_{xy} is the traction in the y-direction across an interface perpendicular to the x-direction. Each component is thus associated with two directions.

24. Inner multiplication and contraction. The quotient law.

If we multiply A_μ by B^ν we obtain sixteen quantities $A_1B^1, A_1B^2, A_2B^1, \ldots$ constituting a mixed tensor. Suppose that we wish to consider the four

* Scalar is a synonym for invariant. I generally use the latter word as the more self-explanatory.

"diagonal" terms A_1B^1, A_2B^2, A_3B^3, A_4B^4; we naturally try to abbreviate these by writing them $A_\mu B^\mu$. But by the summation convention $A_\mu B^\mu$ stands for the sum of the four quantities. The convention is right. We have no use for them individually since they do not form a vector; but the sum is of great importance.

$A_\mu B^\mu$ is called the *inner product* of the two vectors, in contrast to the ordinary or *outer product* $A_\mu B^\nu$.

In rectangular coordinates the inner product coincides with the *scalar-product* defined in the well-known elementary theory of vectors; but the outer product is not the so-called *vector-product* of the elementary theory.

By a similar process we can form from any mixed tensor $A^\tau_{\mu\nu\sigma}$ a "contracted*" tensor $A^\sigma_{\mu\nu\sigma}$, which is two ranks lower since σ has now become a dummy suffix. To prove that $A^\sigma_{\mu\nu\sigma}$ is a tensor, we set $\tau = \sigma$ in (23·3),

$$A'^\sigma_{\mu\nu\sigma} = \frac{\partial x_\alpha}{\partial x_\mu'} \frac{\partial x_\beta}{\partial x_\nu'} \frac{\partial x_\gamma}{\partial x_\sigma'} \frac{\partial x_\sigma'}{\partial x_\delta} A^\delta_{\alpha\beta\gamma}.$$

The substitution operator $\dfrac{\partial x_\gamma}{\partial x_\sigma'} \dfrac{\partial x_\sigma'}{\partial x_\delta}$ changes δ to γ in $A^\delta_{\alpha\beta\gamma}$ by (22·4). Hence

$$A'^\sigma_{\mu\nu\sigma} = \frac{\partial x_\alpha}{\partial x_\mu'} \frac{\partial x_\beta}{\partial x_\nu'} A^\gamma_{\alpha\beta\gamma}.$$

Comparing with the transformation law (23·22) we see that $A^\sigma_{\mu\nu\sigma}$ is a covariant tensor of the second rank. Of course, the dummy suffixes γ and σ are equivalent.

Similarly, setting $\nu = \mu$ in (23·23),

$$A'^\mu_\mu = \frac{\partial x_\alpha}{\partial x_\mu'} \frac{\partial x_\mu'}{\partial x_\beta} A^\beta_\alpha = A^\alpha_\alpha = A^\mu_\mu,$$

that is to say A^μ_μ is unaltered by a transformation of coordinates. Hence it is an invariant.

By the same method we can show that $A_\mu B^\mu$, $A^{\mu\nu}_{\mu\nu}$, $A^\nu_\mu B^\mu_\nu$ are invariants. In general when an upper and lower suffix are the same the corresponding covariant and contravariant qualities cancel out. If all suffixes cancel out in this way, the expression must be invariant. The identified suffixes must be of opposite characters; the expression $A^\tau_{\mu\sigma\sigma}$ is not a tensor, and no interest is attached to it.

We see that the suffixes keep a tally of what we have called the generalised dimensions of the terms in our equations. After cancelling out any suffixes which appear in both upper and lower positions, the remaining suffixes must appear in the same position in each term of an equation. When that is satisfied each term will undergo the same set of operations when a transformation of coordinates is made, and the equation will continue to hold in all

* German, *verjüngt*.

systems of coordinates. This may be compared with the well-known condition that each term must have the same physical dimensions, so that it undergoes multiplication by the same factor when a change of units is made and the equation continues to hold in all systems of units.

Just as we can infer the physical dimensions of some novel entity entering into a physical equation, so we can infer the contravariant and covariant dimensions of an expression whose character was hitherto unknown. For example, if the equation is

$$A(\mu\nu)\, B_{\nu\sigma} = C_{\mu\sigma} \qquad \text{.........................(24·1)},$$

where the nature of $A(\mu\nu)$ is not known initially, we see that $A(\mu\nu)$ must be a tensor of the character A_μ^ν, so as to give

$$A_\mu^\nu B_{\nu\sigma} = C_{\mu\sigma},$$

which makes the covariant dimensions on both sides consistent.

The equation (24·1) may be written symbolically

$$A(\mu\nu) = C_{\mu\sigma}/B_{\nu\sigma},$$

and the conclusion is that not only the product but also the (symbolic) quotient of two tensors is a tensor. Of course, the operation here indicated is not that of ordinary division.

This quotient law is a useful aid in detecting the tensor-character of expressions. It is not claimed that the general argument here given amounts to a strict mathematical proof. In most cases we can supply the proof required by one or more applications of the following rigorous theorem—

A quantity which on inner multiplication by *any* covariant (alternatively, by *any* contravariant) vector always gives a tensor, is itself a tensor.

For suppose that $$A(\mu\nu)\, B^\nu$$

is always a covariant vector for any choice of the contravariant vector B^ν. Then by (23·12)

$$\{A'(\mu\nu)\, B'^\nu\} = \frac{\partial x_\alpha}{\partial x_\mu'} \{A(\alpha\beta)\, B^\beta\} \qquad \text{................(24·2)}.$$

But by (23·11) applied to the reverse transformation from accented to unaccented coordinates

$$B^\beta = \frac{\partial x_\beta}{\partial x_\nu'} B'^\nu.$$

Hence, substituting for B^β in (24·2),

$$B'^\nu \left(A'(\mu\nu) - \frac{\partial x_\alpha}{\partial x_\mu'} \frac{\partial x_\beta}{\partial x_\nu'} A(\alpha\beta) \right) = 0.$$

Since B'^ν is arbitrary the quantity in the bracket must vanish. This shows that $A(\mu\nu)$ is a covariant tensor obeying the transformation law (23·22).

We shall cite this theorem as the "rigorous quotient theorem."

25. The fundamental tensors.

It is convenient to write (22·1) as

$$ds^2 = g_{\mu\nu}(dx)^\mu (dx)^\nu$$

in order to show explicitly the contravariant character of $dx_\mu = (dx)^\mu$. Since ds^2 is independent of the coordinate-system it is an invariant or tensor of zero rank. The equation shows that $g_{\mu\nu}(dx)^\mu$ multiplied by an arbitrarily chosen contravariant vector $(dx)^\nu$ always gives a tensor of zero rank; hence $g_{\mu\nu}(dx)^\mu$ is a vector. Again, we see that $g_{\mu\nu}$ multiplied by an arbitrary contravariant vector $(dx)^\mu$ always gives a vector; hence $g_{\mu\nu}$ is a tensor. This double application of the rigorous quotient theorem shows that $g_{\mu\nu}$ is a tensor; and it is evidently covariant as the notation has anticipated.

Let g stand for the determinant

$$\begin{vmatrix} g_{11} & g_{12} & g_{13} & g_{14} \\ g_{21} & g_{22} & g_{23} & g_{24} \\ g_{31} & g_{32} & g_{33} & g_{34} \\ g_{41} & g_{42} & g_{43} & g_{44} \end{vmatrix}.$$

Let $g^{\mu\nu}$ be defined as the minor of $g_{\mu\nu}$ in this determinant, divided by g *.

Consider the inner product $g_{\mu\sigma}g^{\nu\sigma}$. We see that μ and ν select two rows in the determinant; we have to take each element in turn from the μ row, multiply by the minor of the corresponding element of the ν row, add together, and divide the result by g. This is equivalent to substituting the μ row for the ν row and dividing the resulting determinant by g. If μ is not the same as ν this gives a determinant with two rows identical, and the result is 0. If μ is the same as ν we reproduce the determinant g divided by itself, and the result is 1. We write

$$\left.\begin{aligned} g_\mu^\nu &= g_{\mu\sigma}g^{\nu\sigma} \\ &= 0 \quad \text{if } \mu \neq \nu \\ &= 1 \quad \text{if } \mu = \nu \end{aligned}\right\} \quad \text{.........................(25·1).}$$

Thus g_μ^ν has the same property of a substitution-operator that we found for $\dfrac{\partial x_\mu}{\partial x_\alpha'}\dfrac{\partial x_\alpha'}{\partial x_\nu}$ in (22·4). For example†,

$$g_\mu^\nu A^\mu = A^\nu + 0 + 0 + 0 \quad \text{.....................(25·2).}$$

Note that g_ν^ν has not the same meaning as g_μ^ν with $\mu = \nu$, because a summation is implied. Evidently

$$g_\nu^\nu = 1 + 1 + 1 + 1 = 4 \text{.....................(25·3).}$$

The equation (25·2) shows that g_μ^ν multiplied by any contravariant vector always gives a vector. Hence g_μ^ν is a tensor. It is a very exceptional tensor since its components are the same in all coordinate-systems.

* The notation anticipates the result proved later that $g^{\mu\nu}$ is a contravariant tensor.

† Note that g_μ^ν will act as a substitution-operator on *any* expression and is not restricted to operating on tensors.

Again since $g_{\mu\sigma}g^{\nu\sigma}$ is a tensor we can infer that $g^{\nu\sigma}$ is a tensor. This is proved rigorously by remarking that $g_{\mu\sigma}A^\mu$ is a covariant vector, arbitrary on account of the free choice of A^μ. Multiplying this vector by $g^{\nu\sigma}$ we have

$$g_{\mu\sigma}g^{\nu\sigma}A^\mu = g^\nu_\mu A^\mu = A^\nu,$$

so that the product is always a vector. Hence the rigorous quotient theorem applies.

The tensor character of $g^{\mu\nu}$ may also be demonstrated by a method which shows more clearly the reason for its definition as the minor of $g_{\mu\nu}$ divided by g. Since $g_{\mu\nu}A^\nu$ is a covariant vector, we can denote it by B_μ. Thus

$$g_{11}A^1 + g_{12}A^2 + g_{13}A^3 + g_{14}A^4 = B_1;\ \text{etc.}$$

Solving these four linear equations for A^1, A^2, A^3, A^4 by the usual method of determinants, the result is

$$A^1 = g^{11}B_1 + g^{12}B_2 + g^{13}B_3 + g^{14}B_4;\ \text{etc.},$$

so that

$$A^\mu = g^{\mu\nu}B_\nu.$$

Whence by the rigorous quotient theorem $g^{\mu\nu}$ is a tensor.

We have thus defined three fundamental tensors

$$g_{\mu\nu},\quad g^\nu_\mu,\quad g^{\mu\nu}$$

of covariant, mixed, and contravariant characters respectively.

26. Associated tensors.

We now define the operation of raising or lowering a suffix. Raising the suffix of a vector is defined by the equation

$$A^\mu = g^{\mu\nu}A_\nu,$$

and lowering by the equation

$$A_\mu = g_{\mu\nu}A^\nu.$$

For a more general tensor such as $A^{\gamma\delta}_{\alpha\beta\mu}$, the operation of raising μ is defined in the same way, viz.

$$A^{\gamma\delta\mu}_{\alpha\beta} = g^{\mu\nu}A^{\gamma\delta}_{\alpha\beta\nu} \quad \text{.........................(26·1)},$$

and for lowering

$$A^{\gamma\delta}_{\alpha\beta\mu} = g_{\mu\nu}A^{\gamma\delta\nu}_{\alpha\beta} \quad \text{.........................(26·2)}.$$

These definitions are consistent, since if we raise a suffix and then lower it we reproduce the original tensor. Thus if in (26·1) we multiply by $g_{\mu\sigma}$ in order to lower the suffix on the left, we have

$$\begin{aligned} g_{\mu\sigma}A^{\gamma\delta\mu}_{\alpha\beta} &= g_{\mu\sigma}g^{\mu\nu}A^{\gamma\delta}_{\alpha\beta\nu} \\ &= g^\nu_\sigma A^{\gamma\delta}_{\alpha\beta\nu} \\ &= A^{\gamma\delta}_{\alpha\beta\sigma} \quad \text{by (25·2)}, \end{aligned}$$

which is the rule expressed by (26·2).

It will be noticed that the raising of a suffix ν by means of $g^{\mu\nu}$ is accompanied by the substitution of μ for ν. The whole operation is closely akin to the plain substitution of μ for ν by means of g_μ^ν. Thus

multiplication by $g^{\mu\nu}$ gives substitution with raising,
multiplication by g_μ^ν gives plain substitution,
multiplication by $g_{\mu\nu}$ gives substitution with lowering.

In the case of non-symmetrical tensors it may be necessary to distinguish the place from which the raised suffix has been brought, e.g. to distinguish between $A_\mu{}^\nu$ and $A^\nu{}_\mu$.

It is easily seen that this rule of association between tensors with suffixes in different positions is fulfilled in the case of $g^{\mu\nu}$, g_μ^ν, $g_{\mu\nu}$; in fact the definition of g_μ^ν in (25·1) is a special case of (26·1).

For rectangular coordinates the raising or lowering of a suffix leaves the components unaltered in three-dimensional space*; and it merely reverses the signs of some of the components for Galilean coordinates in four-dimensional space-time. Since the elementary definitions of physical quantities refer to rectangular axes and time, we can generally use any one of the associated tensors to represent a physical entity without infringing pre-relativity definitions. This leads to a somewhat enlarged view of a tensor as having in itself no particular covariant or contravariant character, but having *components* of various degrees of covariance or contravariance represented by the whole system of associated tensors. That is to say, the raising or lowering of suffixes will not be regarded as altering the individuality of the tensor; and reference to a tensor $A_{\mu\nu}$ may (if the context permits) be taken to include the associated tensors A_μ^ν and $A^{\mu\nu}$.

It is useful to notice that dummy suffixes have a certain freedom of movement between the tensor-factors of an expression. Thus

$$A_{\alpha\beta}B^{\alpha\beta} = A^{\alpha\beta}B_{\alpha\beta}, \quad A_{\mu\alpha}B^{\nu\alpha} = A_\mu{}^\alpha B^\nu{}_\alpha \quad \text{.............. (26·3)}.$$

The suffix may be raised in one term provided it is lowered in the other. The proof follows easily from (26·1) and (26·2).

In the elementary vector theory two vectors are said to be *perpendicular* if their scalar-product vanishes; and the square of the *length* of the vector is its scalar-product into itself. Corresponding definitions are adopted in the tensor calculus.

The vectors A_μ and B_μ are said to be *perpendicular* if

$$A_\mu B^\mu = 0 \quad \text{.............................. (26·4)}.$$

If l is the *length* of A_μ (or A^μ)

$$l^2 = A_\mu A^\mu \quad \text{.............................. (26·5)}.$$

A vector is self-perpendicular if its length vanishes.

* If $ds^2 = dx_1^2 + dx_2^2 + dx_3^2$, $g_{\mu\nu} = g^{\mu\nu} = g_\mu^\nu$ so that all three tensors are merely substitution-operators.

The interval is the length of the corresponding displacement dx_μ because

$$ds^2 = g_{\mu\nu}(dx)^\mu . (dx)^\nu$$
$$= (dx)_\nu (dx)^\nu$$

by (26·2). A displacement is thus self-perpendicular when it is along a light-track, $ds = 0$.

If a vector A_μ receives an infinitesimal increment dA_μ perpendicular to itself, its length is unaltered to the first order; for by (26·5)

$$(l + dl)^2 = (A_\mu + dA_\mu)(A^\mu + dA^\mu)$$
$$= A_\mu A^\mu + A^\mu dA_\mu + A_\mu dA^\mu \quad \text{to the first order}$$
$$= l^2 + 2A_\mu dA^\mu \quad \text{by (26·3)},$$

and $A_\mu dA^\mu = 0$ by the condition of perpendicularity (26·4).

In the elementary vector theory, the scalar-product of two vectors is equal to the product of their lengths multiplied by the cosine of the angle between them. Accordingly in the general theory the angle θ between two vectors A_μ and B_μ is defined by

$$\cos\theta = \frac{A_\mu B^\mu}{\sqrt{(A_\alpha A^\alpha)(B_\beta B^\beta)}} \quad \text{....................(26·6).}$$

Clearly the angle so defined is an invariant, and agrees with the usual definition when the coordinates are rectangular. In determining the angle between two intersecting lines it makes no difference whether the world is curved or flat, since only the initial directions are concerned and these in any case lie in the tangent plane. The angle θ (if it is real) has thus the usual geometrical meaning even in non-Euclidean space. It must not, however, be inferred that ordinary angles are invariant for the Lorentz transformation; naturally an angle in three dimensions is invariant only for transformations in three dimensions, and the angle which is invariant for Lorentz transformations is a four-dimensional angle.

From a tensor of even rank we can construct an invariant by bringing half the suffixes to the upper and half to the lower position and contracting. Thus from $A_{\mu\nu\sigma\tau}$ we form $A^{\sigma\tau}_{\mu\nu}$ and contract, obtaining $A = A^{\mu\nu}_{\mu\nu}$. This invariant will be called the *spur**. Another invariant is the square of the length $A_{\mu\nu\sigma\tau}A^{\mu\nu\sigma\tau}$. There may also be intermediate invariants such as $A^{\alpha}_{\mu\nu\alpha}A^{\mu\nu\beta}_{\beta}$.

27. Christoffel's 3-index symbols.

We introduce two expressions (not tensors) of great importance throughout our subsequent work, namely

$$[\mu\nu, \sigma] = \tfrac{1}{2}\left(\frac{\partial g_{\mu\sigma}}{\partial x_\nu} + \frac{\partial g_{\nu\sigma}}{\partial x_\mu} - \frac{\partial g_{\mu\nu}}{\partial x_\sigma}\right) \quad \text{....................(27·1),}$$

$$\{\mu\nu, \sigma\} = \tfrac{1}{2}g^{\sigma\lambda}\left(\frac{\partial g_{\mu\lambda}}{\partial x_\nu} + \frac{\partial g_{\nu\lambda}}{\partial x_\mu} - \frac{\partial g_{\mu\nu}}{\partial x_\lambda}\right) \quad \text{..............(27·2).}$$

* Originally the German word *Spur*.

We have
$$\{\mu\nu, \sigma\} = g^{\sigma\lambda} [\mu\nu, \lambda] \quad \text{......................(27·3)},$$
$$[\mu\nu, \sigma] = g_{\sigma\lambda} \{\mu\nu, \lambda\} \quad \text{......................(27·4)}.$$
The result (27·3) is obvious from the definitions. To prove (27·4), multiply (27·3) by $g_{\sigma\alpha}$; then
$$\begin{aligned} g_{\sigma\alpha} \{\mu\nu, \sigma\} &= g_{\sigma\alpha} g^{\sigma\lambda} [\mu\nu, \lambda] \\ &= g_{\alpha}^{\lambda} [\mu\nu, \lambda] \\ &= [\mu\nu, \alpha], \end{aligned}$$
which is equivalent to (27·4).

Comparing with (26·1) and (26·2) we see that the passage from the "square" to the "curly" symbol, and *vice versa*, is the same process as raising and lowering a suffix. It might be convenient to use a notation in which this was made evident, e.g.
$$\Gamma_{\mu\nu,\sigma} = [\mu\nu, \sigma], \quad \Gamma_{\mu\nu}^{\sigma} = \{\mu\nu, \sigma\},$$
but we shall adhere to the more usual notation.

From (27·1) it is found that
$$[\mu\nu, \sigma] + [\sigma\nu, \mu] = \frac{\partial g_{\mu\sigma}}{\partial x_\nu} \quad \text{....................(27·5)}.$$

There are 40 different 3-index symbols of each kind. It may here be explained that the $g_{\mu\nu}$ are components of a generalised *potential*, and the 3-index symbols components of a generalised *force* in the gravitational theory (see § 55).

28. Equations of a geodesic.

We shall now determine the equations of a geodesic or path between two points for which
$$\int ds \text{ is stationary.}$$
This absolute track is of fundamental importance in dynamics, but at the moment we are concerned with it only as an aid in the development of the tensor calculus*.

Keeping the beginning and end of the path fixed, we give every intermediate point an arbitrary infinitesimal displacement δx_σ so as to deform the path. Since
$$\begin{aligned} ds^2 &= g_{\mu\nu} dx_\mu dx_\nu, \\ 2ds\,\delta(ds) &= dx_\mu dx_\nu \delta g_{\mu\nu} + g_{\mu\nu} dx_\mu \delta(dx_\nu) + g_{\mu\nu} dx_\nu \delta(dx_\mu) \\ &= dx_\mu dx_\nu \frac{\partial g_{\mu\nu}}{\partial x_\sigma} \delta x_\sigma + g_{\mu\nu} dx_\mu d(\delta x_\nu) + g_{\mu\nu} dx_\nu d(\delta x_\mu) \quad \text{...(28·1)}. \end{aligned}$$
The stationary condition is
$$\int \delta(ds) = 0 \quad \text{..........................(28·2)},$$

* Our ultimate goal is equation (29·3). An alternative proof (which does not introduce the calculus of variations) is given in § 31.

which becomes by (28·1)

$$\frac{1}{2}\int\left\{\frac{dx_\mu}{ds}\frac{dx_\nu}{ds}\frac{\partial g_{\mu\nu}}{\partial x_\sigma}\delta x_\sigma + g_{\mu\nu}\frac{dx_\mu}{ds}\frac{d}{ds}(\delta x_\nu) + g_{\mu\nu}\frac{dx_\nu}{ds}\frac{d}{ds}(\delta x_\mu)\right\}ds = 0,$$

or, changing dummy suffixes in the last two terms,

$$\frac{1}{2}\int\left\{\frac{dx_\mu}{ds}\frac{dx_\nu}{ds}\frac{\partial g_{\mu\nu}}{\partial x_\sigma}\delta x_\sigma + \left(g_{\mu\sigma}\frac{dx_\mu}{ds} + g_{\sigma\nu}\frac{dx_\nu}{ds}\right)\frac{d}{ds}(\delta x_\sigma)\right\}ds = 0.$$

Applying the usual method of partial integration, and rejecting the integrated part since δx_σ vanishes at both limits,

$$\frac{1}{2}\int\left\{\frac{dx_\mu}{ds}\frac{dx_\nu}{ds}\frac{\partial g_{\mu\nu}}{\partial x_\sigma} - \frac{d}{ds}\left(g_{\mu\sigma}\frac{dx_\mu}{ds} + g_{\sigma\nu}\frac{dx_\nu}{ds}\right)\right\}\delta x_\sigma ds = 0.$$

This must hold for all values of the arbitrary displacements δx_σ at all points, hence the coefficient in the integrand must vanish at all points on the path. Thus

$$\frac{1}{2}\frac{dx_\mu}{ds}\frac{dx_\nu}{ds}\frac{\partial g_{\mu\nu}}{\partial x_\sigma} - \frac{1}{2}\frac{dg_{\mu\sigma}}{ds}\frac{dx_\mu}{ds} - \frac{1}{2}\frac{dg_{\sigma\nu}}{ds}\frac{dx_\nu}{ds} - \frac{1}{2}g_{\mu\sigma}\frac{d^2x_\mu}{ds^2} - \frac{1}{2}g_{\sigma\nu}\frac{d^2x_\nu}{ds^2} = 0.$$

Now*
$$\frac{dg_{\mu\sigma}}{ds} = \frac{\partial g_{\mu\sigma}}{\partial x_\nu}\frac{dx_\nu}{ds} \quad\text{and}\quad \frac{dg_{\sigma\nu}}{ds} = \frac{\partial g_{\sigma\nu}}{\partial x_\mu}\frac{dx_\mu}{ds}.$$

Also in the last two terms we replace the dummy suffixes μ and ν by ϵ. The equation then becomes

$$\frac{1}{2}\frac{dx_\mu}{ds}\frac{dx_\nu}{ds}\left(\frac{\partial g_{\mu\nu}}{\partial x_\sigma} - \frac{\partial g_{\mu\sigma}}{\partial x_\nu} - \frac{\partial g_{\nu\sigma}}{\partial x_\mu}\right) - g_{\epsilon\sigma}\frac{d^2x_\epsilon}{ds^2} = 0 \quad\text{............(28·3).}$$

We can get rid of the factor $g_{\epsilon\sigma}$ by multiplying through by $g^{\sigma\alpha}$ so as to form the substitution operator g_ϵ^α. Thus

$$\frac{1}{2}\frac{dx_\mu}{ds}\frac{dx_\nu}{ds}g^{\sigma\alpha}\left(\frac{\partial g_{\mu\sigma}}{\partial x_\nu} + \frac{\partial g_{\nu\sigma}}{\partial x_\mu} - \frac{\partial g_{\mu\nu}}{\partial x_\sigma}\right) + \frac{d^2x_\alpha}{ds^2} = 0 \quad\text{.........(28·4),}$$

or, by (27·2)
$$\frac{d^2x_\alpha}{ds^2} + \{\mu\nu, \alpha\}\frac{dx_\mu}{ds}\frac{dx_\nu}{ds} = 0 \quad\text{.....................(28·5).}$$

For $\alpha = 1, 2, 3, 4$ this gives the four equations determining a geodesic.

29. Covariant derivative of a vector.

The derivative of an invariant is a covariant vector (§ 19), but the derivative of a vector is not a tensor. We proceed to find certain tensors which are used in this calculus in place of the ordinary derivatives of vectors.

Since dx_μ is contravariant and ds invariant, a "velocity" dx_μ/ds is a contravariant vector. Hence if A_μ is any covariant vector the inner product

$$A_\mu\frac{dx_\mu}{ds} \text{ is invariant.}$$

* These simple formulae are noteworthy as illustrating the great value of the summation convention. The law of total differentiation for four coordinates becomes *formally* the same as for one coordinate.

The rate of change of this expression per unit interval along any assigned curve must also be independent of the coordinate-system, i.e.

$$\frac{d}{ds}\left(A_\mu \frac{dx_\mu}{ds}\right) \text{ is invariant} \quad\quad (29\cdot1).$$

This assumes that we keep to the same absolute curve however the coordinate-system is varied. The result (29·1) is therefore only of practical use if it is applied to a curve which is defined independently of the coordinate-system. We shall accordingly apply it to a geodesic. Performing the differentiation,

$$\frac{\partial A_\mu}{\partial x_\nu}\frac{dx_\nu}{ds}\cdot\frac{dx_\mu}{ds} + A_\mu \frac{d^2 x_\mu}{ds^2} \text{ is invariant along a geodesic} \quad\quad (29\cdot2).$$

From (28·5) we have that along a geodesic

$$A_\mu \frac{d^2 x_\mu}{ds^2} = A_\alpha \frac{d^2 x_\alpha}{ds^2} = -A_\alpha \{\mu\nu, \alpha\} \frac{dx_\mu}{ds}\frac{dx_\nu}{ds}.$$

Hence (29·2) gives

$$\frac{dx_\mu}{ds}\frac{dx_\nu}{ds}\left(\frac{\partial A_\mu}{\partial x_\nu} - A_\alpha \{\mu\nu, \alpha\}\right) \text{ is invariant.}$$

The result is now general since the curvature (which distinguishes the geodesic) has been eliminated by using the equations (28·5) and only the gradient of the curve (dx_μ/ds and dx_ν/ds) has been left in the expression.

Since dx_μ/ds and dx_ν/ds are contravariant vectors, their co-factor is a covariant tensor of the second rank. We therefore write

$$A_{\mu\nu} = \frac{\partial A_\mu}{\partial x_\nu} - \{\mu\nu, \alpha\} A_\alpha, \quad\quad (29\cdot3),$$

and the tensor $A_{\mu\nu}$ is called the *covariant derivative* of A_μ.

By raising a suffix we obtain two associated tensors $A^\mu{}_\nu$ and $A_\mu{}^\nu$ which must be distinguished since the two suffixes are not symmetrical. The first of these is the most important, and is to be understood when the tensor is written simply as A^μ_ν without distinction of original position.

Since $$A_\sigma = g_{\sigma\epsilon} A^\epsilon,$$

we have by (29·3)

$$\begin{aligned} A_{\sigma\nu} &= \frac{\partial}{\partial x_\nu}(g_{\sigma\epsilon} A^\epsilon) - \{\sigma\nu, \alpha\}(g_{\alpha\epsilon} A^\epsilon) \\ &= g_{\sigma\epsilon}\frac{\partial A^\epsilon}{\partial x_\nu} + A^\epsilon \frac{\partial g_{\sigma\epsilon}}{\partial x_\nu} - [\sigma\nu, \epsilon] A^\epsilon \quad \text{by (27·4)} \\ &= g_{\sigma\epsilon}\frac{\partial A^\epsilon}{\partial x_\nu} + [\epsilon\nu, \sigma] A^\epsilon \quad \text{by (27·5).} \end{aligned}$$

Hence multiplying through by $g^{\mu\sigma}$, and remembering that $g^{\mu\sigma} g_{\sigma\epsilon}$ is a substitution-operator, we have

$$A^\mu{}_\nu = \frac{\partial A^\mu}{\partial x_\nu} + \{\epsilon\nu, \mu\} A^\epsilon \quad\quad (29\cdot4).$$

This is called the covariant derivative of A^μ. The considerable differences between the formulae (29·3) and (29·4) should be carefully noted.

The tensors $A_\mu{}^\nu$ and $A^{\mu\nu}$, obtained from (29·3) and (29·4) by raising the second suffix, are called the *contravariant derivatives* of A_μ and A^μ. We shall not have much occasion to refer to contravariant derivatives.

30. Covariant derivative of a tensor.

The covariant derivatives of tensors of the second rank are formed as follows—

$$A_\sigma^{\mu\nu} = \frac{\partial A^{\mu\nu}}{\partial x_\sigma} + \{\alpha\sigma, \mu\} A^{\alpha\nu} + \{\alpha\sigma, \nu\} A^{\mu\alpha} \quad \text{...........(30·1)},$$

$$A_{\mu\sigma}^\nu = \frac{\partial A_\mu^\nu}{\partial x_\sigma} - \{\mu\sigma, \alpha\} A_\alpha^\nu + \{\alpha\sigma, \nu\} A_\mu^\alpha \quad \text{..............(30·2)},$$

$$A_{\mu\nu\sigma} = \frac{\partial A_{\mu\nu}}{\partial x_\sigma} - \{\mu\sigma, \alpha\} A_{\alpha\nu} - \{\nu\sigma, \alpha\} A_{\mu\alpha} \quad \text{...........(30·3)}.$$

And the general rule for covariant differentiation with respect to x_σ is illustrated by the example

$$A_{\lambda\mu\nu\sigma}^\rho = \frac{\partial}{\partial x_\sigma} A_{\lambda\mu\nu}^\rho - \{\lambda\sigma, \alpha\} A_{\alpha\mu\nu}^\rho - \{\mu\sigma, \alpha\} A_{\lambda\alpha\nu}^\rho - \{\nu\sigma, \alpha\} A_{\lambda\mu\alpha}^\rho + \{\alpha\sigma, \rho\} A_{\lambda\mu\nu}^\alpha \quad \text{........(30·4)}.$$

The above formulae are primarily definitions; but we have to prove that the quantities on the right are actually tensors. This is done by an obvious generalisation of the method of the preceding section. Thus if in place of (29·1) we use

$$\frac{d}{ds}\left(A_{\mu\nu} \frac{dx_\mu}{ds} \frac{dx_\nu}{ds}\right) \text{ is invariant along a geodesic,}$$

we obtain

$$\frac{\partial A_{\mu\nu}}{\partial x_\sigma} \frac{dx_\sigma}{ds} \frac{dx_\mu}{ds} \frac{dx_\nu}{ds} + A_{\mu\nu} \frac{dx_\nu}{ds} \frac{d^2 x_\mu}{ds^2} + A_{\mu\nu} \frac{dx_\mu}{ds} \frac{d^2 x_\nu}{ds^2}.$$

Then substituting for the second derivatives from (28·5) the expression reduces to

$$A_{\mu\nu\sigma} \frac{dx_\mu}{ds} \frac{dx_\nu}{ds} \frac{dx_\sigma}{ds} \text{ is invariant,}$$

showing that $A_{\mu\nu\sigma}$ is a tensor.

The formulae (30·1) and (30·2) are obtained by raising the suffixes ν and μ, the details of the work being the same as in deducing (29·4) from (29·3).

Consider the expression

$$B_{\mu\sigma} C_\nu + B_\mu C_{\nu\sigma},$$

the σ denoting covariant differentiation. By (29·3) this is equal to

$$\left(\frac{\partial B_\mu}{\partial x_\sigma} - \{\mu\sigma, \alpha\} B_\alpha\right) C_\nu + B_\mu \left(\frac{\partial C_\nu}{\partial x_\sigma} - \{\nu\sigma, \alpha\} C_\alpha\right)$$

$$= \frac{\partial}{\partial x_\sigma} (B_\mu C_\nu) - \{\mu\sigma, \alpha\} (B_\alpha C_\nu) - \{\nu\sigma, \alpha\} (B_\mu C_\alpha).$$

But comparing with (30·3) we see that this is the covariant derivative of the tensor of the second rank $(B_\mu C_\nu)$. Hence

$$(B_\mu C_\nu)_\sigma = B_{\mu\sigma} C_\nu + B_\mu C_{\nu\sigma} \quad \text{.....................(30·5).}$$

Thus in covariant differentiation of a product the distributive rule used in ordinary differentiation holds good.

Applying (30·3) to the fundamental tensor, we have

$$\begin{aligned} g_{\mu\nu\sigma} &= \frac{\partial g_{\mu\nu}}{\partial x_\sigma} - \{\mu\sigma, \alpha\}\, g_{\alpha\nu} - \{\nu\sigma, \alpha\}\, g_{\mu\alpha} \\ &= \frac{\partial g_{\mu\nu}}{\partial x_\sigma} - [\mu\sigma, \nu] - [\nu\sigma, \mu] \\ &= 0 \quad \text{by (27·5).} \end{aligned}$$

Hence the covariant derivatives of the fundamental tensors vanish identically, and the fundamental tensors can be treated as *constants* in covariant differentiation. It is thus immaterial whether a suffix is raised before or after the differentiation, as our definitions have already postulated.

If I is an invariant, IA_μ is a covariant vector; hence its covariant derivative is

$$\begin{aligned} (IA_\mu)_\nu &= \frac{\partial}{\partial x_\nu}(IA_\mu) - \{\mu\nu, \alpha\}\, IA_\alpha \\ &= A_\mu \frac{\partial I}{\partial x_\nu} + IA_{\mu\nu}. \end{aligned}$$

But by the rule for differentiating a product (30·5)

$$(IA_\mu)_\nu = I_\nu A_\mu + IA_{\mu\nu},$$

so that

$$I_\nu = \frac{\partial I}{\partial x_\nu}.$$

Hence the covariant derivative of an invariant is the same as its ordinary derivative.

It is, of course, impossible to reserve the notation $A_{\mu\nu}$ exclusively for the covariant derivative of A_μ, and the concluding suffix does not denote differentiation unless expressly stated. In case of doubt we may indicate the covariant and contravariant derivatives by $(A_\mu)_\nu$ and $(A_\mu)^\nu$.

The utility of the covariant derivative arises largely from the fact that, when the $g_{\mu\nu}$ are constants, the 3-index symbols vanish and the covariant derivative reduces to the ordinary derivative. Now in general our physical equations have been stated for the case of Galilean coordinates in which the $g_{\mu\nu}$ are constants; and we may in Galilean equations replace the ordinary derivative by the covariant derivative without altering anything. This is a necessary step in reducing such equations to the general tensor form which holds true for all coordinate-systems.

As an illustration suppose we wish to find the general equation of propagation of a potential with the velocity of light. In Galilean coordinates the equation is of the well-known form

$$\Box\phi \equiv \frac{\partial^2\phi}{\partial t^2} - \frac{\partial^2\phi}{\partial x^2} - \frac{\partial^2\phi}{\partial y^2} - \frac{\partial^2\phi}{\partial z^2} = 0 \quad\text{.................(30·6)}.$$

The Galilean values of $g^{\mu\nu}$ are $g^{44} = 1$, $g^{11} = g^{22} = g^{33} = -1$, and the other components vanish. Hence (30·6) can be written

$$g^{\mu\nu}\frac{\partial^2\phi}{\partial x_\mu \partial x_\nu} = 0 \quad\text{...........................(30·65)}.$$

The potential ϕ being an invariant, its ordinary derivative is a covariant vector $\phi_\mu = \partial\phi/\partial x_\mu$; and since the coordinates are Galilean we may insert the covariant derivative $\phi_{\mu\nu}$ instead of $\partial\phi_\mu/\partial x_\nu$. Hence the equation becomes

$$g^{\mu\nu}\phi_{\mu\nu} = 0 \quad\text{..........................(30·7)}.$$

Up to this point Galilean coordinates are essential; but now, by examining the covariant dimensions of (30·7), we notice that the left-hand side is an invariant, and therefore its value is unchanged by any transformation of coordinates. Hence (30·7) holds for all coordinate-systems, if it holds for any. Using (29·3) we can write it more fully

$$g^{\mu\nu}\left(\frac{\partial^2\phi}{\partial x_\mu \partial x_\nu} - \{\mu\nu, \alpha\}\frac{\partial\phi}{\partial x_\alpha}\right) = 0 \quad\text{.................(30·8)}.$$

This formula may be used for transforming Laplace's equation into curvilinear coordinates, etc.

It must be remembered that a transformation of coordinates does not alter the kind of space. Thus if we know by experiment that a potential ϕ is propagated according to the law (30·6) in Galilean coordinates, it follows rigorously that it is propagated according to the law (30·8) in any system of coordinates in flat space-time; but it does not follow rigorously that it will be propagated according to (30·8) when an irreducible gravitational field is present which alters the kind of space-time. It is, however, a plausible suggestion that (30·8) may be the general law of propagation of ϕ in any kind of space-time; that is the suggestion which the principle of equivalence makes. Like all generalisations which are only tested experimentally in a particular case, it must be received with caution.

The operator $\Box$ will frequently be referred to. In general coordinates it is to be taken as defined by

$$\Box A_{\mu\nu\ldots} = g^{\alpha\beta}(A_{\mu\nu\ldots})_{\alpha\beta} \quad\text{........................(30·9)}.$$

Or we may write it in the form

$$\Box = ((\ldots)_\alpha)^\alpha,$$

i.e. we perform a covariant and contravariant differentiation and contract them.

SUMMARY OF RULES FOR COVARIANT DIFFERENTIATION.

1. To obtain the covariant derivative of any tensor $A^{::::}_{::::}$ with respect to x_σ, we take first the ordinary derivative

$$\frac{\partial}{\partial x_\sigma} A^{::::}_{::::};$$

and for *each* covariant suffix $A^{::::}_{..\mu.}$, we add a term

$$-\{\mu\sigma, \alpha\} A^{::::}_{..\alpha.};$$

and for *each* contravariant suffix $A^{..\mu.}_{::::}$, we add a term

$$+\{\alpha\sigma, \mu\} A^{..\alpha.}_{::::}.$$

2. The covariant derivative of a product is formed by covariant differentiation of each factor in turn, by the same rule as in ordinary differentiation.

3. The fundamental tensor $g_{\mu\nu}$ or $g^{\mu\nu}$ behaves as though it were a constant in covariant differentiation.

4. The covariant derivative of an invariant is its ordinary derivative.

5. In taking second, third or higher derivatives, the order of differentiation is not interchangeable*.

31. Alternative discussion of the covariant derivative.

By (23·22) $$g'_{\mu\nu} = \frac{\partial x_\alpha}{\partial x_\mu'} \frac{\partial x_\beta}{\partial x_\nu'} g_{\alpha\beta}.$$

Hence differentiating

$$\frac{\partial g'_{\mu\nu}}{\partial x_\lambda'} = g_{\alpha\beta} \left\{ \frac{\partial^2 x_\alpha}{\partial x_\lambda' \partial x_\mu'} \frac{\partial x_\beta}{\partial x_\nu'} + \frac{\partial^2 x_\alpha}{\partial x_\lambda' \partial x_\nu'} \frac{\partial x_\beta}{\partial x_\mu'} \right\} + \frac{\partial x_\alpha}{\partial x_\mu'} \frac{\partial x_\beta}{\partial x_\nu'} \frac{\partial x_\gamma}{\partial x_\lambda'} \frac{\partial g_{\alpha\beta}}{\partial x_\gamma} \quad \text{...(31·11)}.$$

Here we have used

$$\frac{\partial g_{\alpha\beta}}{\partial x_\lambda'} = \frac{\partial g_{\alpha\beta}}{\partial x_\gamma} \frac{\partial x_\gamma}{\partial x_\lambda'},$$

and further we have interchanged the dummy suffixes α and β in the second term in the bracket. Similarly

$$\frac{\partial g'_{\nu\lambda}}{\partial x_\mu'} = g_{\alpha\beta} \left\{ \frac{\partial^2 x_\alpha}{\partial x_\mu' \partial x_\nu'} \frac{\partial x_\beta}{\partial x_\lambda'} + \frac{\partial^2 x_\alpha}{\partial x_\mu' \partial x_\lambda'} \frac{\partial x_\beta}{\partial x_\nu'} \right\} + \frac{\partial x_\alpha}{\partial x_\mu'} \frac{\partial x_\beta}{\partial x_\nu'} \frac{\partial x_\gamma}{\partial x_\lambda'} \frac{\partial g_{\beta\gamma}}{\partial x_\alpha} \quad \text{... (31·12)},$$

$$\frac{\partial g'_{\mu\lambda}}{\partial x_\nu'} = g_{\alpha\beta} \left\{ \frac{\partial^2 x_\alpha}{\partial x_\nu' \partial x_\mu'} \frac{\partial x_\beta}{\partial x_\lambda'} + \frac{\partial^2 x_\alpha}{\partial x_\nu' \partial x_\lambda'} \frac{\partial x_\beta}{\partial x_\mu'} \right\} + \frac{\partial x_\alpha}{\partial x_\mu'} \frac{\partial x_\beta}{\partial x_\nu'} \frac{\partial x_\gamma}{\partial x_\lambda'} \frac{\partial g_{\alpha\gamma}}{\partial x_\beta} \quad \text{... (31·13)}.$$

Add (31·12) and (31·13) and subtract (31·11), we obtain by (27·1)

$$[\mu\nu, \lambda]' = g_{\alpha\beta} \frac{\partial^2 x_\alpha}{\partial x_\mu' \partial x_\nu'} \frac{\partial x_\beta}{\partial x_\lambda'} + \frac{\partial x_\alpha}{\partial x_\mu'} \frac{\partial x_\beta}{\partial x_\nu'} \frac{\partial x_\gamma}{\partial x_\lambda'} [\alpha\beta, \gamma] \quad \text{.........(31·2)}.$$

* This is inserted here for completeness; it is discussed later.

Multiply through by $g'^{\lambda\rho}\dfrac{\partial x_\epsilon}{\partial x_\rho'}$, we have by (27·3)

$$\{\mu\nu, \rho\}' \frac{\partial x_\epsilon}{\partial x_\rho'} = g_{\alpha\beta}\frac{\partial^2 x_\alpha}{\partial x_\mu' \partial x_\nu'} \cdot g'^{\lambda\rho}\frac{\partial x_\beta}{\partial x_\lambda'}\frac{\partial x_\epsilon}{\partial x_\rho'} + g'^{\lambda\rho}\frac{\partial x_\gamma}{\partial x_\lambda'}\frac{\partial x_\epsilon}{\partial x_\rho'} \cdot \frac{\partial x_\alpha}{\partial x_\mu'}\frac{\partial x_\beta}{\partial x_\nu'}[\alpha\beta, \gamma]$$

$$= g_{\alpha\beta}g^{\beta\epsilon}\frac{\partial^2 x_\alpha}{\partial x_\mu' \partial x_\nu'} + \frac{\partial x_\alpha}{\partial x_\mu'}\frac{\partial x_\beta}{\partial x_\nu'}g^{\gamma\epsilon}[\alpha\beta, \gamma] \qquad \text{by (23·21)}$$

$$= \frac{\partial^2 x_\epsilon}{\partial x_\mu' \partial x_\nu'} + \frac{\partial x_\alpha}{\partial x_\mu'}\frac{\partial x_\beta}{\partial x_\nu'}\{\alpha\beta, \epsilon\} \qquad \text{.........(31·3)},$$

a formula which determines the second derivative $\partial^2 x_\epsilon/\partial x_\mu' \partial x_\nu'$ in terms of the first derivatives.

By (23·12)

$$A_\mu' = \frac{\partial x_\epsilon}{\partial x_\mu'}A_\epsilon \qquad \text{.........(31·4)}.$$

Hence differentiating

$$\frac{\partial A_\mu'}{\partial x_\nu'} = \frac{\partial^2 x_\epsilon}{\partial x_\mu' \partial x_\nu'}A_\epsilon + \frac{\partial x_\epsilon}{\partial x_\mu'} \cdot \frac{\partial x_\delta}{\partial x_\nu'}\frac{\partial A_\epsilon}{\partial x_\delta}$$

$$= \left(\{\mu\nu, \rho\}'\frac{\partial x_\epsilon}{\partial x_\rho'} - \frac{\partial x_\alpha}{\partial x_\mu'}\frac{\partial x_\beta}{\partial x_\nu'}\{\alpha\beta, \epsilon\}\right)A_\epsilon + \frac{\partial x_\alpha}{\partial x_\mu'}\frac{\partial x_\beta}{\partial x_\nu'}\frac{\partial A_\alpha}{\partial x_\beta} \qquad \text{......(31·5)}$$

by (31·3) and changing the dummy suffixes in the last term.

Also by (23·12)

$$A_\epsilon \frac{\partial x_\epsilon}{\partial x_\rho'} = A_\rho'.$$

Hence (31·5) becomes

$$\frac{\partial A_\mu'}{\partial x_\nu'} - \{\mu\nu, \rho\}' A_\rho' = \frac{\partial x_\alpha}{\partial x_\mu'}\frac{\partial x_\beta}{\partial x_\nu'}\left(\frac{\partial A_\alpha}{\partial x_\beta} - \{\alpha\beta, \epsilon\}A_\epsilon\right) \qquad \text{.........(31·6)},$$

showing that

$$\frac{\partial A_\mu}{\partial x_\nu} - \{\mu\nu, \rho\}A_\rho$$

obeys the law of transformation of a covariant tensor. We thus reach the result (29·3) by an alternative method.

A tensor of the second or higher rank may be taken instead of A_μ in (31·4), and its covariant derivative will be found by the same method.

32. Surface-elements and Stokes's theorem.

Consider the outer product $\Sigma^{\mu\nu}$ of two different displacements dx_μ and δx_ν. The tensor $\Sigma^{\mu\nu}$ will be unsymmetrical in μ and ν. We can decompose any such tensor into the sum of a symmetrical part $\frac{1}{2}(\Sigma^{\mu\nu} + \Sigma^{\nu\mu})$ and an antisymmetrical part $\frac{1}{2}(\Sigma^{\mu\nu} - \Sigma^{\nu\mu})$.

Double* the antisymmetrical part of the product $dx_\mu \delta x_\nu$ is called the *surface-element* contained by the two displacements, and is denoted by $dS^{\mu\nu}$. We have accordingly

$$dS^{\mu\nu} = dx_\mu \delta x_\nu - dx_\nu \delta x_\mu \qquad \text{.........(32·1)}$$

$$= \begin{vmatrix} dx_\mu & dx_\nu \\ \delta x_\mu & \delta x_\nu \end{vmatrix}.$$

* The doubling of the natural expression is avenged by the appearance of the factor $\frac{1}{2}$ in most formulae containing $dS^{\mu\nu}$.

In rectangular coordinates this determinant represents the area of the projection on the $\mu\nu$ plane of the parallelogram contained by the two displacements; thus the components of the tensor are the projections of the parallelogram on the six coordinate planes. In the tensor $dS^{\mu\nu}$ these are repeated twice, once with positive and once with negative sign (corresponding perhaps to the two sides of the surface). The four components dS^{11}, dS^{22}, etc. vanish, as must happen in every antisymmetrical tensor. The appropriateness of the name "surface-element" is evident in rectangular coordinates; the geometrical meaning becomes more obscure in other systems.

The surface-element is always a tensor of the second rank whatever the number of dimensions of space; but in *three* dimensions there is an alternative representation of a surface area by a simple *vector* at right angles to the surface and of length proportional to the area; indeed it is customary in three dimensions to represent any antisymmetrical tensor by an adjoint vector. Happily in four dimensions it is not possible to introduce this source of confusion.

The invariant $\frac{1}{2} A_{\mu\nu} dS^{\mu\nu}$

is called the *flux* of the tensor $A_{\mu\nu}$ through the surface-element. The flux involves only the antisymmetrical part of $A_{\mu\nu}$, since the inner product of a symmetrical and an antisymmetrical tensor evidently vanishes.

Some of the chief antisymmetrical tensors arise from the operation of *curling*. If $K_{\mu\nu}$ is the covariant derivative of K_μ, we find from (29·3) that

$$K_{\mu\nu} - K_{\nu\mu} = \frac{\partial K_\mu}{\partial x_\nu} - \frac{\partial K_\nu}{\partial x_\mu} \quad\quad (32·2)$$

since the 3-index symbols cancel out. Since the left-hand side is a tensor, the right-hand side is also a tensor. The right-hand side will be recognised as the "curl" of elementary vector theory, except that we have apparently reversed the sign. Strictly speaking, however, we should note that the curl in the elementary three-dimensional theory is a vector, whereas our curl is a tensor; and comparison of the sign attributed is impossible.

The result that the covariant curl is the same as the ordinary curl does not apply to contravariant vectors or to tensors of higher rank:

$$K^\mu{}_\nu - K^\nu{}_\mu \neq \frac{\partial K^\mu}{\partial x_\nu} - \frac{\partial K^\nu}{\partial x_\mu}.$$

In tensor notation the famous theorem of Stokes becomes

$$\int K_\mu dx_\mu = -\frac{1}{2}\iint\left(\frac{\partial K_\mu}{\partial x_\nu} - \frac{\partial K_\nu}{\partial x_\mu}\right) dS^{\mu\nu} \quad\quad (32·3),$$

the double integral being taken over any surface bounded by the path of the single integral. The factor $\frac{1}{2}$ is needed because each surface-element occurs twice, e.g. as dS^{12} and $-dS^{21}$. The theorem can be proved as follows—

Since both sides of the equation are invariants it is sufficient to prove the equation for any one system of coordinates. Choose coordinates so that the surface is on one of the fundamental partitions $x_3 = \text{const.}$, $x_4 = \text{const.}$, and so that the contour consists of four parts given successively by $x_1 = \alpha$, $x_2 = \beta$, $x_1 = \gamma$, $x_2 = \delta$; the rest of the mesh-system may be filled up arbitrarily. For an elementary mesh the containing vectors are $(dx_1, 0, 0, 0)$ and $(0, dx_2, 0, 0)$, so that by (32·1)

$$dS^{12} = dx_1 dx_2 = -dS^{21}.$$

Hence the right-hand side of (32·3) becomes

$$-\int_\alpha^\gamma \int_\beta^\delta \left(\frac{\partial K_1}{\partial x_2} - \frac{\partial K_2}{\partial x_1}\right) dx_1 dx_2$$

$$= -\int_\alpha^\gamma \{[K_1]^\delta - [K_1]^\beta\}\, dx_1 + \int_\beta^\delta \{[K_2]^\gamma - [K_2]^\alpha\}\, dx_2,$$

which consists of four terms giving $\int K_\mu dx_\mu$ for the four parts of the contour.

This proof affords a good illustration of the methods of the tensor calculus. The relation to be established is between two quantities which (by examination of their covariant dimensions) are seen to be invariants, viz. $K_\mu (dx)^\mu$ and $(K_{\mu\nu} - K_{\nu\mu})\, dS^{\mu\nu}$, the latter having been simplified by (32·2). Accordingly it is a relation which does not depend on any particular choice of coordinates, although in (32·3) it is expressed as it would appear when referred to a coordinate-system. In proving the relation of the two invariants once for all, we naturally choose for the occasion coordinates which simplify the analysis; and the work is greatly shortened by drawing our curved meshes so that four partition-lines make up the contour.

33. Significance of covariant differentiation.

Suppose that we wish to discuss from the physical point of view how a field of force varies from point to point. If polar coordinates are being used, a change of the r-component does not necessarily indicate a want of uniformity in the field of force; it is at least partly attributable to the inclination between the r-directions at different points. Similarly when rotating axes are used, the rate of change of momentum h is given not by dh_1/dt, etc., but by

$$dh_1/dt - \omega_3 h_2 + \omega_2 h_3, \text{ etc.} \quad \text{.....................(33·1)}.$$

The momentum may be constant even when the time-derivatives of its components are not zero.

We must recognise then that the change of a physical entity is usually regarded as something distinct from the change of the mathematical components into which we resolve it. In the elementary theory a definition of the former change is obtained by identifying it with the change of the components in unaccelerated rectangular coordinates; but this is of no avail in the general case because space-time may be of a kind for which no such coordinates exist. Can we still preserve this notion of a *physical* rate of change in the general case?

Our attention is directed to the rate of change of a physical entity because of its importance in the laws of physics, e.g. force is the time-rate of change of momentum, or the space-rate of change of potential; therefore the rate of change should be expressed by a tensor of some kind in order that it may enter into the general physical laws. Further in order to agree with the customary definition in elementary cases, it must reduce to the rate of change of the rectangular components when the coordinates are Galilean. Both conditions are fulfilled if we define the physical rate of change of the tensor by its covariant derivative.

The covariant derivative $A_{\mu\nu}$ consists of the term $\partial A_\mu/\partial x_\nu$, giving the apparent gradient, from which is subtracted the "spurious change" $\{\mu\nu, \alpha\} A_\alpha$ attributable to the curvilinearity of the coordinate-system. When Cartesian coordinates (rectangular or oblique) are used, the 3-index symbols vanish and there is, as we should expect, no spurious change. For the present we shall call $A_{\mu\nu}$ the rate of *absolute change* of the vector A_μ.

Consider an elementary mesh in the plane of $x_\nu x_\sigma$, the corners being at

$$A\,(x_\nu, x_\sigma), \quad B\,(x_\nu + dx_\nu, x_\sigma), \quad C\,(x_\nu + dx_\nu, x_\sigma + dx_\sigma), \quad D\,(x_\nu, x_\sigma + dx_\sigma).$$

Let us calculate the whole absolute change of the vector-field A_μ as we pass round the circuit $ABCDA$.

(1) From A to B, the absolute change is $A_{\mu\nu} dx_\nu$, calculated for x_σ*.

(2) From B to C, the absolute change is $A_{\mu\sigma} dx_\sigma$, calculated for $x_\nu + dx_\nu$.

(3) From C to D, the absolute change is $-A_{\mu\nu} dx_\nu$, calculated for $x_\sigma + dx_\sigma$.

(4) From D to A, the absolute change is $-A_{\mu\sigma} dx_\sigma$, calculated for x_ν.

Combining (2) and (4) the net result is the difference of the changes $A_{\mu\sigma} dx_\sigma$ at $x_\nu + dx_\nu$ and at x_ν respectively. We might be tempted to set this difference down as

$$\frac{\partial}{\partial x_\nu} (A_{\mu\sigma} dx_\sigma)\, dx_\nu.$$

But as already explained that would give only the difference of the mathematical components and not the "absolute difference." We must take the covariant derivative instead, obtaining (since dx_σ is the same for (2) and (4))

$$A_{\mu\sigma\nu} dx_\sigma dx_\nu.$$

Similarly (3) and (1) give

$$-A_{\mu\nu\sigma} dx_\nu dx_\sigma,$$

so that the total absolute change round the circuit is

$$(A_{\mu\sigma\nu} - A_{\mu\nu\sigma})\, dx_\nu dx_\sigma \quad \text{......................(33·2).}$$

We should naturally expect that on returning to our starting point the absolute change would vanish. How could there have been any absolute change on balance, seeing that the vector is now the same A_μ that we started with? Nevertheless in general $A_{\mu\nu\sigma} \neq A_{\mu\sigma\nu}$, that is to say the order of covariant differentiation is not permutable, and (33·2) does not vanish.

* We suspend the summation convention since dx_ν and dx_σ are edges of a particular mesh. The convention would give correct results; but it goes too fast, and we cannot keep pace with it.

That this result is not unreasonable may be seen by considering a two-dimensional space, the surface of the ocean. If a ship's head is kept straight on the line of its wake, the course is a great circle. Now suppose that the ship sails round a circuit so that the final position and course are the same as at the start. If account is kept of all the successive changes of course, and the angles are added up, these will not give a change zero (or 2π) on balance. For a triangular course the difference is the well-known "spherical excess." Similarly the changes of velocity do not cancel out on balance. Here we have an illustration that the absolute changes of a vector do not cancel out on bringing it back to its initial position.

If the present result sounds self-contradictory, the fault lies with the name "absolute change" which we have tentatively applied to the thing under discussion. The name is illuminating in some respects, because it shows the continuity of covariant differentiation with the conceptions of elementary physics. For instance, no one would hesitate to call (33·1) the absolute rate of change of momentum in contrast to the apparent rate of change dh_1/dt. But having shown the continuity, we find it better to avoid the term in the more general case of non-Euclidean space.

Following Levi-Civita and Weyl we use the term *parallel displacement* for what we have hitherto called displacement without "absolute change." The condition for parallel displacement is that the covariant derivative vanishes.

We have hitherto considered the absolute change necessary in order that the vector may return to its original value, and so be a single-valued function of position. If we do not permit any change *en route*, i.e. if we move the vector by parallel displacement, the same quantity will appear (with reversed sign) as a discrepancy δA_μ between the final and initial vectors. Since these are at the same point the difference of the initial and final vectors can be measured immediately. We have then by (33·2)

$$\delta A_\mu = (A_{\mu\nu\sigma} - A_{\mu\sigma\nu})\, dx_\nu dx_\sigma,$$

which may also be written

$$\delta A_\mu = \tfrac{1}{2}\iint (A_{\mu\nu\sigma} - A_{\mu\sigma\nu})\, dS^{\nu\sigma} \quad \text{.................}(33\cdot3),$$

where the summation convention is now restored. We have only proved this for an infinitesimal circuit occupying a coordinate-mesh, for which $dS^{\nu\sigma}$ has only two non-vanishing components $dx_\nu dx_\sigma$ and $-dx_\nu dx_\sigma$. But the equation is seen to be a tensor-equation, and therefore holds independently of the coordinate-system; thus it applies to circuits of any shape, since we can always choose coordinates for which the circuit becomes a coordinate-mesh. But (33·3) is still restricted to infinitesimal circuits and there is no way of extending it to finite circuits—unlike Stokes's theorem. The reason for this restriction is as follows—

An *isolated vector* A_μ may be taken at the starting point and carried by parallel displacement round the circuit, leading to a determinate value of δA_μ.

In (33·3) this is expressed in terms of derivatives of a *vector-field* A_μ extending throughout the region of integration. For a large circuit this would involve values of A_μ remote from the initial vector, which are obviously irrelevant to the calculation of δA_μ. It is rather remarkable that there should exist such a formula even for an infinitesimal circuit; the fact is that although $A_{\mu\nu\sigma} - A_{\mu\sigma\nu}$ at a point formally refers to a vector-field, its value turns out to depend solely on the isolated vector A_μ (see equation (34·3)).

The contravariant vector dx_μ/ds gives a direction in the four-dimensional world which is interpreted as a velocity from the ordinary point of view which separates space and time. We shall usually call it a "velocity"; its connection with the usual three-dimensional vector (u, v, w) is given by

$$\frac{dx_\mu}{ds} = \beta\,(u, v, w, 1),$$

where β is the FitzGerald factor dt/ds. The length (26·5) of a velocity is always unity.

If we transfer dx_μ/ds continually along itself by parallel displacement we obtain a geodesic. For by (29·4) the condition for parallel displacement is

$$\frac{\partial}{\partial x_\nu}\left(\frac{dx_\mu}{ds}\right) + \{\alpha\nu, \mu\}\frac{dx_\alpha}{ds} = 0.$$

Hence multiplying by dx_ν/ds

$$\frac{d^2 x_\mu}{ds^2} + \{\alpha\nu, \mu\}\frac{dx_\alpha}{ds}\frac{dx_\nu}{ds} = 0 \quad \text{.....................(33·4)},$$

which is the condition for a geodesic (28·5). Thus in the language used at the beginning of this section, a geodesic is a line in four dimensions whose direction undergoes no absolute change.

34. The Riemann-Christoffel tensor.

The second covariant derivative of A_μ is found by inserting in (30·3) the value of $A_{\mu\nu}$ from (29·3). This gives

$$A_{\mu\nu\sigma} = \frac{\partial}{\partial x_\sigma}\left(\frac{\partial A_\mu}{\partial x_\nu} - \{\mu\nu, \alpha\} A_\alpha\right) - \{\mu\sigma, \alpha\}\left(\frac{\partial A_\alpha}{\partial x_\nu} - \{\alpha\nu, \epsilon\} A_\epsilon\right)$$
$$- \{\nu\sigma, \alpha\}\left(\frac{\partial A_\mu}{\partial x_\alpha} - \{\mu\alpha, \epsilon\} A_\epsilon\right)$$
$$= \frac{\partial^2 A_\mu}{\partial x_\sigma \partial x_\nu} - \{\mu\nu, \alpha\}\frac{\partial A_\alpha}{\partial x_\sigma} - \{\mu\sigma, \alpha\}\frac{\partial A_\alpha}{\partial x_\nu} - \{\nu\sigma, \alpha\}\frac{\partial A_\mu}{\partial x_\alpha} + \{\nu\sigma, \alpha\}\{\mu\alpha, \epsilon\} A_\epsilon$$
$$+ \{\mu\sigma, \alpha\}\{\alpha\nu, \epsilon\} A_\epsilon - A_\alpha \frac{\partial}{\partial x_\sigma}\{\mu\nu, \alpha\} \quad \text{.....................(34·1)}.$$

The first five terms are unaltered when ν and σ are interchanged. The last two terms may be written, by changing the dummy suffix α to ϵ in the last term,

$$A_\epsilon\left(\{\mu\sigma, \alpha\}\{\alpha\nu, \epsilon\} - \frac{\partial}{\partial x_\sigma}\{\mu\nu, \epsilon\}\right).$$

Hence

$$A_{\mu\nu\sigma} - A_{\mu\sigma\nu} = A_\epsilon \left(\{\mu\sigma, \alpha\} \{\alpha\nu, \epsilon\} - \frac{\partial}{\partial x_\sigma} \{\mu\nu, \epsilon\} - \{\mu\nu, \alpha\} \{\alpha\sigma, \epsilon\} + \frac{\partial}{\partial x_\nu} \{\mu\sigma, \epsilon\} \right) \quad \text{......(34·2)}.$$

The rigorous quotient theorem shows that the co-factor of A_ϵ must be a tensor. Accordingly we write

$$A_{\mu\nu\sigma} - A_{\mu\sigma\nu} = A_\epsilon B^\epsilon_{\mu\nu\sigma} \quad \text{......(34·3)},$$

where

$$B^\epsilon_{\mu\nu\sigma} = \{\mu\sigma, \alpha\} \{\alpha\nu, \epsilon\} - \{\mu\nu, \alpha\} \{\alpha\sigma, \epsilon\} + \frac{\partial}{\partial x_\nu} \{\mu\sigma, \epsilon\} - \frac{\partial}{\partial x_\sigma} \{\mu\nu, \epsilon\} \quad \text{...(34·4)}.$$

This is called the Riemann-Christoffel tensor. It is only when this tensor vanishes that the order of covariant differentiation is permutable.

The suffix ϵ may be lowered. Thus

$$\begin{aligned} B_{\mu\nu\sigma\rho} &= g_{\rho\epsilon} B^\epsilon_{\mu\nu\sigma} \\ &= \{\mu\sigma, \alpha\} [\alpha\nu, \rho] - \{\mu\nu, \alpha\} [\alpha\sigma, \rho] + \frac{\partial}{\partial x_\nu} [\mu\sigma, \rho] - \frac{\partial}{\partial x_\sigma} [\mu\nu, \rho] \\ &\quad - \{\mu\sigma, \alpha\} \frac{\partial g_{\rho\alpha}}{\partial x_\nu} + \{\mu\nu, \alpha\} \frac{\partial g_{\rho\alpha}}{\partial x_\sigma} \quad \text{......(34·45)}, \end{aligned}$$

where ϵ has been replaced by α in the last two terms,

$$\begin{aligned} &= -\{\mu\sigma, \alpha\} [\rho\nu, \alpha] + \{\mu\nu, \alpha\} [\rho\sigma, \alpha] \\ &\quad + \tfrac{1}{2} \left(\frac{\partial^2 g_{\rho\sigma}}{\partial x_\mu \partial x_\nu} + \frac{\partial^2 g_{\mu\nu}}{\partial x_\rho \partial x_\sigma} - \frac{\partial^2 g_{\mu\sigma}}{\partial x_\rho \partial x_\nu} - \frac{\partial^2 g_{\rho\nu}}{\partial x_\mu \partial x_\sigma} \right) \quad \text{......(34·5)}, \end{aligned}$$

by (27·5) and (27·1).

It will be seen from (34·5) that $B_{\mu\nu\sigma\rho}$, besides being antisymmetrical in ν and σ, is also antisymmetrical in μ and ρ. Also it is symmetrical for the double interchange μ and ν, ρ and σ. It has the further cyclic property

$$B_{\mu\nu\sigma\rho} + B_{\mu\sigma\rho\nu} + B_{\mu\rho\nu\sigma} = 0 \quad \text{......(34·6)},$$

as is easily verified from (34·5).

The general tensor of the fourth rank has 256 different components. Here the double antisymmetry reduces the number (apart from differences of sign) to 6×6. 30 of these are paired because μ, ρ can be interchanged with ν, σ; but the remaining 6 components, in which μ, ρ is the same pair of numbers as ν, σ, are without partners. This leaves 21 different components, between which (34·6) gives only one further relation. We conclude that the Riemann-Christoffel tensor has 20 *independent* components*.

The Riemann-Christoffel tensor is derived solely from the $g_{\mu\nu}$ and therefore belongs to the class of fundamental tensors. Usually we can form from any tensor a series of tensors of continually increasing rank by covariant

* Writing the suffixes in the order $\mu\rho\sigma\nu$ the following scheme gives 21 different components:

1212	1223	1313	1324	1423	2323	2424
1213	1224	1314	1334	1424	2324	2434
1214	1234	1323	1414	1434	2334	3434

with the relation 1234 − 1324 + 1423 = 0.

If we omit those containing the suffix 4, we are left with 6 components in three-dimensional space. In two dimensions there is only the one component 1212.

differentiation. But this process is frustrated in the case of the fundamental tensors because $g_{\mu\nu\sigma}$ vanishes identically. We have got round the gap and reached a fundamental tensor of the fourth rank. The series can now be continued indefinitely by covariant differentiation.

When the Riemann-Christoffel tensor vanishes, the differential equations

$$A_{\mu\nu} \equiv \frac{\partial A_\mu}{\partial x_\nu} - \{\mu\nu, \alpha\} A_\alpha = 0 \quad \text{.....................(34·7)}$$

are integrable. For the integration will be possible if (34·7) makes dA_μ or

$$\frac{\partial A_\mu}{\partial x_\nu} dx_\nu$$

a complete differential, i.e. if

$$\{\mu\nu, \alpha\} A_\alpha dx_\nu$$

is a complete differential. By the usual theory the condition for this is

$$\frac{\partial}{\partial x_\sigma}(\{\mu\nu, \alpha\} A_\alpha) - \frac{\partial}{\partial x_\nu}(\{\mu\sigma, \alpha\} A_\alpha) = 0,$$

or
$$A_\alpha \left(\frac{\partial}{\partial x_\sigma}\{\mu\nu, \alpha\} - \frac{\partial}{\partial x_\nu}\{\mu\sigma, \alpha\}\right) + \{\mu\nu, \alpha\}\frac{\partial A_\alpha}{\partial x_\sigma} - \{\mu\sigma, \alpha\}\frac{\partial A_\alpha}{\partial x_\nu} = 0.$$

Substituting for $\partial A_\alpha/\partial x_\sigma$, $\partial A_\alpha/\partial x_\nu$ from (34·7)

$$A_\alpha \left(\frac{\partial}{\partial x_\sigma}\{\mu\nu, \alpha\} - \frac{\partial}{\partial x_\nu}\{\mu\sigma, \alpha\}\right) + (\{\mu\nu, \alpha\}\{\alpha\sigma, \epsilon\} - \{\mu\sigma, \alpha\}\{\alpha\nu, \epsilon\}) A_\epsilon = 0.$$

Changing the suffix α to ϵ in the first term, the condition becomes

$$A_\epsilon B^\epsilon_{\mu\sigma\nu} = 0.$$

Accordingly when $B^\epsilon_{\mu\sigma\nu}$ vanishes, the differential dA_μ determined by (34·7) will be a complete differential, and

$$\int dA_\mu$$

between any two points will be independent of the path of integration. We can then carry the vector A_μ by parallel displacement to any point obtaining a unique result independent of the route of transfer. If a vector is displaced in this way all over the field, we obtain a *uniform vector-field*.

This construction of a uniform vector-field is only possible when the Riemann-Christoffel tensor vanishes throughout. In other cases the equations have no complete integral, and can only be integrated along a particular route. E.g., we can prescribe a *uniform direction* at all points of a plane, but there is nothing analogous to a uniform direction over the surface of a sphere.

Formulae analogous to (34·3) can be obtained for the second derivatives of a tensor $A_{\ldots\mu\ldots}$ instead of for a vector A_μ. The result is easily found to be

$$A_{\ldots\mu\ldots\nu\sigma} - A_{\ldots\mu\ldots\sigma\nu} = \Sigma B^\epsilon_{\mu\nu\sigma} A_{\ldots\epsilon\ldots} \quad \text{...........(34·8)},$$

the summation being taken over all the suffixes μ of the original tensor.

The corresponding formulae for contravariant tensors follow at once, since the $g^{\mu\nu}$ behave as constants in covariant differentiation, and suffixes may be raised on both sides of (34·8).

35. Miscellaneous formulae.

The following are needed for subsequent use—

Since
$$g_{\mu\nu} g^{\mu\alpha} = 0 \text{ or } 1,$$
$$g^{\mu\alpha} dg_{\mu\nu} + g_{\mu\nu} dg^{\mu\alpha} = 0.$$

Hence
$$g^{\mu\alpha} g^{\nu\beta} dg_{\mu\nu} = - g_{\mu\nu} g^{\nu\beta} dg^{\mu\alpha} = - g^{\beta}_{\mu} dg^{\mu\alpha}$$
$$= - dg^{\alpha\beta} \quad \text{.............(35·11).}$$

Similarly
$$dg_{\alpha\beta} = - g_{\mu\alpha} g_{\nu\beta} dg^{\mu\nu} \quad \text{.............(35·12).}$$

Multiplying by $A^{\alpha\beta}$, we have by the rule for lowering suffixes
$$A^{\alpha\beta} dg_{\alpha\beta} = - (g_{\mu\alpha} g_{\nu\beta} A^{\alpha\beta}) dg^{\mu\nu}$$
$$= - A_{\mu\nu} dg^{\mu\nu} = - A_{\alpha\beta} dg^{\alpha\beta} \quad \text{.............(35·2).}$$

For any tensor $B_{\alpha\beta}$ other than the fundamental tensor the corresponding formula would be
$$A^{\alpha\beta} dB_{\alpha\beta} = A_{\alpha\beta} dB^{\alpha\beta}$$
by (26·3). The exception for $B_{\alpha\beta} = g_{\alpha\beta}$ arises because a change $dg_{\alpha\beta}$ has an additional indirect effect through altering the operation of raising and lowering suffixes.

Again dg is formed by taking the differential of each $g_{\mu\nu}$ and multiplying by its co-factor $g \, . \, g^{\mu\nu}$ in the determinant. Thus
$$\frac{dg}{g} = g^{\mu\nu} dg_{\mu\nu} = - g_{\mu\nu} dg^{\mu\nu} \quad \text{.............(35·3).}$$

The contracted 3-index symbol
$$\{\mu\sigma, \sigma\} = \tfrac{1}{2} g^{\sigma\lambda} \left\{ \frac{\partial g_{\mu\lambda}}{\partial x_\sigma} + \frac{\partial g_{\sigma\lambda}}{\partial x_\mu} - \frac{\partial g_{\mu\sigma}}{\partial x_\lambda} \right\}$$
$$= \tfrac{1}{2} g^{\sigma\lambda} \frac{\partial g_{\sigma\lambda}}{\partial x_\mu}.$$

The other two terms cancel by interchange of the dummy suffixes σ and λ. Hence by (35·3)
$$\{\mu\sigma, \sigma\} = \frac{1}{2g} \frac{\partial g}{\partial x_\mu}$$
$$= \frac{\partial}{\partial x_\mu} \log \sqrt{-g} \quad \text{.............(35·4).}$$

We use $\sqrt{-g}$ because g is always negative for real coordinates.

A possible pitfall in differentiating a summed expression should be noticed. The result of differentiating $a_{\mu\nu} x_\mu x_\nu$ with respect to x_ν is not $a_{\mu\nu} x_\mu$ but $(a_{\mu\nu} + a_{\nu\mu}) x_\mu$. The method of performing such differentiations may be illustrated by the following example. Let
$$h_{\nu\tau} = a_{\mu\nu} a_{\sigma\tau} x_\mu x_\sigma,$$
where $a_{\mu\nu}$ represents constant coefficients. Then
$$\frac{\partial h_{\nu\tau}}{\partial x_\alpha} = a_{\mu\nu} a_{\sigma\tau} \left(\frac{\partial x_\mu}{\partial x_\alpha} x_\sigma + \frac{\partial x_\sigma}{\partial x_\alpha} x_\mu \right)$$
$$= a_{\mu\nu} a_{\sigma\tau} (g^{\mu}_{\alpha} x_\sigma + g^{\sigma}_{\alpha} x_\mu) \quad \text{by (22·3).}$$

Repeating the process,

$$\frac{\partial^2 h_{\nu\tau}}{\partial x_\alpha \partial x_\beta} = a_{\mu\nu} a_{\sigma\tau} (g_\alpha^\mu g_\beta^\sigma + g_\alpha^\sigma g_\beta^\mu)$$
$$= a_{\alpha\nu} a_{\beta\tau} + a_{\beta\nu} a_{\alpha\tau}.$$

Hence changing dummy suffixes

$$\frac{\partial^2}{\partial x_\mu \partial x_\sigma} (a_{\mu\nu} a_{\sigma\tau} x_\mu x_\sigma) = a_{\mu\nu} a_{\sigma\tau} + a_{\sigma\nu} a_{\mu\tau} \quad \text{..............(35·5).}$$

Similarly if $a_{\mu\nu\sigma}$ is symmetrical in its suffixes

$$\frac{\partial^3}{\partial x_\mu \partial x_\nu \partial x_\sigma} (a_{\mu\nu\sigma} x_\mu x_\nu x_\sigma) = 6 a_{\mu\nu\sigma} \quad \text{.................(35·6).}$$

The pitfall arises from repeating a suffix three times in one term. In these formulae the summation applies to the repetition within the bracket, and not to the differentiation.

Summary.

Tensors are quantities obeying certain transformation laws. Their importance lies in the fact that if a tensor equation is found to hold for one system of coordinates, it continues to hold when any transformation of coordinates is made. New tensors are recognised either by investigating their transformation laws directly or by the property that the sum, difference, product or quotient of tensors is a tensor. This is a generalisation of the method of dimensions in physics.

The principal operations of the tensor calculus are addition, multiplication (outer and inner), summation (§ 22), contraction (§ 24), substitution (§ 25), raising and lowering suffixes (§ 26), covariant differentiation (§§ 29, 30). There is no operation of division; but an inconvenient factor $g_{\mu\nu}$ or $g^{\mu\nu}$ can be removed by multiplying through by $g^{\mu\sigma}$ or $g_{\mu\sigma}$ so as to form the substitution-operator. The operation of summation is practically outside our control and always presents itself as a *fait accompli*. The most characteristic process of manipulation in this calculus is the free alteration of dummy suffixes (those appearing twice in a term); it is probably this process which presents most difficulty to the beginner.

Of special interest are the fundamental tensors or world-tensors, of which we have discovered two, viz. $g_{\mu\nu}$ and $B_{\mu\nu\sigma\rho}$. The latter has been expressed in terms of the former and its first and second derivatives. It is through these that the gap between pure geometry and physics is bridged; in particular $g_{\mu\nu}$ relates the observed quantity ds to the mathematical coordinate specification dx_μ.

Since in our work we generally deal with tensors, the reader may be led to overlook the rarity of this property. The juggling tricks which we seem to perform in our manipulations are only possible because the material used is of quite exceptional character.

The further development of the tensor calculus will be resumed in § 48; but a stage has now been reached at which we may begin to apply it to the theory of gravitation.

CHAPTER III

THE LAW OF GRAVITATION

36. The condition for flat space-time. Natural coordinates.

A region of the world is called *flat* or *homaloidal* if it is possible to construct in it a Galilean frame of reference.

It was shown in § 4 that when the $g_{\mu\nu}$ are constants, ds^2 can be reduced to the sum of four squares, and Galilean coordinates can be constructed. Thus an equivalent definition of flat space-time is that it is such that coordinates can be found for which the $g_{\mu\nu}$ are constants.

When the $g_{\mu\nu}$ are constants the 3-index symbols all vanish; but since the 3-index symbols do not form a tensor, they will not in general continue to vanish when other coordinates are substituted in the same flat region. Again, when the $g_{\mu\nu}$ are constants, the Riemann-Christoffel tensor, being composed of products and derivatives of the 3-index symbols, will vanish; and since it is a tensor, it will continue to vanish when any other coordinate-system is substituted in the same region.

Hence the vanishing of the Riemann-Christoffel tensor is a necessary condition for flat space-time.

This condition is also *sufficient*—if the Riemann-Christoffel tensor vanishes space-time must be flat. This can be proved as follows—

We have found (§ 34) that if

$$B^{\epsilon}_{\mu\nu\sigma} = 0 \quad \text{.............................(36·1)},$$

it is possible to construct a uniform vector-field by parallel displacement of a vector all over the region. Let $A^{\mu}_{(\alpha)}$ be four uniform vector-fields given by $\alpha = 1, 2, 3, 4$, so that

$$(A^{\mu}_{(\alpha)})_{\sigma} = 0$$

or by (29·4)

$$\frac{\partial A^{\mu}_{(\alpha)}}{\partial x_{\sigma}} = -\{\epsilon\sigma, \mu\} A^{\epsilon}_{(\alpha)} \quad \text{.......................(36·2)}.$$

Note that α is not a tensor-suffix, but merely distinguishes the four independent vectors.

We shall use these four uniform vector-fields to define a new coordinate-system distinguished by accents. Our unit mesh will be the hyperparallelopiped contained by the four vectors at any point, and the complete mesh-system will be formed by successive parallel displacements of this unit mesh until the whole region is filled. One edge of the unit mesh, given in the old coordinates by

$$dx_{\mu} = A^{\mu}_{(1)},$$

has to become in the new coordinates

$$dx_{\alpha}' = (1, 0, 0, 0).$$

Similarly the second edge, $dx_\mu = A^\mu_{(2)}$, must become $dx_\alpha' = (0, 1, 0, 0)$; etc. This requires the law of transformation

$$dx_\mu = A^\mu_{(\alpha)} dx_\alpha' \quad \text{.............................(36·3)}.$$

Of course, the construction of the accented coordinate-system depends on the possibility of constructing uniform vector-fields, and this depends on (36·1) being satisfied.

Since ds^2 is an invariant

$$\begin{aligned} g'_{\alpha\beta} dx_\alpha' dx_\beta' &= g_{\mu\nu} dx_\mu dx_\nu \\ &= g_{\mu\nu} A^\mu_{(\alpha)} A^\nu_{(\beta)} dx_\alpha' dx_\beta' \quad \text{by (36·3)}. \end{aligned}$$

Hence

$$g'_{\alpha\beta} = g_{\mu\nu} A^\mu_{(\alpha)} A^\nu_{(\beta)}.$$

Accordingly, by differentiation,

$$\begin{aligned} \frac{\partial g'_{\alpha\beta}}{\partial x_\sigma} &= g_{\mu\nu} A^\mu_{(\alpha)} \frac{\partial A^\nu_{(\beta)}}{\partial x_\sigma} + g_{\mu\nu} A^\nu_{(\beta)} \frac{\partial A^\mu_{(\alpha)}}{\partial x_\sigma} + A^\mu_{(\alpha)} A^\nu_{(\beta)} \frac{\partial g_{\mu\nu}}{\partial x_\sigma} \\ &= -g_{\mu\nu} A^\mu_{(\alpha)} A^\epsilon_{(\beta)} \{\epsilon\sigma, \nu\} - g_{\mu\nu} A^\nu_{(\beta)} A^\epsilon_{(\alpha)} \{\epsilon\sigma, \mu\} + A^\mu_{(\alpha)} A^\nu_{(\beta)} \frac{\partial g_{\mu\nu}}{\partial x_\sigma} \end{aligned}$$

by (36·2). By changing dummy suffixes, this becomes

$$\begin{aligned} \frac{\partial g'_{\alpha\beta}}{\partial x_\sigma} &= A^\mu_{(\alpha)} A^\nu_{(\beta)} \left[-g_{\mu\epsilon} \{\nu\sigma, \epsilon\} - g_{\epsilon\nu} \{\mu\sigma, \epsilon\} + \frac{\partial g_{\mu\nu}}{\partial x_\sigma} \right] \\ &= A^\mu_{(\alpha)} A^\nu_{(\beta)} \left[-[\nu\sigma, \mu] - [\mu\sigma, \nu] + \frac{\partial g_{\mu\nu}}{\partial x_\sigma} \right] \\ &= 0 \quad \text{by (27·5)}. \end{aligned}$$

Hence the $g'_{\alpha\beta}$ are constant throughout the region. We have thus constructed a coordinate-system fulfilling the condition that the g's are constant, and it follows that the space-time is flat.

It will be seen that a *uniform* mesh-system, i.e. one in which the unit meshes are connected with one another by parallel displacement, is necessarily a Cartesian system (rectangular or oblique). Uniformity in this sense is impossible in space-time for which the Riemann-Christoffel tensor does not vanish, e.g. there can be no uniform mesh-system on a sphere.

When space-time is not flat we can introduce coordinates which will be approximately Galilean in a small region round a selected point, the $g_{\mu\nu}$ being not constant but stationary there; this amounts to identifying the curved space-time with the osculating flat space-time for a small distance round the point. Expressing the procedure analytically, we choose coordinates such that the 40 derivatives $\partial g_{\mu\nu}/\partial x_\sigma$ vanish *at the selected point*. It is fairly obvious from general considerations that this will always be possible; but the following is a formal proof. Having transferred the origin to the selected point, make the following transformation of coordinates

$$x_\epsilon = g^\mu_\epsilon x_\mu' - \tfrac{1}{2} \{\alpha\beta, \epsilon\}_0 \, g^\mu_\alpha g^\nu_\beta x_\mu' x_\nu' \quad \text{..................(36·4)},$$

where the value of the 3-index symbol at the origin is to be taken. Then at the origin

$$\frac{\partial x_\epsilon}{\partial x_\mu'} = g_\epsilon^\mu \quad \text{.....................................(36·45)},$$

$$\frac{\partial^2 x_\epsilon}{\partial x_\mu' \partial x_\nu'} = -\{\alpha\beta, \epsilon\} g_\alpha^\mu g_\beta^\nu$$

$$= -\{\alpha\beta, \epsilon\} \frac{\partial x_\alpha}{\partial x_\mu'} \frac{\partial x_\beta}{\partial x_\nu'} \qquad \text{by (36·45).}$$

Hence by (31·3) $$\{\mu\nu, \rho\}' \frac{\partial x_\epsilon}{\partial x_\rho'} = 0.$$

But $$\{\mu\nu, \rho\}' \frac{\partial x_\epsilon}{\partial x_\rho'} = \{\mu\nu, \rho\}' g_\epsilon^\rho = \{\mu\nu, \epsilon\}'.$$

Hence in the new coordinates the 3-index symbols vanish at the origin; and it follows by (27·4) and (27·5) that the first derivatives of the $g'_{\mu\nu}$ vanish. This is the preliminary transformation presupposed in § 4.

We pass on to a somewhat more difficult transformation which is important as contributing an insight into the significance of $B^\epsilon_{\mu\nu\sigma}$.

It is not possible to make the second derivatives of the $g_{\mu\nu}$ vanish at the selected point (as well as the first derivatives) unless the Riemann-Christoffel tensor vanishes there; but a great number of other special conditions can be imposed on the 100 second derivatives by choosing the coordinates suitably. Make an additional transformation of the form

$$x_\epsilon = g_\mu^\epsilon x_\mu' + \tfrac{1}{6} a^\epsilon_{\mu\nu\sigma} x_\mu' x_\nu' x_\sigma' \quad \text{.....................(36·5)},$$

where $a^\epsilon_{\mu\nu\sigma}$ represents arbitrary coefficients symmetrical in μ, ν, σ. This new transformation will not affect the first derivatives of the $g_{\mu\nu}$ at the origin, which have already been made to vanish by the previous transformation, but it alters the second derivatives. By differentiating (31·3), viz.

$$\{\mu\nu, \rho\}' \frac{\partial x_\epsilon}{\partial x_\rho'} - \frac{\partial x_\alpha}{\partial x_\mu'} \frac{\partial x_\beta}{\partial x_\nu'} \{\alpha\beta, \epsilon\} = \frac{\partial^2 x_\epsilon}{\partial x_\mu' \partial x_\nu'},$$

we obtain at the origin

$$\frac{\partial}{\partial x_\sigma'} \{\mu\nu, \rho\}' \frac{\partial x_\epsilon}{\partial x_\rho'} - \frac{\partial x_\alpha}{\partial x_\mu'} \frac{\partial x_\beta}{\partial x_\nu'} \frac{\partial x_\gamma}{\partial x_\sigma'} \frac{\partial}{\partial x_\gamma} \{\alpha\beta, \epsilon\} = \frac{\partial^3 x_\epsilon}{\partial x_\mu' \partial x_\nu' \partial x_\sigma'},$$

since the 3-index symbols themselves vanish. Hence by (36·5)*

$$\frac{\partial}{\partial x_\sigma'} \{\mu\nu, \rho\}' . g_\rho^\epsilon - g_\mu^\alpha g_\nu^\beta g_\sigma^\gamma \frac{\partial}{\partial x_\gamma} \{\alpha\beta, \epsilon\} = a^\epsilon_{\mu\nu\sigma},$$

which reduces to $$\frac{\partial}{\partial x_\sigma'} \{\mu\nu, \epsilon\}' - \frac{\partial}{\partial x_\sigma} \{\mu\nu, \epsilon\} = a^\epsilon_{\mu\nu\sigma} \quad \text{.............(36·55).}$$

The transformation (36·5) accordingly increases $\partial \{\mu\nu, \epsilon\}/\partial x_\sigma$ by $a^\epsilon_{\mu\nu\sigma}$.

Owing to the symmetry of $a^\epsilon_{\mu\nu\sigma}$, all three quantities

$$\frac{\partial}{\partial x_\sigma} \{\mu\nu, \epsilon\}, \quad \frac{\partial}{\partial x_\nu} \{\mu\sigma, \epsilon\}, \quad \frac{\partial}{\partial x_\mu} \{\nu\sigma, \epsilon\}$$

* For the disappearance of the factor $\frac{1}{6}$, see (35·6).

are necessarily increased by the same amount. Now the unaltered difference

$$\frac{\partial}{\partial x_\nu}\{\mu\sigma, \epsilon\} - \frac{\partial}{\partial x_\sigma}\{\mu\nu, \epsilon\} = B^\epsilon_{\mu\nu\sigma} \quad \text{..................(36·6)},$$

since the remaining terms of (34·4) vanish in the coordinates here used. We cannot alter any of the components of the Riemann-Christoffel tensor; but, subject to this limitation, the alterations of the derivatives of the 3-index symbols are arbitrary.

The most symmetrical way of imposing further conditions is to make a transformation such that

$$\frac{\partial}{\partial x_\sigma}\{\mu\nu, \epsilon\} + \frac{\partial}{\partial x_\nu}\{\mu\sigma, \epsilon\} + \frac{\partial}{\partial x_\mu}\{\nu\sigma, \epsilon\} = 0 \quad \text{...........(36·7)}.$$

There are 80 different equations of this type, each of which fixes one of the 80 arbitrary coefficients $a^\epsilon_{\mu\nu\sigma}$. In addition there are 20 independent equations of type (36·6) corresponding to the 20 independent components of the Riemann-Christoffel tensor. Thus we have just sufficient equations to determine uniquely the 100 second derivatives of the $g_{\mu\nu}$. Coordinates such that $\partial g_{\mu\nu}/\partial x_\sigma$ is zero and $\partial^2 g_{\mu\nu}/\partial x_\sigma \partial x_\tau$ satisfies (36·7) may be called *canonical coordinates*.

By solving the 100 equations we obtain all the $\partial^2 g_{\mu\nu}/\partial x_\sigma \partial x_\tau$ for canonical coordinates expressed as linear functions of the $B^\epsilon_{\mu\nu\sigma}$.

The two successive transformations which lead to canonical coordinates are combined in the formula

$$x_\epsilon = g^\epsilon_\mu x_\mu' - \tfrac{1}{2}\{\mu\nu, \epsilon\}_0 x_\mu' x_\nu'$$
$$- \frac{1}{18}\left[\frac{\partial}{\partial x_\mu}\{\nu\sigma, \epsilon\} + \frac{\partial}{\partial x_\nu}\{\mu\sigma, \epsilon\} + \frac{\partial}{\partial x_\sigma}\{\mu\nu, \epsilon\}\right]_0 x_\mu' x_\nu' x_\sigma' \quad \text{...(36·8)}.$$

At the origin $\partial x_\epsilon/\partial x_\mu' = g^\epsilon_\mu$, so that the transformation does not alter any tensor at the origin. For example, the law of transformation of $C_{\mu\nu\sigma}$ gives

$$C'_{\mu\nu\sigma} = C_{\alpha\beta\gamma}\frac{\partial x_\alpha}{\partial x_\mu'}\frac{\partial x_\beta}{\partial x_\nu'}\frac{\partial x_\gamma}{\partial x_\sigma'} = C_{\alpha\beta\gamma} g^\alpha_\mu g^\beta_\nu g^\gamma_\sigma$$
$$= C_{\mu\nu\sigma}.$$

The transformation in fact alters the curvature and hypercurvature of the axes passing through the origin, but does not alter the angles of intersection.

Consider any tensor which contains only the $g_{\mu\nu}$ and their first and second derivatives. In canonical coordinates the first derivatives vanish and the second derivatives are linear functions of the $B^\epsilon_{\mu\nu\sigma}$; hence the whole tensor is a function of the $g_{\mu\nu}$ and the $B^\epsilon_{\mu\nu\sigma}$. But neither the tensor itself nor the $g_{\mu\nu}$ and $B^\epsilon_{\mu\nu\sigma}$ have been altered in the reduction to canonical coordinates, hence the same functional relation holds true in the original unrestricted coordinates. We have thus the important result—

The only fundamental tensors which do not contain derivatives of $g_{\mu\nu}$ beyond the second order are functions of $g_{\mu\nu}$ and $B^\epsilon_{\mu\nu\sigma}$.

This shows that our treatment of the tensors describing the character of space-time has been exhaustive as far as the second order. If for suitably chosen coordinates two surfaces have the same $g_{\mu\nu}$ and $B^{\epsilon}_{\mu\nu\sigma}$ at some point, they will be applicable to one another as far as cubes of the coordinates; the two tensors suffice to specify the whole metric round the point to this extent.

Having made the first derivatives vanish, we can by the linear transformation explained in § 4 give the $g_{\mu\nu}$ Galilean values at the selected point. The coordinates so obtained are called *natural coordinates* at the point and quantities referred to these coordinates are said to be expressed in *natural measure*. Natural coordinates are thus equivalent to Galilean coordinates when only the $g_{\mu\nu}$ and their first derivatives are considered; the difference appears when we study phenomena involving the second derivatives.

By making a Lorentz transformation (which leaves the coordinates still a natural system) we can reduce to rest the material located at the point, or an observer supposed to be stationed with his measuring appliances at the point. The *natural measure* is then further particularised as the *proper-measure* of the material, or observer. It may be noticed that the material will be at rest both as regards velocity and acceleration (unless it is acted on by electromagnetic forces) because there is no field of acceleration relative to natural coordinates.

To sum up this discussion of special systems of coordinates.—When the Riemann-Christoffel tensor vanishes, we can adopt Galilean coordinates throughout the region. When it does not vanish we can adopt coordinates which agree with Galilean coordinates at a selected point in the values of the $g_{\mu\nu}$ and their first derivatives but not in the second derivatives; these are called *natural coordinates* at the point. Either Galilean or natural coordinates can be subjected to Lorentz transformations, so that we can select a system with respect to which a particular observer is at rest; this system will be the *proper-coordinates* for that observer. Although we cannot in general make natural coordinates agree with Galilean coordinates in the second derivatives of the $g_{\mu\nu}$, we can impose 80 partially arbitrary conditions on the 100 second derivatives; and when these conditions are selected as in (36·7) the resulting coordinates have been called *canonical*.

There is another way of specialising coordinates which may be mentioned here for completeness. It is always possible to choose coordinates such that the determinant $g=-1$ everywhere (as in Galilean coordinates). This is explained in § 49.

We may also consider another class of specialised coordinates—those which are permissible in special problems. There are certain (non-Euclidean) coordinates found to be most convenient in dealing with the gravitational field of the sun, Einstein's or de Sitter's curved world, and so on. It must be remembered, however, that these refer to idealised problems, and coordinate-systems with simple properties can only be approximately realised in nature.

If possible a *static system* of coordinates is selected, the condition for this being that all the $g_{\mu\nu}$ are independent of one of the coordinates x_4 (which must be of timelike character*). In that case the interval corresponding to any displacement dx_μ is independent of the "time" x_4. Such a system can, of course, only be found if the relative configuration of the attracting masses is maintained unaltered. If in addition it is possible to make $g_{14}, g_{24}, g_{34} = 0$ the time will be reversible, and in particular the forward velocity of light along any track will be equal to the backward velocity; this renders the application of the name "time" to x_4 more just, since one of the alternative conventions of § 11 is satisfied. We shall if possible employ systems which are static and reversible in dealing with large regions of the world; problems in which this simplification is not permissible must generally be left aside as insoluble—e.g. the problem of two attracting bodies. For small regions of the world the greatest simplification is obtained by using natural coordinates.

37. Einstein's law of gravitation.

The contracted Riemann-Christoffel tensor is formed by setting $\epsilon = \sigma$ in $B^\epsilon_{\mu\nu\sigma}$. It is denoted by $G_{\mu\nu}$. Hence by (34·4)

$$G_{\mu\nu} = \{\mu\sigma, \alpha\}\{\alpha\nu, \sigma\} - \{\mu\nu, \alpha\}\{\alpha\sigma, \sigma\} + \frac{\partial}{\partial x_\nu}\{\mu\sigma, \sigma\} - \frac{\partial}{\partial x_\sigma}\{\mu\nu, \sigma\} \quad \text{...(37·1).}$$

The symbols containing a duplicated suffix are simplified by (35·4), viz.

$$\{\mu\sigma, \sigma\} = \frac{\partial}{\partial x_\mu}\log\sqrt{-g}.$$

Hence, with some alterations of dummy suffixes,

$$G_{\mu\nu} = -\frac{\partial}{\partial x_\alpha}\{\mu\nu, \alpha\} + \{\mu\alpha, \beta\}\{\nu\beta, \alpha\} + \frac{\partial^2}{\partial x_\mu \partial x_\nu}\log\sqrt{-g} - \{\mu\nu, \alpha\}\frac{\partial}{\partial x_\alpha}\log\sqrt{-g} \quad \text{......(37·2).}$$

Contraction by setting $\epsilon = \mu$ does not provide an alternative tensor, because

$$B^\mu_{\mu\nu\sigma} = g^{\mu\rho} B_{\mu\nu\sigma\rho} = 0,$$

owing to the antisymmetry of $B_{\mu\nu\sigma\rho}$ in μ and ρ.

The law $$G_{\mu\nu} = 0 \quad \text{..............................(37·3),}$$

in empty space, is chosen by Einstein for his law of gravitation.

We see from (37·2) that $G_{\mu\nu}$ is a symmetrical tensor; consequently the law provides 10 partial differential equations to determine the $g_{\mu\nu}$. It will be found later (§ 52) that there are 4 identical relations between them, so that the number of equations is effectively reduced to 6. The equations are of the second order and involve the second differential coefficients of $g_{\mu\nu}$ linearly. We proved in § 36 that tensors not containing derivatives beyond the second must necessarily be compounded from $g_{\mu\nu}$ and $B^\epsilon_{\mu\nu\sigma}$; so that, unless we are prepared

* dx_4 will be timelike if g_{44} is always positive.

to go beyond the second order, the choice of a law of gravitation is very limited, and we can scarcely avoid relying on the tensor $G_{\mu\nu}$*.

Without introducing higher derivatives, which would seem out of place in this problem, we can suggest as an alternative to (37·3) the law

$$G_{\mu\nu} = \lambda g_{\mu\nu} \quad \text{.............................(37·4)},$$

where λ is a universal constant. There are theoretical grounds for believing that this is actually the correct form; but it is certain that λ must be an extremely small constant, so that in practical applications we still take (37·3) as sufficiently approximate. The introduction of the small constant λ leads to the spherical world of Einstein or de Sitter to which we shall return in Chapter V.

The spur

$$G = g^{\mu\nu} G_{\mu\nu} \quad \text{.............................(37·5)}$$

is called the Gaussian curvature, or simply the *curvature*, of space-time. It must be remembered, however, that the deviation from flatness is described in greater detail by the tensors $G_{\mu\nu}$ and $B_{\mu\nu\sigma\rho}$ (sometimes called *components of curvature*) and the vanishing of G is by no means a sufficient condition for flat space-time.

Einstein's law of gravitation expresses the fact that the geometry of an empty region of the world is not of the most general Riemannian type, but is limited. General Riemannian geometry corresponds to the quadratic form (2·1) with the g's entirely unrestricted functions of the coordinates; Einstein asserts that the natural geometry of an empty region is not of so unlimited a kind, and the possible values of the g's are restricted to those which satisfy the differential equations (37·3). It will be remembered that a field of force arises from the discrepancy between the natural geometry of a coordinate-system and the abstract Galilean geometry attributed to it; thus any law governing a field of force must be a law governing the natural geometry. That is why the law of gravitation must appear as a restriction on the possible natural geometry of the world. The inverse-square law, which is a plausible law of weakening of a supposed absolute force, becomes quite unintelligible (and indeed impossible) when expressed as a restriction on the intrinsic geometry of space-time; we have to substitute some law obeyed by the tensors which describe the world-conditions determining the natural geometry.

38. The gravitational field of an isolated particle.

We have now to determine a particular solution of the equations (37·3). The solution which we shall obtain will ultimately be shown to correspond to the field of an isolated particle continually at rest at the origin; and in seeking a solution we shall be guided by our general idea of the type of solution to be expected for such a particle. This preliminary argument need not be rigorous;

* The law $B_{\mu\nu\sigma\rho}=0$ (giving flat space-time throughout all empty regions) would obviously be too stringent, since it does not admit of the existence of irreducible fields of force.

the final test is whether the formulae suggested by it satisfy the equations to be solved.

In flat space-time the interval, referred to spherical polar coordinates and time, is

$$ds^2 = -dr^2 - r^2 d\theta^2 - r^2 \sin^2\theta\, d\phi^2 + dt^2 \quad\text{.........(38·11).}$$

If we consider what modifications of this can be made without destroying the spherical symmetry in space, the symmetry as regards past and future time, or the static condition, the most general possible form appears to be

$$ds^2 = -U(r)\, dr^2 - V(r)\,(r^2 d\theta^2 + r^2 \sin^2\theta\, d\phi^2) + W(r)\, dt^2 \quad\text{...(38·12),}$$

where U, V, W are arbitrary functions of r. Let

$$r_1^2 = r^2 V(r).$$

Then (38·12) becomes of the form

$$ds^2 = -U_1(r_1)\, dr_1^2 - r_1^2 d\theta^2 - r_1^2 \sin^2\theta\, d\phi^2 + W_1(r_1)\, dt^2 \quad\text{...(38·13),}$$

where U_1 and W_1 are arbitrary functions of r_1. There is no reason to regard r in (38·12) as more immediately the counterpart of r in (38·11) than r_1 is. If the functions U, V, W differ only slightly from unity, both r and r_1 will have approximately the properties of the radius-vector in Euclidean geometry; but no length in non-Euclidean space can have exactly the properties of a Euclidean radius-vector, and it is arbitrary whether we choose r or r_1 as its closest representative. We shall here choose r_1, and accordingly drop the suffix, writing (38·13) in the form

$$ds^2 = -e^\lambda dr^2 - r^2 d\theta^2 - r^2 \sin^2\theta\, d\phi^2 + e^\nu dt^2 \quad\text{...........(38·2),}$$

where λ and ν are functions of r only.

Moreover since the gravitational field (or disturbance of flat space-time) due to a particle diminishes indefinitely as we go to an infinite distance, we must have λ and ν tend to zero as r tends to infinity. Formula (38·2) will then reduce to (38·11) at an infinite distance from the particle.

Our coordinates are

$$x_1 = r, \quad x_2 = \theta, \quad x_3 = \phi, \quad x_4 = t,$$

and the fundamental tensor is by (38·2)

$$g_{11} = -e^\lambda, \quad g_{22} = -r^2, \quad g_{33} = -r^2 \sin^2\theta, \quad g_{44} = e^\nu \quad\text{......(38·31),}$$

and

$$g_{\mu\nu} = 0 \text{ if } \mu \neq \nu.$$

The determinant g reduces to its leading diagonal $g_{11}g_{22}g_{33}g_{44}$. Hence

$$-g = e^{\lambda+\nu} r^4 \sin^2\theta \quad\text{........................(38·32),}$$

and $g^{11} = 1/g_{11}$, etc., so that

$$g^{11} = -e^{-\lambda}, \quad g^{22} = -1/r^2, \quad g^{33} = -1/r^2 \sin^2\theta, \quad g^{44} = e^{-\nu} \quad\text{...(38·33).}$$

Since all the $g^{\mu\nu}$ vanish except when the two suffixes are the same, the summation disappears in the formula for the 3-index symbols (27·2), and

$$\{\mu\nu, \sigma\} = \tfrac{1}{2} g^{\sigma\sigma} \left(\frac{\partial g_{\mu\sigma}}{\partial x_\nu} + \frac{\partial g_{\nu\sigma}}{\partial x_\mu} - \frac{\partial g_{\mu\nu}}{\partial x_\sigma} \right) \quad \text{not summed.}$$

If μ, ν, σ denote *different* suffixes we get the following possible cases (the summation convention being suspended):

$$\left.\begin{aligned}
\{\mu\mu, \mu\} &= \tfrac{1}{2} g^{\mu\mu} \frac{\partial g_{\mu\mu}}{\partial x_\mu} = \tfrac{1}{2} \frac{\partial}{\partial x_\mu} (\log g_{\mu\mu}) \\
\{\mu\mu, \nu\} &= -\tfrac{1}{2} g^{\nu\nu} \frac{\partial g_{\mu\mu}}{\partial x_\nu} \\
\{\mu\nu, \nu\} &= \tfrac{1}{2} g^{\nu\nu} \frac{\partial g_{\nu\nu}}{\partial x_\mu} = \tfrac{1}{2} \frac{\partial}{\partial x_\mu} (\log g_{\nu\nu}) \\
\{\mu\nu, \sigma\} &= 0
\end{aligned}\right\} \quad \text{.............. (38·4).}$$

It is now easy to go systematically through the 40 3-index symbols calculating the values of those which do not vanish. We obtain the following results, the accent denoting differentiation with respect to r:

$$\left.\begin{aligned}
\{11, 1\} &= \tfrac{1}{2}\lambda' \\
\{12, 2\} &= 1/r \\
\{13, 3\} &= 1/r \\
\{14, 4\} &= \tfrac{1}{2}\nu' \\
\{22, 1\} &= -re^{-\lambda} \\
\{23, 3\} &= \cot\theta \\
\{33, 1\} &= -r\sin^2\theta e^{-\lambda} \\
\{33, 2\} &= -\sin\theta\cos\theta \\
\{44, 1\} &= \tfrac{1}{2}e^{\nu-\lambda}\nu'
\end{aligned}\right\} \quad \text{.................... (38·5).}$$

The remaining 31 symbols vanish. Note that $\{21, 2\}$ is the same as $\{12, 2\}$, etc.

These values must be substituted in (37·2). As there may be some pitfalls in carrying this out, we shall first write out the equations (37·2) in full, omitting the terms (223 in number) which now obviously vanish.

$$G_{11} = -\frac{\partial}{\partial r}\{11, 1\} + \{11, 1\}\{11, 1\} + \{12, 2\}\{12, 2\} + \{13, 3\}\{13, 3\} + \{14, 4\}\{14, 4\}$$
$$+ \frac{\partial^2}{\partial r^2}\log\sqrt{-g} - \{11, 1\}\frac{\partial}{\partial r}\log\sqrt{-g},$$

$$G_{22} = -\frac{\partial}{\partial r}\{22, 1\} + 2\{22, 1\}\{21, 2\} + \{23, 3\}\{23, 3\} + \frac{\partial^2}{\partial\theta^2}\log\sqrt{-g}$$
$$- \{22, 1\}\frac{\partial}{\partial r}\log\sqrt{-g},$$

$$G_{33} = -\frac{\partial}{\partial r}\{33, 1\} - \frac{\partial}{\partial\theta}\{33, 2\} + 2\{33, 1\}\{31, 3\} + 2\{33, 2\}\{32, 3\}$$
$$- \{33, 1\}\frac{\partial}{\partial r}\log\sqrt{-g} - \{33, 2\}\frac{\partial}{\partial\theta}\log\sqrt{-g},$$

$$G_{44} = -\frac{\partial}{\partial r}\{44, 1\} + 2\{44, 1\}\{41, 4\} - \{44, 1\}\frac{\partial}{\partial r}\log\sqrt{-g},$$

$$G_{12} = \{13, 3\}\{23, 3\} - \{12, 2\}\frac{\partial}{\partial\theta}\log\sqrt{-g}.$$

The remaining components contain no surviving terms.

Substitute from (38·5) and (38·32) in these, and collect the terms. The equations to be satisfied become

$$G_{11} = \tfrac{1}{2}\nu'' - \tfrac{1}{4}\lambda'\nu' + \tfrac{1}{4}\nu'^2 - \lambda'/r = 0 \quad\text{(38·61)},$$
$$G_{22} = e^{-\lambda}(1 + \tfrac{1}{2}r(\nu' - \lambda')) - 1 = 0 \quad\text{(38·62)},$$
$$G_{33} = \sin^2\theta \,.\, e^{-\lambda}(1 + \tfrac{1}{2}r(\nu' - \lambda')) - \sin^2\theta = 0 \quad\text{(38·63)},$$
$$G_{44} = e^{\nu-\lambda}(-\tfrac{1}{2}\nu'' + \tfrac{1}{4}\lambda'\nu' - \tfrac{1}{4}\nu'^2 - \nu'/r) = 0 \quad\text{(38·64)},$$
$$G_{12} = 0 \qquad\qquad = 0 \quad\text{(38·65)}.$$

We may leave aside (38·63) which is a mere repetition of (38·62); then there are left three equations to be satisfied by λ and ν. From (38·61) and (38·64) we have $\lambda' = -\nu'$. Since λ and ν are to vanish together at $r = \infty$, this requires that

$$\lambda = -\nu.$$

Then (38·62) becomes $\qquad e^{\nu}(1 + r\nu') = 1.$

Set $e^{\nu} = \gamma$, then $\qquad \gamma + r\gamma' = 1.$

Hence, integrating, $\qquad \gamma = 1 - \dfrac{2m}{r} \quad\text{(38·7)},$

where $2m$ is a constant of integration.

It will be found that all three equations are satisfied by this solution. Accordingly, substituting $e^{-\lambda} = e^{\nu} = \gamma$ in (38·2),

$$ds^2 = -\gamma^{-1}dr^2 - r^2 d\theta^2 - r^2 \sin^2\theta \, d\phi^2 + \gamma dt^2 \quad\text{(38·8)},$$

where $\gamma = 1 - 2m/r$, is a particular solution of Einstein's gravitational equations $G_{\mu\nu} = 0$. The solution in this form was first obtained by Schwarzschild.

39. Planetary orbits.

According to (15·7) the track of a particle moving freely in the space-time given by (38·8) is determined by the equations of a geodesic (28·5), viz.

$$\frac{d^2x_\alpha}{ds^2} + \{\mu\nu, \alpha\}\frac{dx_\mu}{ds}\frac{dx_\nu}{ds} = 0 \quad\text{(39·1)}.$$

Taking first $\alpha = 2$, the surviving terms are

$$\frac{d^2x_2}{ds^2} + \{12, 2\}\frac{dx_1}{ds}\frac{dx_2}{ds} + \{21, 2\}\frac{dx_2}{ds}\frac{dx_1}{ds} + \{33, 2\}\frac{dx_3}{ds}\frac{dx_3}{ds} = 0,$$

or using (38·5)

$$\frac{d^2\theta}{ds^2} + \frac{2}{r}\frac{dr}{ds}\frac{d\theta}{ds} - \cos\theta\sin\theta\left(\frac{d\phi}{ds}\right)^2 = 0 \quad\text{(39·2)}.$$

Choose coordinates so that the particle moves initially in the plane $\theta = \tfrac{1}{2}\pi$. Then $d\theta/ds = 0$ and $\cos\theta = 0$ initially, so that $d^2\theta/ds^2 = 0$. The particle therefore continues to move in this plane, and we may simplify the remaining equations by putting $\theta = \tfrac{1}{2}\pi$ throughout. The equations for $\alpha = 1, 3, 4$ are found in like manner, viz.

$$\frac{d^2r}{ds^2} + \tfrac{1}{2}\lambda'\left(\frac{dr}{ds}\right)^2 - re^{-\lambda}\left(\frac{d\phi}{ds}\right)^2 + \tfrac{1}{2}e^{\nu-\lambda}\nu'\left(\frac{dt}{ds}\right)^2 = 0 \quad\text{(39·31)},$$

$$\frac{d^2\phi}{ds^2} + \frac{2}{r}\frac{dr}{ds}\frac{d\phi}{ds} = 0 \quad\text{.......................(39·32)},$$

$$\frac{d^2t}{ds^2} + \nu'\frac{dr}{ds}\frac{dt}{ds} = 0 \quad\text{.......................(39·33)}.$$

The last two equations may be integrated immediately, giving

$$r^2\frac{d\phi}{ds} = h \quad\text{.................................(39·41)},$$

$$\frac{dt}{ds} = ce^{-\nu} = c/\gamma \quad\text{.......................(39·42)},$$

where h and c are constants of integration.

Instead of troubling to integrate (39·31) we can use in place of it (38·8) which plays here the part of an integral of energy. It gives

$$\gamma^{-1}\left(\frac{dr}{ds}\right)^2 + r^2\left(\frac{d\phi}{ds}\right)^2 - \gamma\left(\frac{dt}{ds}\right)^2 = -1 \quad\text{...........(39·43)}.$$

Eliminating dt and ds by means of (39·41) and (39·42)

$$\frac{1}{\gamma}\left(\frac{h}{r^2}\frac{dr}{d\phi}\right)^2 + \frac{h^2}{r^2} - \frac{c^2}{\gamma} = -1 \quad\text{.................(39·44)},$$

whence, multiplying through by γ or $(1-2m/r)$,

$$\left(\frac{h}{r^2}\frac{dr}{d\phi}\right)^2 + \frac{h^2}{r^2} = c^2 - 1 + \frac{2m}{r} + \frac{2m}{r}\cdot\frac{h^2}{r^2},$$

or writing $1/r = u$,

$$\left(\frac{du}{d\phi}\right)^2 + u^2 = \frac{c^2-1}{h^2} + \frac{2m}{h^2}u + 2mu^3 \quad\text{.............(39·5)}.$$

Differentiating with respect to ϕ, and removing the factor $\frac{du}{d\phi}$,

$$\frac{d^2u}{d\phi^2} + u = \frac{m}{h^2} + 3mu^2 \quad\text{.......................(39·61)},$$

with
$$r^2\frac{d\phi}{ds} = h \quad\text{.................................(39·62)}.$$

Compare these with the equations of a Newtonian orbit

$$\frac{d^2u}{d\phi^2} + u = \frac{m}{h^2} \quad\text{.........................(39·71)}$$

with
$$r^2\frac{d\phi}{dt} = h \quad\text{.........................(39·72)}.$$

In (39·61) the ratio of $3mu^2$ to m/h^2 is $3h^2u^2$, or by (39·62)

$$3\left(r\frac{d\phi}{ds}\right)^2.$$

For ordinary speeds this is an extremely small quantity—practically three times the square of the transverse velocity in terms of the velocity of light. For example, this ratio for the earth is ·00000003. In practical cases the extra term in (39·61) will represent an almost inappreciable correction to the Newtonian orbit (39·71).

Again in (39·62) and (39·72) the difference between ds and dt is equally insignificant, even if we were sure of what is meant by dt in the Newtonian theory. The *proper-time* for the body is ds, and it might perhaps be urged that dt in equation (39·72) is intended to refer to this; but on the other hand s cannot be used as a coordinate since ds is not a complete differential, and Newton's "time" is always assumed to be a coordinate.

Thus it appears that a particle moving in the field here discussed will behave as though it were under the influence of the Newtonian force exerted by a particle of gravitational mass m at the origin, the motion agreeing with the Newtonian theory to the order of accuracy for which that theory has been confirmed by observation.

By showing that our solution satisfies $G_{\mu\nu} = 0$, we have proved that it describes a possible state of the world which might be met with in nature under suitable conditions. By deducing the orbit of a particle, we have discovered how that state of the world would be recognised observationally if it did exist. In this way we conclude that the space-time field represented by (38·8) is the one which accompanies (or "is due to") a particle of mass m at the origin.

The gravitational mass m is the measure adopted in the Newtonian theory of the power of the particle in causing a field of acceleration around it, the units being here chosen so that the velocity of light and the constant of gravitation are both unity. It should be noticed that we have as yet given no reason to expect that m in the present chapter has anything to do with the m introduced in § 12 to measure the inertial properties of the particle.

For a circular orbit the Newtonian theory gives

$$m = \omega^2 r^3 = v^2 r,$$

the constant of gravitation being unity. Applying this to the earth, $v = 30$ km. per sec. $= 10^{-4}$ in terms of the velocity of light, and $r = 1{\cdot}5 \,.\, 10^8$ km. Hence the mass m of the sun is approximately 1·5 kilometres. The mass of the earth is 1/300,000th of this, or about 5 millimetres*.

More accurately, the mass of the sun, $1{\cdot}99 \,.\, 10^{33}$ grams, becomes in gravitational units 1·47 kilometres; and other masses are converted in a like proportion.

* Objection is sometimes taken to the use of a centimetre as a unit of gravitational (i.e. gravitation-exerting) mass; but the same objection would apply to the use of a gram, since the gram is properly a measure of a different property of the particle, viz. its *inertia*. Our constant of integration m is clearly a length and the reader may, if he wishes to make this clear, call it the gravitational radius instead of the gravitational mass. But when it is realised that the gravitational radius in centimetres, the inertia in grams, and the energy in ergs, are merely measure-numbers in different codes of the *same* intrinsic quality of the particle, it seems unduly pedantic to insist on the older discrimination of these units which grew up on the assumption that they measured qualities which were radically different.

40. The advance of perihelion.

The equation (39·5) for the orbit of a planet can be integrated in terms of elliptic functions; but we obtain the astronomical results more directly by a method of successive approximation. We proceed from equation (39·61)

$$\frac{d^2u}{d\phi^2} + u = \frac{m}{h^2} + 3mu^2 \quad \text{.........................(40·1).}$$

Neglecting the small term $3mu^2$, the solution is

$$u = \frac{m}{h^2}(1 + e\cos(\phi - \varpi)) \quad \text{....................(40·2),}$$

as in Newtonian dynamics. The constants of integration, e and ϖ, are the eccentricity and longitude of perihelion.

Substitute this first approximation in the small term $3mu^2$, then (40·1) becomes

$$\frac{d^2u}{d\phi^2} + u = \frac{m}{h^2} + 3\frac{m^3}{h^4} + 6\frac{m^3}{h^4}e\cos(\phi - \varpi) + \frac{3}{2}\frac{m^3}{h^4}e^2(1 + \cos 2(\phi - \varpi)) \quad \text{......(40·3).}$$

Of the additional terms the only one which can produce an effect within the range of observation is the term in $\cos(\phi - \varpi)$; this is of the right period to produce a continually increasing effect by resonance. Remembering that the particular integral of

$$\frac{d^2u}{d\phi^2} + u = A\cos\phi$$

is

$$u = \tfrac{1}{2}A\phi\sin\phi,$$

this term gives a part of u

$$u_1 = 3\frac{m^3}{h^4}e\phi\sin(\phi - \varpi) \quad \text{....................(40·4),}$$

which must be added to the complementary integral (40·2). Thus the second approximation is

$$u = \frac{m}{h^2}\left(1 + e\cos(\phi - \varpi) + 3\frac{m^2}{h^2}e\phi\sin(\phi - \varpi)\right)$$

$$= \frac{m}{h^2}(1 + e\cos(\phi - \varpi - \delta\varpi)),$$

where

$$\delta\varpi = 3\frac{m^2}{h^2}\phi \quad \text{............................(40·5),}$$

and $(\delta\varpi)^2$ is neglected.

Whilst the planet moves through 1 revolution, the perihelion ϖ advances a fraction of a revolution equal to

$$\frac{\delta\varpi}{\phi} = \frac{3m^2}{h^2} = \frac{3m}{a(1 - e^2)} \quad \text{.........................(40·6),}$$

using the well-known equation of areas $h^2 = ml = ma(1 - e^2)$.

Another form is obtained by using Kepler's third law,

$$m = \left(\frac{2\pi}{T}\right)^2 a^3,$$

giving
$$\frac{\delta\varpi}{\phi} = \frac{12\pi^2 a^2}{c^2 T^2 (1-e^2)} \quad\text{.........................(40·7)},$$

where T is the period, and the velocity of light c has been reinstated.

This advance of the perihelion is appreciable in the case of the planet Mercury, and the predicted value is confirmed by observation.

For a circular orbit we put dr/ds, $d^2r/ds^2 = 0$, so that (39·31) becomes

$$-re^{-\lambda}\left(\frac{d\phi}{ds}\right)^2 + \tfrac{1}{2}e^{\nu-\lambda}\nu'\left(\frac{dt}{ds}\right)^2 = 0.$$

Whence
$$\left(\frac{d\phi}{dt}\right)^2 = \tfrac{1}{2}e^{\nu}\nu'/r = \tfrac{1}{2}\gamma'/r$$
$$= m/r^3,$$

so that Kepler's third law is *accurately* fulfilled. This result has no observational significance, being merely a property of the particular definition of r here adopted. Slightly different coordinate-systems exist which might with equal right claim to correspond to polar coordinates in flat space-time; and for these Kepler's third law would no longer be exact.

We have to be on our guard against results of this latter kind which would only be of interest if the radius-vector were a directly measured quantity instead of a conventional coordinate. The advance of perihelion is a phenomenon of a different category. Clearly the number of years required for an eccentric orbit to make a complete revolution returning to its original position is capable of observational test, unaffected by any convention used in defining the exact length of the radius-vector.

For the four inner planets the following table gives the corrections to the centennial motion of perihelion predicted by Einstein's theory:

	$\delta\varpi$	$e\delta\varpi$
Mercury	+42″·9	+8″·82
Venus	+ 8·6	+ 0·05
Earth	+ 3·8	+ 0·07
Mars	+ 1·35	+ 0·13

The product $e\delta\varpi$ is a better measure of the observable effect to be looked for, and the correction is only appreciable in the case of Mercury. After applying these corrections to $e\delta\varpi$, the following discrepancies between theory and observation remain in the secular changes of the elements of the inner planets, i and Ω being the inclination and the longitude of the node:

	$e\delta\varpi$	δe	$\sin i\delta\Omega$	δi
Mercury	−0″·58 ± 0″·29	−0″·88 ± 0″·33	+0″·46 ± 0″·34	+0″·38 ± 0″·54
Venus	− 0·11 ± 0·17	+ 0·21 ± 0·21	+ 0·53 ± 0·12	+ 0·38 ± 0·22
Earth	0·00 ± 0·09	+ 0·02 ± 0·07	... ...	− 0·22 ± 0·18
Mars	+ 0·51 ± 0·23	+ 0·29 ± 0·18	− 0·11 ± 0·15	− 0·01 ± 0·13

The probable errors here given include errors of observation, and also errors in the theory due to uncertainty of the masses of the planets. The positive sign indicates excess of observed motion over theoretical motion*.

Einstein's correction to the perihelion of Mercury has removed the principal discordance in the table, which on the Newtonian theory was nearly 30 times the probable error. Of the 15 residuals 8 exceed the probable error, and 3 exceed twice the probable error—as nearly as possible the proper proportion. But whereas we should expect the greatest residual to be about 3 times the probable error, the residual of the node of Venus is rather excessive at $4\frac{1}{2}$ times the probable error, and may perhaps be a genuine discordance. Einstein's theory throws no light on the cause of this discordance.

41. The deflection of light.

For motion with the speed of light $ds = 0$, so that by (39·62) $h = \infty$, and the orbit (39·61) reduces to

$$\frac{d^2u}{d\phi^2} + u = 3mu^2 \quad \text{.........................(41·1).}$$

The track of a light-pulse is also given by a geodesic with $ds = 0$ according to (15·8). Accordingly the orbit (41·1) gives the path of a ray of light.

We integrate by successive approximation. Neglecting $3mu^2$ the solution of the approximate equation

$$\frac{d^2u}{d\phi^2} + u = 0$$

is the straight line

$$u = \frac{\cos\phi}{R} \quad \text{..........................(41·2).}$$

Substituting this in the small term $3mu^2$, we have

$$\frac{d^2u}{d\phi^2} + u = \frac{3m}{R^2}\cos^2\phi.$$

A particular integral of this equation is

$$u_1 = \frac{m}{R^2}(\cos^2\phi + 2\sin^2\phi),$$

so that the complete second approximation is

$$u = \frac{\cos\phi}{R} + \frac{m}{R^2}(\cos^2\phi + 2\sin^2\phi) \quad \text{..............(41·3).}$$

Multiply through by rR,

$$R = r\cos\phi + \frac{m}{R}(r\cos^2\phi + 2r\sin^2\phi),$$

or in rectangular coordinates, $x = r\cos\phi$, $y = r\sin\phi$,

$$x = R - \frac{m}{R}\frac{x^2 + 2y^2}{\sqrt{(x^2 + y^2)}} \quad \text{........................(41·4).}$$

* Newcomb, *Astronomical Constants*. His results have been slightly corrected by using a modern value of the constant of precession in the above table; see de Sitter, *Monthly Notices*, vol. 76, p. 728.

The second term measures the very slight deviation from the straight line $x = R$. The asymptotes are found by taking y very large compared with x. The equation then becomes

$$x = R - \frac{m}{R}(\pm 2y),$$

and the small angle between the asymptotes is (in circular measure)

$$\frac{4m}{R}.$$

For a ray grazing the sun's limb, $m = 1{\cdot}47$ km., $R = 697{,}000$ km., so that the deflection should be $1''{\cdot}75$ The observed values obtained by the British eclipse expeditions in 1919 were

Sobral expedition	$1''{\cdot}98 \pm 0''{\cdot}12$
Principe expedition	$1''{\cdot}61 \pm 0''{\cdot}30$

It has been explained in *Space, Time and Gravitation* that this deflection is double that which might have been predicted on the Newtonian theory. In this connection the following paradox has been remarked. Since the curvature of the light-track is doubled, the acceleration of the light at each point is double the Newtonian acceleration; whereas for a slowly moving object the acceleration is practically the same as the Newtonian acceleration. To a man in a lift descending with acceleration m/r^2 the tracks of ordinary particles will appear to be straight lines; but it looks as though it would require an acceleration $2m/r^2$ to straighten out the light-tracks. Does not this contradict the principle of equivalence?

The fallacy lies in a confusion between two meanings of the word "curvature." The *coordinate* curvature obtained from the equation of the track (41·4) is not the *geodesic* curvature. The latter is the curvature with which the local observer—the man in the lift—is concerned. Consider the curved light-track traversing the hummock corresponding to the sun's field; its curvature can be reckoned by projecting it either on the base of the hummock or on the tangent plane at any point. The curvatures of the two projections will generally be different. The projection into Euclidean coordinates (x, y) used in (41·4) is the projection on the base of the hummock; in applying the principle of equivalence the projection is on the tangent plane, since we consider a region of the curved world so small that it cannot be discriminated from its tangent plane.

42. Displacement of the Fraunhofer lines.

Consider a number of similar atoms vibrating at different points in the region. Let the atoms be momentarily at rest in our coordinate-system (r, θ, ϕ, t). The test of similarity of the atoms is that corresponding intervals should be equal, and accordingly the *interval* of vibration of all the atoms will be the same.

Since the atoms are at rest we set $dr, d\theta, d\phi = 0$ in (38·8), so that

$$ds^2 = \gamma dt^2 \quad \text{..............................(42·1)}.$$

Accordingly the *times* of vibration of the differently placed atoms will be inversely proportional to $\sqrt{\gamma}$.

Our system of coordinates is a *static system*, that is to say the $g_{\mu\nu}$ do not change with the time. (An arbitrary coordinate-system has not generally this property; and further when we have to take account of two or more attracting bodies, it is in most cases impossible to find a strictly static system of coordinates.) Taking an observer at rest in the system (r, θ, ϕ, t) a wave emitted by one of the atoms will reach him at a certain time δt after it leaves the atom; and owing to the static condition this time-lag remains constant for subsequent waves. Consequently the waves are received at the same time-periods as they are emitted. We are therefore able to compare the time-periods dt of the different atoms, by comparing the periods of the waves received from them, and can verify experimentally their dependence on the value of $\sqrt{\gamma}$ at the place where they were emitted. Naturally the most hopeful test is the comparison of the waves received from a solar and a terrestrial atom whose periods should be in the ratio 1·00000212 : 1. For wave-length 4000 Å, this amounts to a relative displacement of 0·0082 Å of the respective spectral lines. This displacement is believed to have been verified observationally, but the test is difficult and perhaps uncertain. The theory has been strikingly confirmed in the spectrum of the Companion of Sirius where the predicted displacement was 30 times larger.

The quantity dt is merely an auxiliary quantity introduced through the equation (38·8) which defines it. The fact that it is carried to us unchanged by light-waves is not of any physical interest, since dt was *defined* in such a way that this must happen. The absolute quantity ds, the interval of the vibration, is not carried to us unchanged, but becomes gradually modified as the waves take their course through the non-Euclidean space-time. It is in transmission through the solar system that the absolute difference is introduced into the waves, which the experiment hopes to detect.

The argument refers to *similar* atoms and the question remains whether, for example, a hydrogen atom on the sun is truly similar to a hydrogen atom on the earth. Strictly speaking it cannot be exactly similar because it is in a different kind of space-time, in which it would be impossible to make a finite structure exactly similar to one existing in the space-time near the earth. But if the interval of vibration of the hydrogen atom is modified by the kind of space-time in which it lies, the difference must be dependent on some invariant of the space-time. The simplest invariant which differs at the sun and the earth is the square of the length of the Riemann-Christoffel tensor, viz.

$$B^{\epsilon}_{\mu\nu\sigma} B^{\mu\nu\sigma}_{\epsilon}.$$

The value of this can be calculated from (38·8) by the method used in that section for calculating the $G_{\mu\nu}$. The result is

$$48 \frac{m^2}{r^6}.$$

By consideration of dimensions it seems clear that the proportionate change of ds would be of the order

$$\frac{\sigma^4 m^2}{r^6},$$

where σ is the radius of the atom; there does not seem to be any other length concerned. For a comparison of solar and terrestrial atoms this would be about 10^{-100}. In any case it seems impossible to construct from the invariants of space-time a term which would compensate the predicted shift of the spectral lines, which is proportional to m/r.

43. Isotropic coordinates.

We can transform the expression for the interval (38·8) by making the substitution

$$r = \left(1 + \frac{m}{2r_1}\right)^2 r_1 \quad \text{.............................(43·1)},$$

so that

$$dr = \left(1 - \frac{m^2}{4r_1^2}\right) dr_1,$$

$$\gamma = \left(1 - \frac{m}{2r_1}\right)^2 \Big/ \left(1 + \frac{m}{2r_1}\right)^2$$

Then (38·8) becomes

$$ds^2 = -(1 + m/2r_1)^4 (dr_1^2 + r_1^2 d\theta^2 + r_1^2 \sin^2\theta\, d\phi^2) + \frac{(1 - m/2r_1)^2}{(1 + m/2r_1)^2} dt^2 \quad \text{...(43·2)}.$$

The coordinates (r_1, θ, ϕ) are called *isotropic* polar coordinates. The corresponding isotropic rectangular coordinates are obtained by putting

$$x = r_1 \sin\theta \cos\phi, \quad y = r_1 \sin\theta \sin\phi, \quad z = r_1 \cos\theta,$$

giving

$$ds^2 = -(1 + m/2r_1)^4 (dx^2 + dy^2 + dz^2) + \frac{(1 - m/2r_1)^2}{(1 + m/2r_1)^2} dt^2 \quad \text{...(43·3)},$$

with

$$r_1 = \sqrt{(x^2 + y^2 + z^2)}.$$

This system has some advantages. For example, to obtain the motion of a light-pulse we set $ds = 0$ in (43·3). This gives

$$\left(\frac{dx}{dt}\right)^2 + \left(\frac{dy}{dt}\right)^2 + \left(\frac{dz}{dt}\right)^2 = \frac{(1 - m/2r_1)^2}{(1 + m/2r_1)^6}.$$

At a distance r_1 from the origin the velocity of light is accordingly

$$\frac{(1 - m/2r_1)}{(1 + m/2r_1)^3} \quad \text{.............................(43·4)}$$

in all directions. For the original coordinates of (38·8) the velocity of light is not the same for the radial and transverse directions.

Again in the isotropic system the coordinate length $(\sqrt{(dx^2 + dy^2 + dz^2)})$ of a small rod which is rigid (ds = constant) does not alter when the orientation of the rod is altered. This system of coordinates is naturally arrived at when we partition space by rigid scales or by light-triangulations in a small region, e.g. in terrestrial measurements. Since the ultimate measurements involved in any observation are carried out in a terrestrial laboratory we ought, strictly speaking, always to employ the isotropic system which conforms to assumptions made in those measurements*. But on the earth the quantity m/r is negligibly small, so that the two systems coalesce with one another and with Euclidean coordinates. Non-Euclidean geometry is only required in the theoretical part of the investigation—the laws of planetary motion and propagation of light through regions where m/r is not negligible; as soon as the light-waves have been safely steered into the terrestrial observatory, the need for non-Euclidean geometry is at an end, and the difference between the isotropic and non-isotropic systems practically disappears.

In either system the forward velocity of light along any line is equal to the backward velocity. Consequently the coordinate t conforms to the convention (§ 11) that simultaneity may be determined by means of light-signals. If we have a clock at A and send a light-signal at time t_A which reaches B and is immediately reflected so as to return to A at time t_A', the time of arrival at B will be $\frac{1}{2}(t_A + t_A')$ just as in the special relativity theory. But the alternative convention, that simultaneity can be determined by slow transport of chronometers, breaks down when there is a gravitational field. This is evident from § 42, since the time-rate of a clock will depend on its position in the field. In any case slow transport of a clock is unrealisable because of the acceleration which all objects must submit to.

The isotropic system could have been found directly by seeking particular solutions of Einstein's equations having the form (38·12), or

$$ds^2 = -e^\lambda dr^2 - e^\mu (r^2 d\theta^2 + r^2 \sin^2\theta\, d\phi^2) + e^\nu dt^2,$$

where λ, μ, ν are functions of r. By the method of § 38, we find

$$\left.\begin{aligned}
G_{11} &= \mu'' + \tfrac{1}{2}\nu'' + \frac{2}{r}\mu' - \frac{1}{r}\lambda' + \tfrac{1}{2}\mu'^2 - \tfrac{1}{2}\lambda'\mu' - \tfrac{1}{4}\lambda'\nu' + \tfrac{1}{4}\nu'^2 \\
G_{22} &= e^{\mu-\lambda}\left[1 + 2r\mu' + \tfrac{1}{2}r(\nu' - \lambda') + \tfrac{1}{2}r^2\mu'' + \tfrac{1}{2}r^2\mu'(\mu' + \tfrac{1}{2}\nu' - \tfrac{1}{2}\lambda')\right] - 1 \\
G_{33} &= G_{22}\sin^2\theta \\
G_{44} &= -e^{\nu-\lambda}\left[\tfrac{1}{2}\nu'' + \frac{1}{r}\nu' + \tfrac{1}{2}\nu'\mu' - \tfrac{1}{4}\lambda'\nu' + \tfrac{1}{4}\nu'^2\right]
\end{aligned}\right\} \qquad \text{......(43·5).}$$

The others are zero.

Owing to an identical relation between G_{11}, G_{22} and G_{44}, the vanishing of this tensor gives only two equations to determine the three unknowns λ, μ, ν. There exists therefore an infinite series of particular solutions, differing according to the third equation between λ, μ, ν which is at our disposal. The two solutions hitherto considered are obtained by taking $\mu = 0$, and $\lambda = \mu$, respectively. The same series of solutions is obtained in a simpler way by substituting arbitrary functions of r instead of r in (38·8).

* But the terrestrial laboratory is falling freely towards the sun, and is therefore accelerated relatively to the coordinates (x, y, z, t).

The possibility of substituting any function of r for r without destroying the spherical symmetry is obvious from the fact that a coordinate is merely an identification-number; but analytically this possibility is bound up with the existence of an identical relation between G_{11}, G_{22} and G_{44}, which makes the equations too few to determine a unique solution.

This introduces us to a theorem of great consequence in our later work. If Einstein's ten equations $G_{\mu\nu} = 0$ were all independent, the ten $g_{\mu\nu}$ would be uniquely determined by them (the boundary conditions being specified). The expression for ds^2 would be unique and no transformation of coordinates would be possible. Since we know that we can transform coordinates as we please, there must exist identical relations between the ten $G_{\mu\nu}$; and these will be found in § 52.

44. Problem of two bodies—Motion of the moon.

The field described by the $g_{\mu\nu}$ may be (artificially) divided into a *field of pure inertia* represented by the Galilean values, and a *field of force* represented by the deviations of the $g_{\mu\nu}$ from the Galilean values. It is not possible to superpose the fields of force due to two attracting particles; because the sum of the two solutions will not satisfy $G_{\mu\nu} = 0$, these equations being non-linear in the $g_{\mu\nu}$.

No solution of Einstein's equations has yet been found for a field with two singularities or particles. The simplest case to be examined would be that of two equal particles revolving in circular orbits round their centre of mass. Apparently there should exist a statical solution with two equal singularities; but the conditions at infinity would differ from those adopted for a single particle since the axes corresponding to the static solution constitute what is called a rotating system. The solution has not been found, and it is even possible that no such statical solution exists. I do not think it has yet been proved that two bodies can revolve without radiation of energy by gravitational waves. In discussions of this radiation problem there is a tendency to beg the question; it is not sufficient to constrain the particles to revolve uniformly, then calculate the resulting gravitational waves, and verify that the radiation of gravitational energy across an infinite sphere is zero. That shows that a statical solution is not obviously inconsistent with itself, but does not demonstrate its possibility.

The problem of two bodies on Einstein's theory remains an outstanding challenge to mathematicians—like the problem of three bodies on Newton's theory.

For practical purposes methods of approximation will suffice. We shall consider the problem of the field due to the combined attractions of the earth and sun, and apply it to find the modifications of the moon's orbit required by the new law of gravitation. The problem has been treated in considerable detail by de Sitter*. We shall not here attempt a complete survey of the

* *Monthly Notices*, vol. 77, p. 155.

problem; but we shall seek out the largest effects to be looked for in refined observations. There are three sources of fresh perturbations:

(1) The sun's attraction is not accurately given by Newton's law, and the solar perturbations of the moon's orbit will require corrections.

(2) Cross-terms between the sun's and the earth's fields of force will arise, since these are not additive.

(3) The earth's field is altered and would *inter alia* give rise to a motion of the lunar perigee analogous to the motion of Mercury's perihelion. It is easily calculated that this is far too small to be detected.

If Ω_S, Ω_E are the Newtonian potentials of the sun and earth, the leading terms of (1), (2), (3) will be relatively of order of magnitude

$$\Omega_S^2, \quad \Omega_S\Omega_E, \quad \Omega_E.$$

For the moon $\Omega_S = 750\Omega_E$. We may therefore confine attention to terms of type (1). If these prove to be too small to be detected, the others will presumably be not worth pursuing.

We were able to work out the planetary orbits from Einstein's law independently of the Newtonian theory; but in the problem of the moon's motion we must concentrate attention on the *difference* between Einstein's and Newton's formulae if we are to avoid repeating the whole labour of the classical lunar theory. In order to make this comparison we transform (39·31) and (39·32) so that t is used as the independent variable.

$$\frac{d^2}{ds^2} = \left(\frac{dt}{ds}\right)^2 \frac{d^2}{dt^2} + \frac{dt}{ds}\frac{d}{dt}\left(\frac{dt}{ds}\right)\frac{d}{dt}$$

$$= \left(\frac{dt}{ds}\right)^2 \left(\frac{d^2}{dt^2} + \lambda' \frac{dr}{dt}\frac{d}{dt}\right) \qquad \text{by (39·42).}$$

Hence the equations (39·31) and (39·32) become

$$\frac{d^2r}{dt^2} + \tfrac{3}{2}\lambda' \left(\frac{dr}{dt}\right)^2 - re^{-\lambda}\left(\frac{d\phi}{dt}\right)^2 + \tfrac{1}{2}e^{2\nu}\nu' = 0,$$

$$\frac{d^2\phi}{dt^2} + \lambda' \frac{dr}{dt}\frac{d\phi}{dt} + \frac{2}{r}\frac{dr}{dt}\frac{d\phi}{dt} = 0.$$

Whence

$$\left.\begin{aligned} \frac{d^2r}{dt^2} - r\left(\frac{d\phi}{dt}\right)^2 + \frac{m}{r^2} &= R \\ r\left(\frac{d^2\phi}{dt^2} - \frac{2}{r}\frac{dr}{dt}\frac{d\phi}{dt}\right) &= \Phi \end{aligned}\right\} \qquad \text{(44·1)},$$

where

$$\left.\begin{aligned} R &= -\tfrac{3}{2}\lambda' u^2 - \frac{2m}{r^2}v^2 + \frac{2m^2}{r^3} \\ \Phi &= -\lambda' uv \end{aligned}\right\} \qquad \text{(44·21)}$$

and

$$u = dr/dt, \quad v = rd\phi/dt.$$

Equations (44·1) show that R and Φ are the radial and transverse perturbing forces which Einstein's theory adds to the classical dynamics. To a sufficient approximation $\lambda' = -2m/r^2$, so that

$$\left.\begin{aligned} R &= \frac{m}{r^2}(3u^2 - 2v^2) + \frac{2m^2}{r^3} \\ \Phi &= \frac{m}{r^2} \cdot 2uv \end{aligned}\right\} \qquad (44{\cdot}22).$$

In three-dimensional problems the perturbing forces become

$$\left.\begin{aligned} R &= \frac{m}{r^2}(3u^2 - 2v^2 - 2w^2) + \frac{2m^2}{r^3} \\ \Phi &= \frac{m}{r^2} \cdot 2uv \\ Z &= \frac{m}{r^2} \cdot 2uw \end{aligned}\right\} \qquad (44{\cdot}23).$$

It must be pointed out that these perturbing forces are Einstein's corrections to the law of central force m/r^2, where r is *the coordinate used in our previous work.* Whether these forces represent the actual differences between Einstein's and Newton's laws depends on what Newton's r is supposed to signify. De Sitter, making a slightly different choice of r, obtains different expressions for R, Φ*. One cannot say that one set of perturbing forces rather than the other represents the difference from the older theory, because the older theory was not sufficiently explicit. The classical lunar theory has been worked out on the basis of the law m/r^2; the ambiguous quantity r occurs in the results, and according as we have assigned to it one meaning or another, so we shall have to apply different corrections to those results. But the final comparison with observation does not depend on the choice of the intermediary quantity r.

Take fixed rectangular axes referred to the ecliptic with the sun as origin, and let

$(a, 0, 0)$ be the coordinates of the earth at the instant considered,

(x, y, z) the coordinates of the moon relative to the earth.

Taking the earth's orbit to be circular and treating the mass of the moon as infinitesimal, the earth's velocity will be $(0, v, 0)$, where $v^2 = m/a$.

To find the difference of the forces R, Φ, Z on the moon and on the earth, we differentiate (44·23) and set

$$\delta r = x, \quad \delta(u, v, w) = (dx/dt, dy/dt, dz/dt),$$

and, after the differentiation,

$$r = a, \quad (u, v, w) = (0, v, 0).$$

* *Monthly Notices*, vol. 76, p. 723, equations (53).

The result will give the perturbing forces on the moon's motion relative to the earth, viz.

$$\left.\begin{aligned} \delta R = X &= \frac{4mx}{a^3} v^2 - \frac{6m^2 x}{a^4} - \frac{4m}{a^2} v \frac{dy}{dt} = -\frac{2m^2 x}{a^4} - \frac{4m}{a^2} v \frac{dy}{dt} \\ \delta \Phi = Y &= \frac{2m}{a^2} v \frac{dx}{dt} \\ Z &= 0 \end{aligned}\right\} \quad \ldots(44{\cdot}3).$$

We shall omit the term $-2m^2x/a^4$ in X. It can be verified that it gives no important observable effects. It produces only an apparent distortion of the orbit attributable to our use of non-isotropic coordinates (§ 43). Transforming to new axes (ξ, η) rotated through an angle θ with respect to (x, y) the remaining forces become

$$\left.\begin{aligned} \Xi &= \frac{m}{a^2} v \left(-2\cos\theta\sin\theta\frac{d\xi}{dt} - (4\cos^2\theta + 2\sin^2\theta)\frac{d\eta}{dt}\right) \\ \mathrm{H} &= \frac{m}{a^2} v \left(\;\; 2\cos\theta\sin\theta\frac{d\eta}{dt} + (4\sin^2\theta + 2\cos^2\theta)\frac{d\xi}{dt}\right) \end{aligned}\right\} \quad \ldots(44{\cdot}4).$$

We keep the axes (ξ, η) permanently fixed; the angle θ which gives the direction of the sun (the old axis of x) will change uniformly, and in the long run take all values with equal frequency independently of the moon's position in its orbit. We can only hope to observe the secular effects of the small forces Ξ, H, accumulated through a long period of time. Accordingly, averaging the trigonometrical functions, the secular terms are

$$\left.\begin{aligned} \Xi &= -3\frac{m}{a^2} v \frac{d\eta}{dt} = -2\omega\frac{d\eta}{dt} \\ \mathrm{H} &= \;\;\; 3\frac{m}{a^2} v \frac{d\xi}{dt} = \;\;\; 2\omega\frac{d\xi}{dt} \end{aligned}\right\} \quad \ldots\ldots\ldots\ldots\ldots\ldots\ldots(44{\cdot}5),$$

where $\omega = \frac{3}{2} mv/a^2$ ....(44·6).

If (F_ξ, F_η) is the Newtonian force, the equations of motion including these secular perturbing forces will be

$$\frac{d^2\xi}{dt^2} + 2\omega\frac{d\eta}{dt} = F_\xi, \qquad \frac{d^2\eta}{dt^2} - 2\omega\frac{d\xi}{dt} = F_\eta \quad \ldots\ldots\ldots\ldots(44{\cdot}7).$$

It is easily seen that ω is a very small quantity, so that ω^2 is negligible The equations (44·7) are then recognised as the Newtonian equations referred to axes rotating with angular velocity $-\omega$. Thus if we take the Newtonian orbit and give it an angular velocity $+\omega$, the result will be the solution of (44·7). The leading correction to the lunar theory obtained from Einstein's equations is a precessional effect, indicating that the classical results refer to a frame of reference advancing with angular velocity ω compared with the general inertial frame of the solar system.

From this cause the moon's node and perigee will advance with velocity ω. If Ω is the earth's angular velocity

$$\frac{\omega}{\Omega} = \frac{3}{2}\frac{m}{a} = \tfrac{3}{2}\,.\,10^{-8}.$$

Hence the advance of perigee and node in a century is

$$3\pi \,.\, 10^{-6} \text{ radians} = 1''\cdot 94.$$

We may notice the very simple theoretical relation that Einstein's corrections cause an advance of the moon's perigee which is *one half* the advance of the earth's perihelion.

Neither the lunar theory nor the observations are as yet carried quite far enough to take account of this small effect; but it is only a little below the limit of detection. The result agrees with de Sitter's value except in the second decimal place which is only approximate.

There are well-known irregular fluctuations in the moon's longitude which attain rather large values; but it is generally considered that these are not of a type which can be explained by any amendment of gravitational theory and their origin must be looked for in other directions. At any rate Einstein's theory throws no light on them.

The advance of 1″·94 per century has not exclusive reference to the moon; in fact the elements of the moon's orbit do not appear in (44·6). It represents a property of the space surrounding the earth—a precession of the inertial frame in this region relative to the general inertial frame of the sidereal system. If the earth's rotation could be accurately measured by Foucault's pendulum or by gyrostatic experiments, the result would differ from the rotation relative to the fixed stars by this amount. This result seems to have been first pointed out by J. A. Schouten. One of the difficulties most often urged against the relativity theory is that the earth's rotation relative to the mean of the fixed stars appears to be an absolute quantity determinable by dynamical experiments on the earth*; it is therefore of interest to find that these two rotations are not exactly the same, and the earth's rotation relative to the stellar system (supposed to agree with the general inertial frame of the universe) *cannot be determined except by astronomical observations.*

The argument of the relativist is that the observed effect on Foucault's pendulum can be accounted for indifferently by a field of force or by rotation. The anti-relativist replies that the field of force is clearly a mathematical fiction, and the only possible *physical cause* must be absolute rotation. It is pointed out to him that nothing essential is gained by choosing the so-called non-rotating axes, because in any case the main part of the field of force remains, viz. terrestrial gravitation. He retorts that with his non-rotating axes he has succeeded in making the field of force vanish at infinity, so that the residuum is accounted for as a local disturbance by the earth; whereas, if axes fixed in the earth are admitted, the corresponding field of force becomes larger and larger as we recede from the earth, so that the relativist demands enormous forces in distant parts for which no physical cause can be assigned. Suppose, however, that the earth's rotation were much slower than it is now,

* *Space, Time and Gravitation*, p. 152.

and that Foucault's experiment had indicated a rotation of only $-1''\cdot 94$ per century. Our two disputants on the cloud-bound planet would no doubt carry on a long argument as to whether this was essentially an absolute rotation of the earth in space, the irony of the situation being that the earth all the while was non-rotating in the anti-relativist's sense, and the proposed transformation to allow for the Foucault rotation would actually have the effect of introducing the enormous field of force in distant parts of space which was so much objected to. When the origin of the $1''\cdot 94$ has been traced as in the foregoing investigation, the anti-relativist who has been arguing that the observed effect is definitely caused by rotation, must change his position and maintain that it is definitely due to a gravitational perturbation exerted by the sun on Foucault's pendulum; the relativist holds to his view that the two causes are not distinguishable.

45. Solution for a particle in a curved world.

In later work Einstein has adopted the more general equations (37·4)

$$G_{\mu\nu} = \alpha g_{\mu\nu} \quad \text{.............................(45·1).}$$

In this case we must modify (38·61), etc. by inserting $\alpha g_{\mu\nu}$ on the right. We then obtain

$$\tfrac{1}{2}\nu'' - \tfrac{1}{4}\lambda'\nu' + \tfrac{1}{4}\nu'^2 - \lambda'/r = -\alpha e^{\lambda} \quad \text{..............(45·21),}$$

$$e^{-\lambda}(1 + \tfrac{1}{2}r(\nu' - \lambda')) - 1 = -\alpha r^2 \quad \text{..............(45·22),}$$

$$e^{\nu-\lambda}(-\tfrac{1}{2}\nu'' + \tfrac{1}{4}\lambda'\nu' - \tfrac{1}{4}\nu'^2 - \nu'/r) = \alpha e^{\nu} \quad \text{...........(45·23).}$$

From (45·21) and (45·23), $\lambda' = -\nu'$, so that we may take $\lambda = -\nu$. An additive constant would merely amount to an alteration of the unit of time. Equation (45·22) then becomes

$$e^{\nu}(1 + r\nu') = 1 - \alpha r^2.$$

Let $e^{\nu} = \gamma$; then

$$\gamma + r\gamma' = 1 - \alpha r^2$$

which on integration gives

$$\gamma = 1 - \frac{2m}{r} - \tfrac{1}{3}\alpha r^2 \quad \text{.......................(45·3).}$$

The only change is the substitution of this new value of γ in (38·8).

By recalculating the few steps from (39·44) to (39·61) we obtain the equation of the orbit

$$\frac{d^2u}{d\phi^2} + u = \frac{m}{h^2} + 3mu^2 - \frac{1}{3}\frac{\alpha}{h^2}u^{-3} \quad \text{................(45·4).}$$

The effect of the new term in α is to give an additional motion of perihelion

$$\frac{\delta\varpi}{\phi} = \frac{1}{2}\frac{\alpha h^6}{m^4} = \frac{1}{2}\frac{\alpha a^3}{m}(1 - e^2)^3 \quad \text{.....................(45·5).}$$

At a place where γ vanishes there is an impassable barrier, since any change dr corresponds to an infinite distance ids surveyed by measuring-rods. The two positive roots of the cubic (45·3) are approximately

$$r = 2m \text{ and } r = \sqrt{(3/\alpha)}.$$

The first root would represent the boundary of the particle—if a genuine particle could exist—and give it the appearance of impenetrability. The second barrier is at a very great distance and may be described as the *horizon* of the world.

It is clear that the latter barrier (or illusion of a barrier) cannot be at a less distance than the most remote celestial objects observed, say 10^{25} cm. This makes α less than 10^{-50} (cm.)$^{-2}$. Inserting this value (in 45·5) we find that the additional motion of perihelion will be well below the limit of observational detection for all planets in the solar system*.

If in (45·3) we set $m = 0$, we abolish the particle at the origin and obtain the solution for an entirely empty world

$$ds^2 = -(1 - \tfrac{1}{3}\alpha r^2)^{-1} dr^2 - r^2 d\theta^2 - r^2 \sin^2\theta \, d\phi^2 + (1 - \tfrac{1}{3}\alpha r^2)\, dt^2 \quad \text{.....(45·6)}.$$

This will be further discussed in Chapter V.

46. Transition to continuous matter.

In the Newtonian theory of attractions the potential Ω in empty space satisfies the equation

$$\nabla^2 \Omega = 0,$$

of which the elementary solution is $\Omega = m/r$; then by a well-known procedure we are able to deduce that in continuous matter

$$\nabla^2 \Omega = -4\pi\rho \quad \text{.....(46·1)}.$$

We can apply the same principle to Einstein's potentials $g_{\mu\nu}$, which in empty space satisfy the equations $G_{\mu\nu} = 0$. The elementary solution has been found, and it remains to deduce the modification of the equations in continuous matter. The logical aspects of the transition from discrete particles to continuous density need not be discussed here, since they are the same for both theories.

When the square of m/r is neglected, the isotropic solution (43·3) for a particle continually at rest becomes†

$$ds^2 = -\left(1 + \frac{2m}{r}\right)(dx^2 + dy^2 + dz^2) + \left(1 - \frac{2m}{r}\right) dt^2 \quad \text{......(46·15)}.$$

The particle need not be at the origin provided that r is the distance from the particle to the point considered.

Summing the fields of force of a number of particles, we obtain

$$ds^2 = -(1 + 2\Omega)(dx^2 + dy^2 + dz^2) + (1 - 2\Omega)\, dt^2 \quad \text{......(46·2)},$$

* This could scarcely have been asserted a few years ago, when it was not known that the stars extended much beyond 1000 parsecs distance. A horizon distant 700 parsecs corresponds to a centennial motion of about 1″ in the earth's perihelion, and greater motion for the more distant planets in direct proportion to their periods.

† This approximation though sufficient for the present purpose is not good enough for a discussion of the perihelion of Mercury. The term in m^2/r^2 in the coefficient of dt^2 would have to be retained.

where

$$\Omega = \Sigma \frac{m}{r} = \text{Newtonian potential at the point considered.}$$

The inaccuracy in neglecting the interference of the fields of the particles is of the same order as that due to the neglect of m^2/r^2, if the number of particles is not unduly large.

Now calculate the $G_{\mu\nu}$ for the expression (46·2). We have

$$G_{\mu\nu} = g^{\tau\rho} B_{\mu\nu\tau\rho} = \tfrac{1}{2} g^{\sigma\rho} \left(\frac{\partial^2 g_{\mu\nu}}{\partial x_\rho \partial x_\sigma} + \frac{\partial^2 g_{\rho\sigma}}{\partial x_\mu \partial x_\nu} - \frac{\partial^2 g_{\mu\sigma}}{\partial x_\rho \partial x_\nu} - \frac{\partial^2 g_{\rho\nu}}{\partial x_\mu \partial x_\sigma} \right) \quad \ldots(46\cdot3)$$

by (34·5). The non-linear terms are left out because they would involve Ω^2 which is of the order $(m/r)^2$ already neglected.

The only terms which survive are those in which the g's have like suffixes. Consider the last three terms in the bracket; for G_{11} they become

$$\frac{1}{2} \left(g^{11} \frac{\partial^2 g_{11}}{\partial x_1^2} + g^{22} \frac{\partial^2 g_{22}}{\partial x_1^2} + g^{33} \frac{\partial^2 g_{33}}{\partial x_1^2} + g^{44} \frac{\partial^2 g_{44}}{\partial x_1^2} - g^{11} \frac{\partial^2 g_{11}}{\partial x_1^2} - g^{11} \frac{\partial^2 g_{11}}{\partial x_1^2} \right).$$

Substituting for the g's from (46·2) we find that the result vanishes (neglecting Ω^2). For G_{44} the result vanishes for a different reason, viz. because Ω does not contain $x_4 (= t)$. Hence

$$G_{\mu\nu} = \tfrac{1}{2} g^{\sigma\rho} \frac{\partial^2 g_{\mu\nu}}{\partial x_\sigma \partial x_\rho} = \tfrac{1}{2} \Box g_{\mu\nu} \quad \text{as in (30·65)} \ldots(46\cdot4).$$

Since time is not involved $\qquad \Box = -\nabla^2,$

$$G_{11},\ G_{22},\ G_{33},\ G_{44} = -\tfrac{1}{2} \nabla^2 (g_{11},\ g_{22},\ g_{33},\ g_{44})$$
$$= \nabla^2 \Omega \qquad \text{by (46·2)}.$$

Hence, making at this point the transition to continuous matter,

$$G_{11},\ G_{22},\ G_{33},\ G_{44} = -4\pi\rho \qquad \text{by (46·1)} \ldots\ldots\ldots\ldots\ldots(46\cdot5).$$

Also

$$G = g^{\mu\nu} G_{\mu\nu} = -G_{11} - G_{22} - G_{33} + G_{44}$$
$$= 8\pi\rho$$

to the same approximation.

Consider the tensor defined by

$$-8\pi T_{\mu\nu} = G_{\mu\nu} - \tfrac{1}{2} g_{\mu\nu} G \ldots\ldots\ldots\ldots\ldots\ldots\ldots\ldots(46\cdot6).$$

We readily find $\qquad T_{\mu\nu} = 0$, except $T_{44} = \rho$,

and raising the suffixes

$$T^{\mu\nu} = 0, \text{ except } T^{44} = \rho \ldots\ldots\ldots\ldots\ldots\ldots\ldots\ldots(46\cdot7),$$

since the $g^{\mu\nu}$ are Galilean to the order of approximation required.

Consider the expression

$$\rho_0 \frac{dx_\mu}{ds} \frac{dx_\nu}{ds},$$

where dx_μ/ds refers to the motion of the matter, and ρ_0 is the proper-density (an invariant). The matter is at rest in the coordinates hitherto used, and consequently

$$\frac{dx_1}{ds},\ \frac{dx_2}{ds},\ \frac{dx_3}{ds} = 0, \quad \frac{dx_4}{ds} = 1,$$

so that all components of the expression vanish, except the component $\mu, \nu = 4$ which is equal to ρ_0. Accordingly in these coordinates

$$T^{\mu\nu} = \rho_0 \frac{dx_\mu}{ds}\frac{dx_\nu}{ds} \quad \text{.........................(46·8)},$$

since the density ρ in (46·7) is clearly the proper-density.

Now (46·8) is a tensor equation*, and since it has been verified for one set of coordinates it is true for all coordinate-systems. Equations (46·6) and (46·8) together give the extension of Einstein's law of gravitation for a region containing continuous matter of proper-density ρ_0 and velocity dx_μ/ds.

The question remains whether the neglect of m^2 causes any inaccuracy in these equations. In passing to continuous matter we diminish m for each particle indefinitely, but increase the number of particles in a given volume. To avoid increasing the number of particles we may diminish the volume, so that the formulae (46·5) will be true for the limiting case of a point inside a very small portion of continuous matter. Will the addition of surrounding matter in large quantities make any difference? This can contribute nothing directly to the tensor $G_{\mu\nu}$, since so far as this surrounding matter is concerned the point is in empty space; but Einstein's equations are non-linear and we must consider the possible cross-terms.

Draw a small sphere surrounding the point P which is being considered. Let $g_{\mu\nu} = \delta_{\mu\nu} + h_{\mu\nu} + h'_{\mu\nu}$, where $\delta_{\mu\nu}$ represents the Galilean values, and $h_{\mu\nu}$ and $h'_{\mu\nu}$ represent the fields of force contributed independently by the matter internal to and external to the sphere. By § 36 we can choose coordinates such that at P $h'_{\mu\nu}$ and its first derivatives vanish; and by the symmetry of the sphere the first derivatives of $h_{\mu\nu}$ vanish, whilst $h_{\mu\nu}$ itself tends to zero for an infinitely small sphere. Hence the cross-terms which are of the form

$$h'_{\sigma\tau}\frac{\partial^2 h_{\mu\nu}}{\partial x_\lambda \partial x_\rho}, \quad \frac{\partial h'_{\sigma\tau}}{\partial x_\lambda}\frac{\partial h_{\mu\nu}}{\partial x_\rho}, \quad \text{and} \quad h_{\sigma\tau}\frac{\partial^2 h'_{\mu\nu}}{\partial x_\lambda \partial x_\rho}$$

will all vanish at P. Accordingly with these limitations there are no cross-terms, and the sum of the two solutions $h_{\mu\nu}$ and $h'_{\mu\nu}$ is also a solution of the accurate equations. Hence the values (46·5) remain true. It will be seen that the limitation is that the coordinates must be "natural coordinates" at the point P. We have already paid heed to this in taking ρ to be the proper-density.

We have assumed that the matter at P is not accelerated with respect to these natural axes at P. (The original particles had to be *continually* at rest, otherwise the solution (46·15) does not apply.) If it were accelerated there would have to be a stress causing the acceleration. We shall find later that a stress contributes additional terms to the $G_{\mu\nu}$. The formulae (46·5) apply only strictly when there is no stress and the continuous medium is specified by one variable only, viz. the density.

* When an equation is stated to be a tensor equation, the reader is expected to verify that the covariant dimensions of both sides are the same.

The reader may feel that there is still some doubt as to the rigour of this justification of the neglect of m^2*. Lest he attach too great importance to the matter, we may state at once that the subsequent developments will not be based on this investigation. In the next chapter we shall arrive at the same formulae by a different line of argument, and proceed in the reverse direction from the laws of continuous matter to the particular case of an isolated particle.

The equation (46·2) is a useful expression for the gravitational field due to a static distribution of mass. It is only a first approximation correct to the order m/r, but *no second approximation exists* except in the case of a solitary particle. This is because when more than one particle is present accelerations necessarily occur, so that there cannot be an exact solution of Einstein's equations corresponding to a number of particles continually at rest. It follows that any constraint which could keep them at rest must necessarily be of such a nature as to contribute a gravitational field on its own account.

It will be useful to give the values of $G_{\mu\nu} - \frac{1}{2}g_{\mu\nu}G$ corresponding to the symmetrical formula for the interval (38·2). By varying λ and ν this can represent any distribution of continuous matter with spherical symmetry. We have

$$G = -e^{-\lambda}(\nu'' - \tfrac{1}{2}\lambda'\nu' + \tfrac{1}{2}\nu'^2 + 2(\nu' - \lambda')/r + 2(1 - e^{\lambda})/r^2)$$

$$\left.\begin{aligned}
G_{11} - \tfrac{1}{2}g_{11}G &= -\nu'/r - (1 - e^{\lambda})/r^2 \\
G_{22} - \tfrac{1}{2}g_{22}G &= -r^2 e^{-\lambda}(\tfrac{1}{2}\nu'' - \tfrac{1}{4}\lambda'\nu' + \tfrac{1}{4}\nu'^2 + \tfrac{1}{2}(\nu' - \lambda')/r) \\
G_{33} - \tfrac{1}{2}g_{33}G &= -r^2 \sin^2\theta e^{-\lambda}(\tfrac{1}{2}\nu'' - \tfrac{1}{4}\lambda'\nu' + \tfrac{1}{4}\nu'^2 + \tfrac{1}{2}(\nu' - \lambda')/r) \\
G_{44} - \tfrac{1}{2}g_{44}G &= e^{\nu-\lambda}(-\lambda'/r + (1 - e^{\lambda})/r^2)
\end{aligned}\right\} \quad (46{\cdot}9).$$

47. Experiment and deductive theory.

So far as I am aware, the following is a complete list of the postulates which have been introduced into our mathematical theory up to the present stage:

1. The fundamental hypothesis of § 1.
2. The interval depends on a quadratic function of four coordinate-differences (§ 2).
3. The path of a freely moving particle is in all circumstances a geodesic (§ 15).
4. The track of a light-wave is a geodesic with $ds = 0$ (§ 15).
5. The law of gravitation for empty space is $G_{\mu\nu} = 0$, or more probably $G_{\mu\nu} = \lambda g_{\mu\nu}$, where λ is a very small constant (§ 37).

* To illustrate the difficulty, what exactly does ρ_0 mean, assuming that it is not *defined* by (46·6) and (46·7)? If the particles do not interfere with each other's fields, ρ_0 is Σm per unit volume; but if we take account of the interference, m is undefined—it is the constant of integration of an equation which does not apply. Mathematically, we cannot say what m would have been if the other particles had been removed; the question is nonsensical. Physically we could no doubt say what would have been the masses of the atoms if widely separated from one another, and compare them with the gravitational power of the atoms under actual conditions; but that involves laws of atomic structure which are quite outside the scope of the argument.

No. 4 includes the identification of the velocity of light with the fundamental velocity, which was originally introduced as a separate postulate in § 6.

In the mathematical theory we have two objects before us—to examine how we may test the truth of these postulates, and to discover how the laws which they express originate in the structure of the world. We cannot neglect either of these aims; and perhaps an ideal logical discussion would be divided into two parts, the one showing the gradual ascent from experimental evidence to the finally adopted specification of the structure of the world, the other starting with this specification and deducing all observational phenomena. The latter part is specially attractive to the mathematician for the proof may be made rigorous; whereas at each stage in the ascent some new inference or generalisation is introduced which, however plausible, can scarcely be considered incontrovertible. We can show that a certain structure will explain all the phenomena; we cannot show that nothing else will.

We may put to the experiments three questions in *crescendo.* Do they verify? Do they suggest? Do they (within certain limitations) compel the laws we adopt? It is when the last question is put that the difficulty arises for there are always limitations which will embarrass the mathematician who wishes to keep strictly to rigorous inference. What, for example, does experiment enable us to assert with regard to the gravitational field of a particle (the other four postulates being granted)? Firstly, we are probably justified in assuming that the interval can be expressed in the form (38·2), and experiment shows that λ and ν tend to zero at great distances. Provided that e^λ and e^ν are simple functions it will be possible to expand the coefficients in the form

$$ds^2 = -\left(1 + \frac{a_1}{r} + \frac{a_2}{r^2} + \ldots\right)^{-1} dr^2 - r^2\,d\theta^2 - r^2 \sin^2\theta\,d\phi^2 + \left(1 + \frac{b_1}{r} + \frac{b_2}{r^2} + \frac{b_3}{r^3} + \ldots\right) dt^2.$$

Now reference to §§ 39, 40, 41 enables us to decide the following points:

(1) The Newtonian law of gravitation shows that $b_1 = -2m$.

(2) The observed deflection of light then shows that $a_1 = -2m$.

(3) The motion of perihelion of Mercury then shows that $b_2 = 0$.

The last two coefficients are not determined experimentally with any high accuracy; and we have no experimental knowledge of the higher coefficients. If the higher coefficients are zero we can proceed to deduce that this field satisfies $G_{\mu\nu} = 0$.

If small concessions are made, the case for the law $G_{\mu\nu} = 0$ can be strengthened. Thus if only one linear constant m is involved in the specification of the field, b_3 must contain m^3, and the corresponding term is of order $(m/r)^3$, an extremely small quantity. Whatever the higher coefficients may be, $G_{\mu\nu}$ will then vanish to a very high order of approximation.

Turning to the other object of our inquiry, we have yet to explain how these five laws originate in the structure of the world. In the next chapter we shall be concerned mainly with Nos. 3 and 5, which are not independent of one another. They will be replaced by a broader principle which contains them both and is of a more axiomatic character. No. 4 will be traced to its origin in the electromagnetic theory of Chapter VI. Finally a synthesis of these together with Nos. 1 and 2 will be attempted in the closing chapter.

The following forward references will enable the reader to trace exactly what becomes of these postulates in the subsequent advance towards more primitive conceptions:

Nos. 1 and 2 are not further considered until § 97.

No. 3 is obtained directly from the law of gravitation in § 56.

No. 4 is obtained from the electromagnetic equations in § 74. These are traced to their origin in § 96.

No. 5 is obtained from the principle of identification in § 54, and more completely from the principle of measurement in § 66. The possibility of alternative laws is discussed in § 62.

In the last century the ideal explanation of the phenomena of nature consisted in the construction of a mechanical model, which would act in the way observed. Whatever may be the practical helpfulness of a model, it is no longer recognised as contributing in any way to an ultimate explanation. A little later, the standpoint was reached that on carrying the analysis as far as possible we must ultimately come to a set of differential equations of which further explanation is impossible. We can then trace the *modus operandi*, but as regards ultimate causes we have to confess that "things happen so, because the world was made in that way." But in the kinetic theory of gases and in thermodynamics we have laws which can be explained much more satisfactorily. The principal laws of gases hold, not because a gas is made "that way," but because it is made "just anyhow." This is perhaps not to be taken quite literally; but if we could see that there was the same inevitability in Maxwell's laws and in the law of gravitation that there is in the laws of gases, we should have reached an explanation far more complete than an ultimate arbitrary differential equation. This suggests striving for an ideal—to show, not that the laws of nature come from a special construction of the ultimate basis of everything, but that the same laws of nature would prevail for the widest possible variety of structure of that basis. The complete ideal is probably unattainable and certainly unattained; nevertheless we shall be influenced by it in our discussion, and it appears that considerable progress in this direction is possible.

CHAPTER IV

RELATIVITY MECHANICS

48. The antisymmetrical tensor of the fourth rank.

A tensor $A_{\mu\nu}$ is said to be antisymmetrical if

$$A_{\nu\mu} = -A_{\mu\nu}.$$

It follows that $A_{11} = -A_{11}$, so that A_{11}, A_{22}, A_{33}, A_{44} must all be zero.

Consider a tensor of the fourth rank $E^{\alpha\beta\gamma\delta}$ which is antisymmetrical for all pairs of suffixes. Any component with two suffixes alike must be zero, since by the rule of antisymmetry $E^{\alpha\beta 11} = -E^{\alpha\beta 11}$. In the surviving components, α, β, γ, δ, being all different, must stand for the numbers 1, 2, 3, 4 in arbitrary order. We can pass from any of these components to E^{1234} by a series of interchanges of the suffixes in pairs, and each interchange merely reverses the sign. Writing E for E^{1234}, all the 256 components have one or other of the values

$$+E, \quad 0, \quad -E.$$

We shall write

$$E^{\alpha\beta\gamma\delta} = E \,.\, \epsilon_{\alpha\beta\gamma\delta} \qquad \text{.........................(48·1)},$$

where

$\epsilon_{\alpha\beta\gamma\delta} = 0$, when the suffixes are not all different,

$\phantom{\epsilon_{\alpha\beta\gamma\delta}} = +1$, when they can be brought to the order 1, 2, 3, 4 by an even number of interchanges,

$\phantom{\epsilon_{\alpha\beta\gamma\delta}} = -1$, when an odd number of interchanges is needed.

It will appear later that E is not an invariant; consequently $\epsilon_{\alpha\beta\gamma\delta}$ is not a tensor.

The coefficient $\epsilon_{\alpha\beta\gamma\delta}$ is particularly useful for dealing with determinants. If $|k_{\mu\nu}|$ denotes the determinant formed with the elements $k_{\mu\nu}$ (which need not form a tensor), we have

$$4! \times |k_{\mu\nu}| = \epsilon_{\alpha\beta\gamma\delta}\, \epsilon_{\epsilon\zeta\eta\theta}\, k_{\alpha\epsilon}\, k_{\beta\zeta}\, k_{\gamma\eta}\, k_{\delta\theta} \qquad \text{..............(48·2)},$$

because the terms of the determinant are obtained by selecting four elements, one from each row (α, β, γ, δ, all different) and also from each column (ϵ, ζ, η, θ, all different) and affixing the + or − sign to the product according as the order of the columns is brought into the order of the rows by an even or odd number of interchanges. The factor 4! appears because every possible permutation of the *same* four elements is included separately in the summation on the right.

It is possible by corresponding formulae to define and manipulate determinants in three dimensions (with 64 elements arranged in a cube) or in four dimensions.

Note that

$$\epsilon_{\alpha\beta\gamma\delta}\, \epsilon_{\alpha\beta\gamma\delta} = 4! \qquad \text{...........................(48·31)}.$$

The determinants with which we are most concerned are the fundamental determinant g and the Jacobian of a transformation

$$J = \frac{\partial (x_1', x_2', x_3', x_4')}{\partial (x_1, x_2, x_3, x_4)}.$$

By (48·2)

$$4!\, g = \epsilon_{\alpha\beta\gamma\delta}\, \epsilon_{\epsilon\zeta\eta\theta}\, g_{\alpha\epsilon}\, g_{\beta\zeta}\, g_{\gamma\eta}\, g_{\delta\theta} \quad \text{.................(48·32)},$$

$$4!\, J = \epsilon_{\alpha\beta\gamma\delta}\, \epsilon_{\epsilon\zeta\eta\theta} \frac{\partial x_\epsilon'}{\partial x_\alpha} \frac{\partial x_\zeta'}{\partial x_\beta} \frac{\partial x_\eta'}{\partial x_\gamma} \frac{\partial x_\theta'}{\partial x_\delta} \quad \text{..............(48·33).}$$

To illustrate the manipulations we shall prove that*

$$g = J^2 g'.$$

By (48·32) and (48·33)

$$(4!)^3 J^2 g' = \epsilon_{\alpha\beta\gamma\delta}\, \epsilon_{\epsilon\zeta\eta\theta}\, g'_{\alpha\epsilon}\, g'_{\beta\zeta}\, g'_{\gamma\eta}\, g'_{\delta\theta} \,.\, \epsilon_{\iota\kappa\lambda\mu}\, \epsilon_{\nu\xi o\varpi} \frac{\partial x_\nu'}{\partial x_\iota} \frac{\partial x_\xi'}{\partial x_\kappa} \frac{\partial x_o'}{\partial x_\lambda} \frac{\partial x_\varpi'}{\partial x_\mu}$$

$$.\, \epsilon_{\rho\sigma\tau\upsilon}\, \epsilon_{\phi\chi\psi\omega} \frac{\partial x_\phi'}{\partial x_\rho} \frac{\partial x_\chi'}{\partial x_\sigma} \frac{\partial x_\psi'}{\partial x_\tau} \frac{\partial x_\omega'}{\partial x_\upsilon} \quad \text{......(48·41).}$$

There are about 280 billion terms on the right, and we proceed to rearrange those which do not vanish.

For non-vanishing terms the letters ν, ξ, o, ϖ denote the same suffixes as α, β, γ, δ, but (usually) in a different order. Permute the four factors in which they occur so that they come into the same order; the suffixes of the denominators will then come into a new order, say, i, k, l, m. Thus

$$\frac{\partial x_\nu'}{\partial x_\iota} \frac{\partial x_\xi'}{\partial x_\kappa} \frac{\partial x_o'}{\partial x_\lambda} \frac{\partial x_\varpi'}{\partial x_\mu} = \frac{\partial x_\alpha'}{\partial x_i} \frac{\partial x_\beta'}{\partial x_k} \frac{\partial x_\gamma'}{\partial x_l} \frac{\partial x_\delta'}{\partial x_m} \quad \text{...........(48·42).}$$

Since the number of interchanges of the denominators is the same as the number of interchanges of the numerators

$$\frac{\epsilon_{\nu\xi o\varpi}}{\epsilon_{\alpha\beta\gamma\delta}} = \pm 1 = \frac{\epsilon_{\iota\kappa\lambda\mu}}{\epsilon_{iklm}} \quad \text{.........................(48·43),}$$

so that the result of the transposition is

$$\epsilon_{\alpha\beta\gamma\delta}\, \epsilon_{\iota\kappa\lambda\mu} \frac{\partial x_\nu'}{\partial x_\iota} \frac{\partial x_\xi'}{\partial x_\kappa} \frac{\partial x_o'}{\partial x_\lambda} \frac{\partial x_\varpi'}{\partial x_\mu} = \epsilon_{\nu\xi o\varpi}\, \epsilon_{iklm} \frac{\partial x_\alpha'}{\partial x_i} \frac{\partial x_\beta'}{\partial x_k} \frac{\partial x_\gamma'}{\partial x_l} \frac{\partial x_\delta'}{\partial x_m} \quad \text{...(48·5).}$$

Making a similar transposition of the last four terms, (48·41) becomes

$$(4!)^3 J^2 g' = g'_{\alpha\epsilon}\, g'_{\beta\zeta}\, g'_{\gamma\eta}\, g'_{\delta\theta} \,.\, \frac{\partial x_\alpha'}{\partial x_i} \frac{\partial x_\beta'}{\partial x_k} \frac{\partial x_\gamma'}{\partial x_l} \frac{\partial x_\delta'}{\partial x_m} \,.\, \frac{\partial x_\epsilon'}{\partial x_r} \frac{\partial x_\zeta'}{\partial x_s} \frac{\partial x_\eta'}{\partial x_t} \frac{\partial x_\theta'}{\partial x_u}$$

$$.\, \epsilon_{iklm}\, \epsilon_{\nu\xi o\varpi}\, \epsilon_{\nu\xi o\varpi}\, \epsilon_{rstu}\, \epsilon_{\phi\chi\psi\omega}\, \epsilon_{\phi\chi\psi\omega}.$$

But by (23·22)

$$g'_{\alpha\epsilon} \frac{\partial x_\alpha'}{\partial x_i} \frac{\partial x_\epsilon'}{\partial x_r} = g_{ir}.$$

Hence

$$(4!)^3 J^2 g' = (4!)^2\, \epsilon_{iklm}\, \epsilon_{rstu}\, g_{ir}\, g_{ks}\, g_{lt}\, g_{mu}$$

$$= (4!)^3 g,$$

which proves the theorem.

* A shorter proof is given at the end of this section.

To calculate them we use the series of § 54, with the scheme

$$\begin{pmatrix} f & H & f \\ 4 & 4 & 4 \\ 0 & r & 1 \end{pmatrix},$$

which gives

$$\sum_i \frac{\binom{3}{\iota}\binom{r}{\iota}}{\binom{7-\iota}{\iota}} \{(f, H)^{1+\iota}, f\}^{r-\iota}$$

$$= \sum_\iota \frac{\binom{4-r}{\iota}\binom{1}{\iota}}{\binom{9-2r-\iota}{\iota}} \{(f, f)^{r+\iota}, H\}^{1-\iota}.$$

If $r = 1$ we find

$$\{(f, H), f\} + \tfrac{3}{6}\{(f, H)^2, f\}^0 = \{(f, f), H\} + \tfrac{3}{6}\{(f, f)^2, H\}^0,$$

$$(t, f) + \tfrac{1}{12} if^2 = \tfrac{1}{2} H^2,$$

since $$(f, H)^2 = \tfrac{1}{6} if,$$

hence $$(t, f) = \tfrac{1}{2} H^2 - \tfrac{1}{12} if^2.$$

If $r = 2$, $$(t, f)^2 + \{(f, H)^2, f\} + \tfrac{3}{10}\{(fH)^3, f\}^0$$

$$= \tfrac{1}{2}(H, H)^2,$$

and since $$(f, H)^2 = \tfrac{1}{6} if, \quad (fH)^3 = 0,$$

this gives $$(t, f)^2 = 0.$$

From $r = 3$,

$$(t, f)^3 + \tfrac{3}{2}\{(f, H)^2, f\}^2 + \tfrac{9}{10}\{(f, H)^3, f\} + \tfrac{1}{4}(f, H)^4 . f$$

$$= \{(f, f)^3, H\} + \tfrac{1}{2}(f, f)^4 . H,$$

or $$(t, f)^3 + \tfrac{1}{4} iH + \tfrac{1}{4} jf = \tfrac{1}{2} iH,$$

on putting in the values for the transvectants of f and H.

Thus

$$(t, f)^3 = \tfrac{1}{4}(iH - jf).$$

The series does not apply when $r = 4$ because then $r + 1 > 4$, but it is easy to calculate

$$(t, f)^4$$

directly or as in § 94.

For

$$\{(ah)\,a_x^3h_x^3,\ b_x^4\}^4 = (ah)\,(ab)^3\,b_xh_x^3 + \lambda\,\{(ah)^2\,a_x^2h_x^2,\ b_x^4\}^3$$
$$+\,\mu\,\{(ah)^3\,a_xh_x,\ b_x^4\}^2 + \nu\,\{(ah)^4,\ b_x^4\},$$

while

$$(ab)^3\,(ah)\,b_x^3h_x^3 = \{(ab)^3\,a_xb_x,\ h_x^4\} = 0,$$
$$(ah)^2\,a_x^2h_x^2 = \tfrac{1}{6}if;\quad (f, f)^3 = 0,$$
$$(ah)^3\,a_xh_x = 0,$$

hence

$$(t, f)^4 = 0.$$

Hence there are no irreducible forms of degree four. In fact there are no more irreducible forms, because the next in order of degree would be of the form

$$(f^{\alpha}H^{\beta}t^{\gamma}, f)^{\rho}.$$

Now since f, H, t are each of order four at least, irreducible forms can only arise from products containing each of the three by itself, and all these, viz.

$$(f, f)^{\rho},\quad (H, f)^{\rho},\quad (t, f)^{\rho}$$

have been already considered. Hence every invariant or covariant of the quartic is a rational integral function of f, H, t, i and j.

90. Quintic. To illustrate still further the method of this chapter we shall apply it to some extent to the binary quintic. The covariants of degree two are

$$(f, f)^2 = H \text{ and } (f, f)^4 = i.$$

The products of powers of f, H and i which are of degree two are f^2, H and i, and to find the covariants of degree three we have to consider transvectants of these three forms with f.

The transvectants arising from f^2 may be neglected, and hence we are left with

$$(H, f)^1,\ (H, f)^2,\ (H, f)^3,\ (H, f)^4,\ (H, f)^5,$$
$$(i, f)^1,\ (i, f)^2$$

as the only possible irreducible covariants of degree three.

Now of these

$$(H, f)^2 = \{(ab)^2\,a_x^3b_x^3,\ c_x^5\}^2$$

and involves a term

$$(ab)^2 (ac)^2 a_x b_x^3 c_x^3,$$

$$\therefore (H, f)^2 = (ab)^2 (ac)^2 a_x b_x^3 c_x^3 + \lambda \{(ab)^4 a_x b_x, c_x^5\}^0$$

$$= \tfrac{1}{3} a_x b_x c_x \{(ab)^2 (ac)^2 b_x^2 c_x^2 + (ba)^2 (bc)^2 c_x^2 a_x^2 + (ca)^2 (cb)^2 a_x^2 b_x^2\} + \lambda if$$

$$= \tfrac{1}{3} a_x b_x c_x \tfrac{1}{2} \{(ab)^4 c_x^4 + (bc)^4 a_x^4 + (ca)^4 b_x^4\} + \lambda if$$

$$= (\lambda + \tfrac{1}{2})\, if,$$

so that $(H, f)^2$ is reducible.

Again $$(H, f)^3 = \{(ab)^2 a_x^3 b_x^3, c_x^5\}^3$$

and contains the term

$$(ab)^2 (ac)^3 b_x^3 c_x^2.$$

This term can be transformed so as to contain $(ac)^4$ and hence must be a multiple of

$$(ab)^4 (ac)\, b_x c_x^4$$

since the letters are equivalent.

$$\therefore (H, f)^3 = \lambda (ab)^4 (ac)\, b_x c_x^4 + \mu \{(ab)^4 a_x b_x, c_x^5\}$$

$$= (\lambda + \mu)(i, f).$$

Further

$$(H, f)^4 = (ab)^2 (bc)^2 (ac)^2 a_x b_x c_x + \lambda (i, f)^2$$

and

$$(H, f)^5 = (ab)^2 (bc)^3 (ac)^2 a_x$$

$$= \tfrac{1}{3} (bc)^2 (ca)^2 (ab)^2 \{(bc)\, a_x + (ca)\, b_x + (ab)\, c_x\} = 0.$$

Finally $(i, f)^1$ is an irreducible form and

$$(i, f)^2 = (ab)^4 (ac)(bc)\, c_x^3$$

$$= -\tfrac{1}{3} (bc)(ca)(ab) \{(ab)^3 c_x^3 + (bc)^3 a_x^3 + (ca)^3 b_x^3\}$$

$$= -(bc)^2 (ca)^2 (ab)^2 a_x b_x c_x \quad (\S\, 22).$$

Hence the only irreducible covariants of degree three are

$$(H, f) = t, \ (f, i)$$

and $$(f, i)^2 = -(bc)^2 (ca)^2 (ab)^2 a_x b_x c_x = -j.$$

The reader may now find the irreducible forms of degree four and verify the result by reference to the chapter on the quintic. It will be seen at once that the method leads to much labour, that the reduction processes are not easy to discover, and, when we mention that for the quintic we have to proceed step by step until we get to degree 18 before the irreducible system is obtained, the impracticability of these methods in dealing with forms of order greater than four will be at once admitted.

91. Further Theory of the Cubic. Syzygy among the irreducible forms. There is an identical relation connecting the irreducible concomitants of the binary cubic—the simplest example of what is known as a syzygy among the covariants of a single binary form.

In fact since t is the Jacobian of f and H we have, § 78,

$$-2t^2 = (f, f)^2 H^2 + (H, H)^2 f^2 - 2(H, f)^2 Hf.$$

Now

$$(f, f)^2 = H$$

$$(H, f)^2 = (ah)^2 a_x = 0$$

$$(H, H)^2 = \{(ab)^2 a_x b_x, (cd)^2 c_x d_x\}^2 = (ab)^2 (cd)^2 (ac)(bd) = \Delta,$$

for although there are four terms in the transvectant they are identical in value, and we have

$$-2t^2 = H^3 + \Delta f^2,$$

the relation required.

92. Since every covariant of the cubic is a rational integral function of f, H, t and Δ it follows that all expressions derived by convolution from products of powers of f, H and t can be expressed in terms of f, H, t and Δ.

The form of the expression can be easily inferred by consideration of its degree and order, but the actual determination of the coefficients may be a troublesome process.

As an example consider the Hessian of t, *i.e.* $(t, t)^2$. It is a covariant of degree six and order two since t is of degree three, and the only product of f, H, t, Δ fulfilling these conditions is $H\Delta$. We at once see that $(t, t)^2$ is a numerical multiple of $H\Delta$.

The reader may calculate the actual value directly by using the series of § 54—we give an alternative process.

Let $$J = (t, f),$$

then $$-2J^2 = (t, t)^2 f^2 + (f, f)^2 t^2 - 2(f, t)^2 ft.$$

But $\quad (t, f) = \frac{1}{2} H^2, \quad (f, f)^2 = H, \quad (f, t)^2 = 0$

and $\quad t^2 = -\frac{1}{2} H^3 - \frac{1}{2} \Delta f^2;$

therefore $\quad -\frac{1}{2} H^4 = (t, t)^2 f^2 + H(-\frac{1}{2} H^3 - \frac{1}{2} \Delta f^2);$

that is $\quad (t, t)^2 f^2 = \frac{1}{2} \Delta f^2 H$

or $\quad (t, t)^2 = \frac{1}{2} \Delta H.$

Again, consider the symbolical product

$$(ab)^2 (ac) (bd) (cd) c_x d_x$$

which represents a covariant of degree four and order two.

Since there is no product

$$f^\alpha H^\beta t^\gamma \Delta^\delta$$

of this degree and order the covariant in question must vanish identically.

To verify this we remark that, on interchanging a, b and c, d, the expression

$$(ab)^2 (ac) (bd) (cd) c_x d_x$$

changes sign; hence it vanishes.

Ex. (i). Calculate the following transvectants in terms of f, H, t, Δ, viz.

$$(H, H), \ (H, H)^2, \ (t, t), \ (t, t)^2, \ (t, t)^3, \ (H, t), \ (H, t)^2.$$

(The only one presenting any difficulty is (H, t) and this is the Jacobian of a Jacobian; its value is $-\frac{1}{2} \Delta f$.)

Ex. (ii). Shew that any symbolical product involving the factor $(ah)^2$ vanishes identically.

Ex. (iii). Shew that

$$(H^3, f^2)^5 = 0, \ (H^3, t^2)^5 = 0, \ (H^3, f^2) = H^2 f (H, f) = -H^2 ft.$$

93. Further Theory of the Quartic. As in the case of the cubic, the square of the covariant t can be expressed rationally in terms of the remaining forms.

In fact we have, § 78,

$$-2t^2 = (f, f)^2 H^2 - 2(H, f)^2 Hf + (H, H)^2 f^2,$$

while $\quad (f, f)^2 = H, \ (H, f)^2 = \frac{1}{6} if,$

so that it only remains to calculate $(H, H)^2$.

Now using the series of § 54 with the scheme

$$\begin{pmatrix} f, & f, & H \\ 4, & 4, & 4 \\ 0, & 2, & 2 \end{pmatrix}$$

we have

$$\sum_i \frac{\binom{2}{i}\binom{2}{i}}{\binom{5-i}{i}} \{(f, f)^{2+i}, H\}^{2-i} = \sum_i \frac{\binom{2}{i}\binom{2}{i}}{\binom{5-i}{i}} \{(f, H)^{2+i}, f\}^{2-i}$$

or

$$(H, H)^2 + \{(f, f)^3, H\} + \tfrac{1}{3}(f, f)^4 . H$$
$$= \{(f, H)^2, f\}^2 + \{(f, H)^3, f\} + \tfrac{1}{3}(f, H)^4 . f,$$

that is $$(H, H)^2 + \tfrac{1}{3} iH = \tfrac{1}{6} i (f, f)^2 + \tfrac{1}{3} jf,$$

since $$(f, H)^3 = 0 \text{ etc.}$$

Hence $$(H, H)^2 = \tfrac{1}{3} jf - \tfrac{1}{6} iH.$$

Consequently

$$-2t^2 = H^3 - \tfrac{1}{3} if . Hf + (\tfrac{1}{3} jf - \tfrac{1}{6} iH) f^2$$
$$= H^3 - \tfrac{1}{2} iHf^2 + \tfrac{1}{3} jf^3,$$

which is the syzygy required.

94. We shall illustrate the reduction of covariants of the quartic by calculating the values of the transvectants of

$$f, H, t,$$

taken two together.

The transvectants of f with itself, H and t have already been found.

As regards the transvectant

$$(H, H)^r$$

we remark that it vanishes when r is odd and

$$(H, H)^2 = \tfrac{1}{3} jf - \tfrac{1}{6} iH.$$

There only remains $(H, H)^4$.

This is equal to

$$\{(ab)^2 a_x^2 b_x^2, h_x^4\}^4 = (ab)^2 (ah)^2 (bh)^2 = \{(ah)^2 a_x^2 h_x^2, b_x^4\}^4$$
$$= \{(H, f)^2, f\}^4 = \tfrac{1}{6} i (f, f)^4 = \tfrac{1}{6} i^2.$$

To calculate the transvectants

$$(t, H)^r, \quad r \not> 3,$$

apply the series of § 54 with the scheme

$$\begin{pmatrix} H, & f, & H \\ 4, & 4, & 4 \\ 0, & r, & 1 \end{pmatrix}$$

and we have

$$\Sigma \frac{\binom{3}{i}\binom{r}{i}}{\binom{7-i}{i}} \{(H, f)^{1+i}, H\}^{r-i} = \Sigma \frac{\binom{4-r}{i}\binom{1}{i}}{\binom{9-2r-i}{i}} \{(H, H)^{r+i}, f\}^{1-i}.$$

On taking $r = 1, 2, 3$ successively and putting in the values of the transvectants $(H, H)^{r+i}$ we find

$$(t, H) = \tfrac{1}{6} f(iH - jf)$$

$$(t, H)^2 = 0$$

$$(t, H)^3 = \tfrac{1}{4} jH - \tfrac{1}{24} i^2 f.$$

For $(t, H)^4$, the scheme

$$\begin{pmatrix} H, & f, & H \\ 4, & 4, & 4 \\ 1, & 3, & 1 \end{pmatrix}$$

must be used; or else as in the case of $(t, f)^4$ it is easy to see that

$$(t, H)^4 = 0.$$

To find $(t, t)^2$ we apply the series with the scheme

$$\begin{pmatrix} f, & H, & t \\ 4, & 4, & 6 \\ 0, & 2, & 1 \end{pmatrix}$$

which gives

$$\{(f, H), t\}^2 + \{(f, H)^2, t\} + \tfrac{3}{10}(f, H)^3 t = \{(f, t)^2, H\} + \tfrac{2}{3}(f, t)^3 H$$

or

$$(t, t)^2 + \frac{i}{6}(f, t) = \frac{2}{3}(f, t)^3 \,.\, H.$$

On substituting the values for (t, f) and $(t, f)^3$ we find

$$(t, t)^2 = \tfrac{1}{6}(jfH - \tfrac{1}{2} iH^2 - \tfrac{1}{12} i^2 f^2).$$

For $(t, t)^4$ we use the scheme

$$\begin{pmatrix} f, & H, & t \\ 4, & 4, & 6 \\ 1, & 3, & 1 \end{pmatrix}$$

leading to

$$(t, t)^4 + \frac{i}{6}(f, t)^3 = \{(t, f)^3, H\}^2$$

$$= \frac{i}{4}(H, H)^2 - \frac{j}{4}(f, H)^2$$

and hence $$(t, t)^4 = 0.$$

Finally

$$(t, t)^6 = \{(t, f)^3, H\}^4 = \tfrac{1}{4} i (H, H)^4 - \tfrac{1}{4} j (f, H)^4$$

$$= \frac{1}{4}\left(\frac{i^3}{6} - j^2\right).$$

Ex. (i). Deduce the value of $(t, t)^2$ from the relation

$$-2\{(t, f)\}^2 = (t, t)^2 f^2 + (f, f)^2 t^2 - 2(f, t)^2 ft.$$

Ex. (ii). Apply Gordan's series to calculate $(t, t)^2$ for the cubic.

Ex. (iii). Prove that

$$\{(t, f)^3, H\}^2 = \{(t, H)^3, f\}^2.$$

Ex. (iv). Prove that for the quartic

$$\{(H, H)^2, H\} = +\tfrac{1}{3} jt$$

$$\{(H, H)^2, H\}^4 = \tfrac{1}{3} j^2 - \tfrac{1}{36} i^3.$$

Ex. (v). Prove that the Hessian of the Hessian of the Hessian of a quartic f is

$$-\tfrac{1}{216} i^2 j f + \tfrac{1}{9} H (j^2 - \tfrac{1}{24} i^3).$$

Ex. (vi). Calculate the values of $\{(t, t)^2, t\}^r$ for $r = 1, 2, 3, 4, 5, 6$.

Ex. (vii). If a quartic f be the product of a cubic by one of the linear factors of its Hessian, then

$$(f, f)^4 = 0.$$

CHAPTER VI.

GORDAN'S THEOREM.

95. WE have already referred to Gordan's theorem which asserts the existence of a finite complete system of covariants for any binary form, and, in fact, we have illustrated the truth of the theorem in obtaining the complete systems for the quadratic, cubic, and quartic. Our previous method is of little practical utility in dealing with forms of order greater than four, but a comparison between it and the procedure of Gordan may not be without value as a primary indication of the salient features of the latter. In the last chapter covariants were classified according to their degree and we shewed how to obtain those of degree m by transvection from those of less degree. In Gordan's investigation covariants are classified according to their grade—the grade being a definite even number associated with any symbolical product, § 61, and all covariants of grade $2r$ are obtained by transvection from those of inferior grade together with some of grade $2r$.

The advantage of using the grade is that no covariant can be of grade greater than n; accordingly, the number of steps in the process is small, whereas there being no limit, *à priori*, to the degree of an irreducible covariant, and the actual degree reached by irreducible forms of quintics, etc., being very high, the number of steps in the other process is uncertain and at the best large. As will be seen later, on the other hand, the transition from grade $2r-2$ to grade $2r$ is commonly much more difficult than that from one degree to the next higher.

Several preliminary propositions are necessary before we can undertake the actual proof of the existence of the complete system; these we now proceed to explain.

96. The first lemma required belongs to that branch of the theory of numbers known as Diophantine Equations.

For the sake of clearness we shall begin by giving an illustration.

Consider the homogeneous linear equation

$$2x + 5y = 3z;$$

it is easy to see that the number of solutions in positive integers is infinite.

Moreover, if

$$x = p, \quad y = q, \quad z = r;$$

and

$$x = p', \quad y = q', \quad z = r',$$

be two solutions, then

$$x = p + p', \quad y = q + q', \quad z = r + r'$$

is also a solution.

We shall call this latter the sum of the former two solutions, and when any solution can be written as the sum of two smaller solutions (throughout we deal only with solutions in positive integers), it is said to be reducible. Otherwise a solution is irreducible, and the important fact for us is that the number of irreducible solutions is always finite.

Thus for the equation above the only irreducible solutions are

$$x = 3,\ y = 0,\ z = 2;\ x = 0,\ y = 3,\ z = 5;$$
$$x = 1,\ y = 2,\ z = 4;\ x = 2,\ y = 1,\ z = 3.$$

In fact if $x > 3$, then $z > 2$, and the solution can be reduced by means of $x = 3$, $y = 0$, $z = 2$; whereas if $y > 3$, then $z > 5$, and the solution can be reduced by means of $x = 0$, $y = 3$, $z = 5$; thus in an irreducible solution neither x nor y can exceed 3, and, as the number of remaining possibilities is finite, the irreducible solutions can be easily found by trial.

By continually reducing a given solution, say $x = p$, $y = q$, $z = r$, we can express it in terms of the irreducible solutions, that is in the form

$$\left.\begin{aligned} p &= 3\lambda + \nu + 2\rho \\ q &= 3\mu + 2\nu + \rho \\ r &= 2\lambda + 5\mu + 4\nu + 3\rho \end{aligned}\right\} \quad \ldots\ldots\ldots\ldots\ldots\text{(A)},$$

where λ, μ, ν, ρ are positive integers:

e.g. take the solution

$$x = 50, \quad y = 7, \quad z = 45,$$

reducing by means of

$$x = 3, \quad y = 0, \quad z = 2,$$

$$x = 16 \, . \, 3 + 2, \quad y = 7, \quad z = 16 \, . \, 2 + 13\,;$$

then reducing

$$x' = 2, \quad y' = 7, \quad z' = 13$$

by means of

$$x = 0, \quad y = 3, \quad z = 5,$$

$$x' = 2, \quad y' = 2 \, . \, 3 + 1, \quad z' = 2 \, . \, 5 + 3,$$

and the remaining part

$$x'' = 2, \quad y'' = 1, \quad z'' = 3$$

is irreducible.

Hence in this case we have

$$\lambda = 16, \quad \mu = 2, \quad \nu = 0, \quad \rho = 1.$$

Of course if we substitute the expressions in (A) for x, y, z the equation is satisfied identically; the important point is that *every* positive integral solution can be written in the form there indicated.

97. The idea of reducibility can be extended at once to any number of linear homogeneous equations, for the sum of two solutions is always a solution, and we may enunciate our first lemma as follows:

The number of irreducible solutions in positive integers of a system of homogeneous linear equations is finite.

Consider first a single equation,

$$a_1 x_1 + a_2 x_2 + \ldots + a_m x_m = b_1 y_1 + b_2 y_2 + \ldots + b_n y_n,$$

connecting the x's and y's, where the coefficients a, b are positive integers.

If the two solutions

$$x_1 = \xi_1, \; x_2 = \xi_2, \; \ldots \; x_m = \xi_m, \; y_1 = \eta_1, \; y_2 = \eta_2, \; \ldots \; y_n = \eta_n\,;$$

$$x_1 = \xi_1', \; x_2 = \xi_2', \; \ldots \; x_m = \xi_m', \; y_1 = \eta_1', \; y_2 = \eta_2', \; \ldots \; y_n = \eta_n',$$

be typified by

$$x = \xi, \quad y = \eta\,;$$

$$x = \xi', \quad y = \eta',$$

respectively, then

$$x = \xi + \xi', \quad y = \eta + \eta'$$

also typifies a solution and this latter is reducible.

First the equation has mn solutions of the type

$$x_r = b_s, \quad y_s = a_r$$

with the rest of the variables zero.

Next suppose that in a solution one of the x's (say x_1) is greater than

$$b_1 + b_2 + \ldots + b_n,$$

then the right-hand side of the equation must be greater than

$$a_1 (b_1 + b_2 + \ldots + b_n),$$

i.e. $$b_1 (y_1 - a_1) + b_2 (y_2 - a_1) + \ldots + b_n (y_n - a_1) > 0,$$

so that at least one y must be greater than a_1. Let $y_r > a_1$, then the solution in question is reducible by means of the solution

$$x_1 = b_r, \quad y_r = a_1$$

with the other variables zero.

Hence $(b_1 + b_2 + \ldots + b_n)$ is an upper limit to the value of any x in an irreducible solution, similarly $(a_1 + a_2 + \ldots + a_m)$ is an upper limit to the value of any y; but the number of solutions reducible or irreducible subject to these restrictions is manifestly finite, therefore *à fortiori* the number of irreducible solutions is finite.

If the irreducible solutions be typified by

$$x = \alpha_1, \quad y = \beta_1;$$
$$x = \alpha_2, \quad y = \beta_2;$$
$$\vdots \qquad \vdots$$
$$x = \alpha_\rho, \quad y = \beta_\rho,$$

then by continued reduction any solution can be expressed in the typical form

$$x = t_1 \alpha_1 + t_2 \alpha_2 + \ldots + t_\rho \alpha_\rho$$
$$y = t_1 \beta_1 + t_2 \beta_2 + \ldots + t_\rho \beta_\rho,$$

where the t's are all positive integers.

Suppose now we have a second equation of the same nature between the variables; then on replacing the x's and y's by their values in terms of the t's the first equation will be satisfied identically and the second equation will become a linear equation between the t's with integral coefficients.

Hence by the above reasoning every solution of the equation among the t's may be written in the typical form

$$t = T_1\gamma_1 + T_2\gamma_2 + \dots + T_\sigma\gamma_\sigma,$$

where $$t = \gamma_1,\ t = \gamma_2,\ \dots\ t = \gamma_\sigma$$

typify the irreducible solutions, and the T's are all positive integers.

Now substitute these values for the t's in the expressions for the x's and y's and we find at once that

$$x = \kappa_1 T_1 + \kappa_2 T_2 + \dots + \kappa_\sigma T_\sigma$$

$$y = \lambda_1 T_1 + \lambda_2 T_2 + \dots + \lambda_\sigma T_\sigma,$$

where the κ's and λ's are fixed positive integers.

Thus the only possible irreducible solutions of the two equations are those typified by

$$x = \kappa_1,\ y = \lambda_1;\ x = \kappa_2,\ y = \lambda_2;\ \dots\ x = \kappa_\sigma,\ y = \lambda_\sigma,$$

for every other solution can be expressed as a linear combination of these.

If we had a third equation, on substituting for the x's and y's their values in terms of the T's the first two equations would be satisfied identically, and the third would become a linear equation among the T's. Then this equation in turn has only a finite number of irreducible solutions, and hence, reasoning exactly as before, we should find that the three equations given have only a finite number of irreducible solutions. The process can be manifestly extended to any number of equations, and hence our theorem is established. A formal proof by induction from $(r-1)$ equations to r equations could of course be easily given.

Ex. To find the irreducible solutions of the two equations

$$\left.\begin{aligned} 2x + 3y &= z + w \\ x + w &= y + z \end{aligned}\right\}.$$

The irreducible solutions of the second equation are easily found since no letter can exceed 2; they are

$$x=1,\ y=1;\ x=1,\ z=1;\ y=1,\ w=1;\ z=1,\ w=1;$$

variables not mentioned in a solution being zero.

Hence the general solution of the second equation is

$$x=a+b,\ y=a+c,\ z=b+d,\ w=c+a,$$

where a, b, c, d are positive integers.

The first equation now becomes

$$5a+b+2c=2d$$

and for an irreducible solution a, b, c cannot exceed 2.

On trial we find the following irreducible solution

$$a=2,\ d=5;\ b=2,\ d=1;\ c=1,\ d=1;\ a=1,\ b=1,\ d=3.$$

Hence the general solution is

$$a=2\alpha+\delta,\ b=2\beta+\delta,\ c=\gamma,\ d=5\alpha+\beta+\gamma+3\delta.$$

These values for a, b, c, d give

$$\left.\begin{aligned} x&=2\alpha+2\beta \qquad +2\delta\\ y&=2\alpha \qquad +\gamma+\delta\\ z&=5\alpha+3\beta+\gamma+4\delta\\ w&=5\alpha+\beta+2\gamma+3\delta\end{aligned}\right\}$$

as the general solution of the two equations.

The only possible irreducible solutions are accordingly

$$\left.\begin{aligned} x=2,\quad y=2,\quad z=5,\quad w=5\\ x=2,\quad y=0,\quad z=3,\quad w=1\\ x=0,\quad y=1,\quad z=1,\quad w=2\\ x=2,\quad y=1,\quad z=4,\quad w=3\end{aligned}\right\}$$

Of these the first is the sum of the third and fourth, while the fourth is the sum of the second and third, so the only irreducible solutions are the second and third. In other words any solution of the two equations may be written in the form

$$x=2p,\ y=q,\ z=3p+q,\ w=p+2q,$$

where p, q are positive integers.

Ex. (i). Prove that the equation $7x+4y=3z$ has four irreducible solutions and that every solution of the two equations $7x+4y=3z$, $z+5w=2y$ can be written in the form

$$x=2a+c,\ y=7a+15b+11c,\ z=14a+20b+17c,\ w=2b+c.$$

Ex. (ii). Find the number of irreducible solutions of the equation

$$a_1x_1+a_2x_2+\dots+a_nx_n=2x,$$

the a's being positive integers.

Ans. If all the a's are even there are n irreducible solutions, if r of the a's are odd there are $n+\dfrac{r(r-1)}{2}$ solutions. In an irreducible solution at the most only two of the letters on the left are different from zero.

Ex. (iii). Prove directly that in an irreducible solution of the two equations

$$a_1x_1+a_2x_2+\dots+a_mx_m=x+z$$

$$b_1y_1+b_2y_2+\dots+b_ny_n=y+z$$

x is less than the greatest a, and y is less than the greatest b.

98. System of forms derived by transvection from two given systems. Consider two systems of binary forms in the same variables x_1, x_2, viz.

$$A_1,\ A_2,\ \dots\ A_m, \text{ of orders } a_1,\ a_2,\ \dots\ a_m \text{ respectively}$$

and $B_1,\ B_2,\ \dots\ B_n$, of orders $b_1,\ b_2,\ \dots\ b_n$ respectively;

we suppose each form written symbolically and denote by U, V two products of the types

$$A_1^{\alpha_1}A_2^{\alpha_2}\dots A_m^{\alpha_m},\quad B_1^{\beta_1}B_2^{\beta_2}\dots B_n^{\beta_n}$$

wherein all the exponents are either zero or positive integers.

The system C is said to be derived from the systems A and B by transvection when it includes all terms in all transvectants of the form

$$(U,\ V)^{\gamma}.$$

It is clear that some of the members of the system C are reducible, that is they can be expressed as rational integral functions of simpler members of that system—in fact, if

$$U = U_1U_2,\ V = V_1V_2,\ \gamma = \gamma_1 + \gamma_2\text{*},$$

then there are many terms in the transvectant

$$(U,\ V)^{\gamma}$$

which are products of two terms, one belonging to the transvectant

$$(U_1,\ V_1)^{\gamma_1}$$

and the other to the transvectant

$$(U_2,\ V_2)^{\gamma_2}.$$

99. We can now enunciate and prove our first theorem, viz.

The number of transvectants of the form $(U,\ V)^{\gamma}$ *which do not contain reducible terms is finite.*

* It is of course assumed that γ_1, γ_2 are such that these transvectants are possible, *e.g.* γ_1 must not exceed the order of U_1.

For suppose that any term of the transvectant

$$(U, V)^\gamma$$

contains ρ symbols of the A's not in combination with a symbol of the B's and σ symbols of the B's not in combination with a symbol of the A's, then we have

$$\left.\begin{array}{l} a_1\alpha_1 + a_2\alpha_2 + \dots + a_m\alpha_m = \rho + \gamma \\ b_1\beta_1 + b_2\beta_2 + \dots + b_n\beta_n = \sigma + \gamma \end{array}\right\} \dots\dots\dots\dots\dots (\text{I})$$

because each side of the first equation, for example, represents the total number of the symbols of the forms A which occur in the product U.

Now to each positive integral solution of the above equations in α, β, ρ, σ, γ there correspond definite products U, V and a definite value of γ and hence a unique transvectant. But as we have already remarked if the solution corresponding to $(U, V)^\gamma$ be the sum of those corresponding to $(U_1, V_1)^{\gamma_1}$ and $(U_2, V_2)^{\gamma_2}$, then $(U, V)^\gamma$ certainly contains reducible terms. Hence transvectants corresponding to reducible solutions always contain reducible terms and inasmuch as the number of irreducible solutions has been proved to be finite it follows that the number of transvectants not containing reducible terms is finite.

100. In actually finding the transvectants which do not contain reducible terms we may use the equations (I), but it is generally easier to proceed directly.

Suppose that the system A contains the single form $f = a_x^5$ and the system B the single form $i = b_x^2$, then we have to consider transvectants

$$(f^\alpha, i^\beta)^\gamma.$$

If $\gamma > 2\beta$ this transvectant vanishes, if $\gamma < 2\beta - 1$ it contains terms of $i(f^\alpha, i^{\beta-1})^\gamma$; hence for an irreducible transvectant we must have $\gamma = 2\beta - 1$ or 2β. In the same way

$$\gamma \not> 5\alpha \text{ and } \not< 5\alpha - 4.$$

Again if $\alpha > 2$, then for an irreducible transvectant $\gamma > 10$ and hence $\beta > 4$, so that some terms may be reduced by means of $(f^2, i^5)^{10}$. Thus we need only consider $\alpha = 0$, $\alpha = 1$ and $\alpha = 2$.

For $\alpha = 0$ we have i.

For $\alpha = 1$ we have f, (f, i), $(f, i)^2$, $(f, i^2)^3$, $(f, i^2)^4$, $(f, i^3)^5$.

For $\alpha = 2$ we have $(f^2, i^3)^6$, $(f^2, i^4)^7$, $(f^2, i^4)^8$, $(f^2, i^5)^9$, $(f^2, i^5)^{10}$.

Now $(f^2, i^3)^6$ contains terms which are products of a term of $(f, i^2)^4$ and $(f, i)^2$ and a like argument applies to

$$(f^2, i^4)^7,\ (f^2, i^4)^8,\ (f^2, i^5)^9,$$

so the only transvectants not containing reducible terms are

$$f,\ i,\ (f, i),\ (f, i)^2,\ (f, i^2)^3,\ (f, i^2)^4,\ (f, i^3)^5,\ (f^2, i^5)^{10}.$$

Ex. (i). If f be any form of order $2n+1$, then the transvectants

$$(f^\alpha, i^\beta)^r$$

which do not contain reducible terms are $2n+4$ in number.

Ex. (ii). Find the corresponding result when f is a form of even order.

Ex. (iii). The only transvectants

$$f_1^{\alpha_1}(f_2^{\alpha_2}, i^\beta)^r,$$

where $f_1 = a_x^4$ and $f_2 = b_x^3$, which do not contain reducible terms are

$$f_1,\ f_2,\ i,\ (f_1, i),\ (f_1, i)^2,\ (f_1, i^2)^3,\ (f_1, i^2)^4,$$
$$(f_2, i),\ (f_2, i)^2,\ (f_2, i^2)^3,\ (f_2^2, i^3)^6.$$

101. Definition. The system of forms A is said to be *complete* when any expression derived by convolution from a product U of powers of the forms A is itself a rational integral function of the A's.

Thus for example the system of forms

$$f = a_x^3 = b_x^3 = \dots$$
$$H = (ab)^2 a_x b_x$$
$$t = (ab)^2 (ca) b_x c_x^2$$
$$\Delta = (ab)^2 (cd)^2 (ac)(bd)$$

is complete because any expression derived in the above manner is a covariant of f and therefore a rational integral function of f, H, t and Δ. Again the system H and Δ included in the above is itself complete.

More generally the system A is said to be *relatively complete* for the modulus G consisting of a number of symbolical determinants when any expression derived by convolution from a product U is a rational integral function of the A's together with terms involving the modulus G.

Thus the system consisting of a single form

$$f = a_x^n = b_x^n = \dots$$

is relatively complete for the modulus $(ab)^2$, since any expression derived by convolution from a power of f can be transformed so that a factor $(ab)^2$ occurs in it.

Again for a quartic the system

$$f = a_x^4, \quad H = (ab)^2 a_x^2 b_x^2, \quad t = (ab)^2 (ca)\, a_x b_x^2 c_x^3,$$
$$j = (bc)^2 (ca)^2 (ab)^2$$

is relatively complete for the modulus $(ab)^4$, for all covariants of f are rational integral functions of

$$f, H, t, j, i,$$

where $i = (ab)^4$.

We may extend our definition of relative completeness still further: a system A is said to be complete relatively for several moduli G_1, G_2 ... when any expression derived by convolution from a product U is a rational integral function of the A's together with terms involving one at least of the moduli G_1, G_2 ....

It will be seen later (or it can be verified without difficulty) that in connection with any quantic $a_x^n = b_x^n$ ... the single form

$$H = (ab)^2 a_x^{n-2} b_x^{n-2}$$

is relatively complete with respect to the modulus $(ab)^4$ except when $n = 4$.

If $n = 4$ the complete system worked out for the form H shews that any expression derived by convolution from a power of H is a rational integral function of H together with terms involving i or j.

That is H is relatively complete for the two moduli $(ab)^4$ and $(bc)^2 (ca)^2 (ab)^2$.

It will be noticed that a complete system is relatively complete for any modulus or systems of moduli.

102. The system C derived by transvection from two given systems contains an infinite number of forms, but it is said to be a finite system when all its members can be expressed as rational integral functions of a certain finite number of them. More generally it is said to be relatively finite for a given modulus G

when every member of C can be expressed as a rational integral function of a certain finite number of them together with terms all of which involve the modulus G.

For example the number of covariants of a binary cubic is infinite but inasmuch as every one is a rational integral function of f, H, t and Δ the system of forms is said to be finite.

Again it will be seen later that every covariant of the binary n-ic

$$f = a_x^n = b_x^n = c_x^n$$

can be expressed in terms of f, H, t, where

$$H = (ab)^2 a_x^{n-2} b_x^{n-2}$$

$$t = (ab)^2 (ca) a_x^{n-3} b_x^{n-2} c_x^{n-1}$$

together with terms involving the factor $(ab)^4$. We should state this fact thus—The system of covariants of a binary n-ic is relatively finite for the modulus $(ab)^4$.

103. Theorem. *If the systems of forms A and B are both finite and complete, then the system derived from them by transvection is finite and complete.*

(*a*) The system is finite.

In the proof of this theorem we shall consider the transvectants

$$(U, V)^\gamma$$

in a certain order defined as follows:—

(i) Transvectants are taken in order of ascending total degree of the product UV in the coefficients of the forms involved in A and B.

(ii) Those for which the total degree is the same are taken in ascending order of indices.

Further than this the order is immaterial.

With this convention let T and T' be any two terms of the transvectant

$$(U, V)^\gamma,$$

then
$$(T - T') = \Sigma (\overline{U}, \overline{V})^{\gamma'}$$

where $\gamma' < \gamma$ and $\overline{U}$, $\overline{V}$ are derived by convolution from U, V respectively.

But since the systems A and B are complete

$$\overline{U} = F(A),$$
$$\overline{V} = \Phi(B),$$

where $F(A)$ is a rational integral function of the A's, that is, an aggregate of products of the type U, and $\Phi(B)$ a similar function of the B's.

Thus $$(\overline{U}, \overline{V})^{\gamma'}$$

can be expressed as the sum of a number of transvectants in each of which the index is less than γ. By hypothesis all such transvectants have been examined before the one now under consideration and hence if all the C's derived from previously considered transvectants can be expressed in terms of

$$C_1, C_2, \ldots C_r,$$

then all C's up to and including those derived from

$$(U, V)^\gamma$$

can be expressed in terms of

$$C_1, C_2, \ldots C_r, T,$$

where T is any term of the last transvectant.

But if the transvectant

$$(U, V)^\gamma$$

contain a reducible term, say $T = T_1 T_2$, then inasmuch as T_1, T_2 must both arise from transvectants previously considered no term T need be added to the system

$$C_1, C_2, \ldots C_r.$$

Thus in gradually building up a system of C's in terms of which all C's can be expressed we need only add a new member when we come to a transvectant containing no reducible term and then we need add only one new member. But the number of transvectants containing no reducible term is finite and hence a finite number of C's can be chosen such that every other is a rational integral function of these, that is the system C is finite.

Remark. A set of C's in terms of which all others can be expressed rationally and integrally can be chosen in various ways, for any term may be selected from each transvectant containing no reducible terms. Further since the difference of two terms of a transvectant can be expressed by means of terms of transvectants previously considered we may, instead of choosing a single term from any transvectant, take an aggregate of any number of such terms or even the transvectant itself, and it will still be true that every member of C can be expressed as a rational integral function of the members of our finite system.

(b) The finite system so constructed is complete.

Let $$C_1, C_2, \dots C_r$$

be the finite system, then we have to prove that an expression $\overline{W}$ derived by convolution from any product of the form

$$W = C_1^{\gamma_1} C_2^{\gamma_2} \dots C_r^{\gamma_r}$$

is a rational integral function of $C_1, C_2, \dots C_r$.

Suppose that $\overline{W}$ contains ρ determinantal factors in which a symbol belonging to a form A occurs in combination with a symbol belonging to a form B.

Then $\overline{W}$ is a term in a transvectant

$$(\overline{U}, \overline{V})^\rho,$$

where $\overline{U}$ contains only symbols of the A's and $\overline{V}$ only symbols of the B's, so that $\overline{U}$ is derived by convolution from a product U of the A's and $\overline{V}$ is derived by convolution from a product V of the B's.

Thus $$\overline{W} = (\overline{U}, \overline{V})^\rho + \Sigma (\overline{\overline{U}} \overline{\overline{V}})^{\rho'},$$

where $\rho' < \rho$ and $\overline{\overline{U}} \overline{\overline{V}}$ are derived from $\overline{U}$, $\overline{V}$ by convolution and therefore ultimately from U, V.

Now $$\overline{U} = F(A),$$

$$\overline{V} = \Phi(B),$$

accordingly $\overline{W}$ can be expressed as an aggregate of transvectants of the form

$$(U, V)^\nu.$$

But we have just proved that every term of such a transvectant is a rational integral function of the C's and consequently $\overline{W}$ is also a rational integral function of them.

Hence the system is not only finite but complete.

104. Theorem. *If a finite system of forms A, all the members of which are covariants of a binary form f, include f and be relatively complete for the modulus H; if, further, a finite system B be relatively complete for the modulus G, and include one form B_1 whose only determinantal factors are H, then the system C derived by transvection from A and B is relatively finite and complete for the modulus G.*

As an example of the theorem let A consist of

$$f = a_x^3 = b_x^3,$$

and B of the two forms

$$H = (ab)^2 a_x b_x, \quad \Delta = (ab)^2 (ac) (bd) (cd)^2.$$

Then A is relatively complete for the modulus $(ab)^2$, § 88, and B is absolutely complete, being the complete system of the Hessian of the cubic; hence according to the theorem the system derived by transvection should be absolutely complete. This is obviously true, for the new system contains f, H, t, Δ, where

$$t = (f, H) = -(ab)^2 (ac) b_x c_x^2,$$

and every possible member of the derived system is a covariant of f, therefore they are all rational integral functions of f, H, t, Δ, which constitute the complete system of the cubic.

105. Lemma. *If P be derived by convolution from a power of f any term in the transvectant*

$$(P, V)^\rho$$

can be expressed as an aggregate of transvectants of the type

$$(U, V)^\sigma$$

in which the degree of U is at most equal to that of P.

(Throughout we shall use U, V as typical symbols for products of powers of the forms of A and B respectively.)

This statement is manifestly true when the degree of P is zero; assuming it true when the degree of P is less than r we shall establish it when the degree is r.

In fact if T be a term in

$$(P, V)^\rho,$$

$$T = (P, V)^\rho + \Sigma (\overline{P}, \overline{V})^{\rho'};$$

and since P, $\bar{P}$ are derived by convolution from a power of the form f which is contained in A,

$$P = F(A) + HW^*,$$
$$\bar{P} = F'(A) + HW',$$

while $$\bar{V} = \Phi(B) + GZ \equiv \Phi(B), \mod G.$$

Hence T can be expressed as the sum of three parts;

(i) transvectants of the type $\{F(A), \Phi(B)\}^\sigma$ the degree of $F(A)$ being r;

(ii) transvectants of the type $(Q, V)^\tau$, where Q is of the same degree as P and contains the factor H;

(iii) terms containing the factor G.

Now Q can be derived by convolution from

$$B_1 f^s,$$

where s is less than r the degree of P; therefore any term in

$$(Q, V)^\tau$$

can be derived by convolution from

$$f^s B_1 V,$$

and is expressible in the form

$$\Sigma(P', \overline{B_1 V}),$$

where P' is derived by convolution from f^s and is of degree less than P. But by hypothesis every term in these transvectants can be expressed as an aggregate

$$\Sigma(U, V)^\sigma, \mod G,$$

for $$\overline{B_1 V} \equiv \Phi(B), \mod G,$$

where the degree of U is at most equal to s and therefore less than r.

On referring to the expression for T we see that T can be written in the form

$$\Sigma(U, V)^\sigma, \mod G;$$

consequently the statement in the lemma can be completely established by induction.

* HW simply means a symbolical product containing the factor H.

COR. *If the product P contain the factor H, then any term in*

$$(P, V)^\rho$$

can be expressed in the form

$$\Sigma (U, V)^\sigma,$$

where the degree of U is less than that of P.

For P is now of the form Q just discussed, and any term in a transvectant

$$(Q, V)^\rho$$

can be expressed as a sum

$$\Sigma (U, V)^\sigma$$

in which the degree of U is at most equal to s which is less than the degree of P.

106. The proof of the theorem is now the same in principle as that in § 103.

The transvectants are considered in the following order.

(i) In order of ascending degree of UV in the coefficients of f.

(ii) Those for which the degree of UV is the same are taken in order of ascending degree of U.

(iii) Transvectants for which these two degrees are the same are taken in order of ascending index.

Further than this the order is immaterial.

If T and T' be two terms in

$$(U, V)^\nu,$$

then
$$T' - T = \Sigma (\bar{U}, \bar{V})^{\nu'},$$
where
$$\nu' < \nu.$$
But
$$\bar{U} = F(A) + HW,$$
$$\bar{V} = \Phi(B) + GW';$$
therefore
$$T' - T \equiv \Sigma \{F(A),\ \Phi(B)\}^{\nu'} + \Sigma \{HW,\ \Phi(B)\}^{\nu'},\ \bmod G.$$

Transvectants of the type

$$\{F(A), \Phi(B)\}^{\nu'}$$

have been previously considered, for the degree of $F(A)$ is the same as that of U and $\nu' < \nu$; further by the lemma transvectants of the type

$$\{HW, \Phi(B)\}^{\nu'}$$

can be expressed in the form

$$\Sigma\,(U', V')^{\sigma}$$

where the degree of U' is less than that of HW, *i.e.* less than that of U.

Thus $T' - T$ can be written

$$\Sigma\,(U'', V'')^{\nu'} + \Sigma\,(U', V')^{\sigma}, \text{ mod } G,$$

where the degree of U'' is the same as that of U and $\nu' < \nu$, while the degree of U' is less than that of U.

Hence if all terms of transvectants considered previously to

$$(U, V)^{\nu}$$

can be expressed rationally and integrally in terms of

$$C_1, C_2, \ldots C_p$$

(except for terms involving G); then all terms of transvectants up to and including

$$(U, V)^{\nu}$$

can be expressed in the form

$$F(C_1, C_2, \ldots C_p, T), \text{ mod } G,$$

where T is any term of the last transvectant.

If the transvectant

$$(U, V)^{\nu}$$

contain a reducible term we may suppose it to be T, and since $T = T_1 T_2$ where T_1, T_2 are terms of former transvectants, there is no need to add the term T to $C_1, C_2, \ldots C_p$.

It follows that in constructing a system of C's in terms of which all C's can be expressed we have to add a new member only when we come to a transvectant containing no reducible terms and then one only. The number of transvectants containing no irreducible terms is finite, § 99; hence if $C_1, C_2, \ldots C_q$ be a series of terms one from each of this finite number of transvectants, any other member of the system C derived by transvection from A and B can be expressed as a rational integral function of $C_1, C_2, \ldots C_q$ together with terms involving the factor G; in other words, the system C is relatively finite for the modulus G.

Next the system

$$C_1, C_2, \ldots C_q$$

is relatively complete with respect to the modulus G.

For any term T derived by convolution from

$$W = C_1^{\gamma_1} C_2^{\gamma_2} \ldots C_q^{\gamma_q}$$

may be regarded as a term in a transvectant

$$(\bar{U}, \bar{V})^\rho,$$

where $\bar{U}$ is derived by convolution from a product of the A's and $\bar{V}$ from a product of the B's.

Hence T can be expressed as an aggregate of transvectants

$$(\bar{U}, \bar{V})^\sigma;$$

while $\bar{U} = P$ can be derived by convolution from a power of f and

$$\bar{V} \equiv \Phi(B), \text{ mod } G;$$

therefore

$$\begin{aligned} T &\equiv \Sigma \{P, \Phi(B)\}^\sigma, \text{ mod } G, \\ &\equiv \Sigma (P, V)^\sigma, \text{ mod } G, \\ &\equiv \Sigma (U, V)^{\sigma'}, \text{ mod } G. \quad \text{(Lemma.)} \end{aligned}$$

Consequently, as has just been proved,

$$T = F(C_1, C_2, \ldots C_q), \text{ mod } G,$$

and the system is complete.

107. COR. I. *If the system B is absolutely complete, then the system derived by transvection from A and B is absolutely complete.*

COR. II. If the system B is complete for two moduli G and G' and contains a form whose only determinantal factors are H, then the derived system is complete for the two moduli G and G'.

To prove this we have only to write

$$\bar{B} \equiv F(B), \text{ modd } (G, G')$$

instead of

$$\bar{B} \equiv F(B), \text{ mod } (G)$$

at every stage of the foregoing proof.

108. Gordan's Theorem. These long preliminary explanations are now at an end and the actual proof of the theorem does not present much difficulty.

Every covariant of a binary form

$$f = a_x^n = b_x^n = \text{etc.}$$

is either a power of f or else contains a factor $(ab)^2$, and hence the form f itself is a complete system with respect to the modulus $(ab)^2$.

Assuming now that a system of covariants containing f and relatively complete for the modulus $(ab)^{2k}$ can be found we shall shew how to construct a system also containing f and relatively complete for the modulus $(ab)^{2k+2}$. The system relatively complete mod $(ab)^{2k}$ is called A_{k-1}, and since every covariant can be derived from f by convolution it is a rational integral function of the forms in A_{k-1} except for terms involving the factor $(ab)^{2k}$.

To construct the system A_k when A_{k-1} is known we make use of the theorem of § 104.

We must therefore begin by constructing a system B_{k-1} possessing the following properties:

(i) it contains the form $(ab)^{2k} a_x^{n-2k} b_x^{n-2k}$,

(ii) it is relatively complete for the modulus $(ab)^{2k+2}$.

Then the system derived from A_{k-1} and B_{k-1} by transvection will be finite and complete with respect to the modulus $(ab)^{2k+2}$, and as it obviously contains f which is contained in A_{k-1} it is the system A_k required.

109. Accordingly we have now to shew how to construct the system B_{k-1}.

There are three cases.

I. If $2k < \frac{n}{2}$ then any form derived by convolution from a power of $H_k = (ab)^{2k} a_x^{n-2k} b_x^{n-2k}$ is of grade $(2k+1)$ at least and therefore of grade $(2k+2)$ since all symbols are now equivalent.

Hence H_k is itself relatively complete for the modulus $(ab)^{2k+2}$ and in this case the system B_k consists of the single form

$$(ab)^{2k} a_x^{n-2k} b_x^{n-2k}. \quad (\S\ 74.)$$

II. If $2k > \frac{n}{2}$ then $H_k = (ab)^{2k} a_x^{n-2k} b_x^{n-2k}$ is of order less than n, say m.

Now we suppose that the complete system of covariants for a form of order $< n$ is known and we derive a system from H_k on the model of the complete system of a_x^m as explained in §§ 79, 80.

Neglecting terms containing $(ab)^{2k+2}$ we can replace each copied form by a single term; the system so derived is complete for the modulus $(ab)^{2k+2}$ and is therefore the system B_{k-1} required.

III. If $2k = \frac{n}{2}$—a case which can only arise when n is a multiple of 4—we have a rather different state of things.

Here the form $H_k = (ab)^{2k} a_x^{n-2k} b_x^{n-2k}$ is relatively complete for the two moduli

$$(ab)^{2k+2}, \quad (ab)^{2k} (bc)^{2k} (ca)^{2k},$$

the latter being an invariant J, and hence by Cor. II. § 107 the system derived by transvection from A_{k-1} and B_{k-1} is relatively complete for the moduli $(ab)^{2k+2}$ and J; calling this system C_k for a moment we have

$$\bar{C}_k \equiv F(C_k) + J \,.\, P_1, \quad \text{mod } (ab)^{2k+2},$$

where P_1 is a covariant of degree less than $\bar{C}_k$.

Further since P_1 can be derived by convolution from f which is contained in C_k, we have

$$P_1 \equiv F_1(C_k) + J \,.\, P_2, \quad \text{mod } (ab)^{2k+2},$$

where P_2 is a covariant of degree less than P_1.

Proceeding in this way we see that $\bar{C}_k$ is a rational integral function of J and the forms in C_k together with terms involving the factor $(ab)^{2k+2}$.

Hence if we add J to the system C_k and call the total system A_k it follows at once that A_k is relatively complete for the modulus $(ab)^{2k+2}$.

Therefore in every case, given the complete system mod $(ab)^{2k}$ we can construct that mod $(ab)^{2k+2}$; but the system A_0 is f, thence we find the system A_1, then from that the system A_2 and so on, in fact we can construct the system A_k relatively complete for the modulus $(ab)^{2k+2}$.

110. Consider now a little more closely what happens when we come to the end of the sequence of moduli $(ab)^2, (ab)^4, (ab)^6 \ldots$, and first let n be even and equal to $2g$.

Then the system A_{g-1} is relatively complete for the modulus $(ab)^{2g}$, and the system B_{g-1} consists of the single invariant $(ab)^{2g}$ so that it is of course absolutely complete.

Hence the system derived from A_{g-1} and B_{g-1} by transvection is absolutely complete and it contains f, therefore it is the complete system of invariants and covariants; further since B_{g-1} consists of a single invariant the complete system A_g consists of A_{g-1} and that invariant $(ab)^{2g}$.

Secondly let n be odd and equal to $2g+1$, then the system A_{g-1} can be constructed and it both contains f and is relatively complete for the modulus $(ab)^{2g}$.

The system B_{g-1} is derived from the quadratic

$$(ab)^{2g} a_x b_x$$

by the same convolutions as the complete system of the quadratic $\alpha_x^2 = \beta_x^2$ is found from this form. This complete system being α_x^2 and $(\alpha\beta)^2$ the system B_{g-1} consists of

$$(ab)^{2g} a_x b_x, \quad (ab)^{2g} (ac)(bd)(cd)^{2g}.$$

This system is relatively complete for the modulus $(ab)^{2g+1}$ by § 109 II, and this being a vanishing invariant it follows that B_{g-1} is absolutely complete.

Hence the system derived from A_{g-1} and B_{g-1} contains f and is absolutely complete, that is it constitutes the complete system of f.

To recapitulate—the complete system mod $(ab)^2$ can be written down at once, then from that we deduce the complete system mod $(ab)^4$ and proceeding step by step we can finally construct an absolutely complete system as the last step in our series.

We have therefore proved that the complete system is finite, for all the systems $A_1, A_2, \ldots$ are finite, and we have shewn how to construct it on the assumption that the systems for forms of lower orders are known—the proof is thus inductive in its nature.

111. We shall illustrate the above process by applying it to the quadratic, cubic, and quartic.

(i) *Quadratic.* The system A_0 is

$$f = a_x^2 = b_x^2$$

and the system B_0 is $(ab)^2$, hence the complete system is

$$a_x^2, \ (ab)^2.$$

(ii) *Cubic.* Here A_0 is

$$f = a_x^3 = b_x^3 = \text{etc.}$$

and B_1 is $$(ab)^2 a_x b_x, \ (ab)^2 (ac)(bd)(cd)^2,$$
in fact $$H, \ (H, H)^2.$$

This system B_1 is absolutely complete, therefore the system derived by transvection is the complete system.

It consists of

$$f, H, (H, H)^2 = \Delta \text{ and } (f^\alpha, H^\beta)^\gamma.$$

Proceeding as in § 88 we can shew that the only irreducible transvectant is (f, H).

(iii) *Quartic.* Here A_0 is

$$f = a_x^4 = b_x^4 \ldots,$$

B_0 is $$H = (ab)^2 a_x^2 b_x^2,$$

and this is complete modd $(ab)^4$ and $(ab)^2 (bc)^2 (ca)^2$.

The system derived by transvection is

$$(f^\alpha, H^\beta)^\gamma.$$

If $\gamma \geqslant 2$ this has a term containing the factor $(ab)^2 (ac)^2$ which is congruent to zero modd $(ab)^4$, $(ab)^2 (bc)^2 (ca)^2$.

Hence we need take only $\gamma = 1$ and thence only $\alpha = 1$, $\beta = 1$, and we find that

$$f, H, (f, H)$$

is relatively complete modd $(ab)^4$, $(ab)^2 (bc)^2 (ca)^2$.

Therefore $$f, H, (f, H), (ab)^2 (bc)^2 (ca)^2$$

is complete mod $(ab)^4$ and is the system A_1.

Then B_1 being the invariant $i = (ab)^4$ we have for the complete system

$$f, H, t = (f, H), \ i = (ab)^4, \ j = (ab)^2 (bc)^2 (ca)^2.$$

112. We shall now apply the principles of §§ 73, 76 to the deduction of a complete system mod $(ab)^{\frac{n}{2}}$ for the binary form of order n.

The system A_0 consists of

$$f = a_x^n = b_x^n = \text{etc.}$$

and B_0 of $$H = (ab)^2 a_x^{n-2} b_x^{n-2} \ldots.$$

The system A_1 is derived by transvection from A_0 and B_0.

Now $$(f^{\alpha}, H^{\beta})^{\gamma}$$

has a term containing the factor $(ab)^2(ac)^2$ if $\gamma > 1$, and since such a term is

$$\equiv 0 \bmod (ab)^4$$

the transvectant may be rejected.

If $\gamma = 1$ the transvectant contains reducible terms unless $\alpha = \beta = 1$, and hence A_1 consists of

$$f, H, (f, H) = t.$$

The system B_1 is

$$(ab)^4 a_x^{n-4} b_x^{n-4}$$

and A_2 is derived by transvection from A_1 and B_1.

If the index of a transvectant be greater than two it contains a term having a factor $(ab)^4(ac)^3$ and this is

$$\equiv 0 \bmod (ab)^6. \quad (\S\ 70.)$$

We need only consider the cases in which the index is $\leqslant 2$, and since the order of each form in A_1 is certainly greater than 2 (in fact $\frac{n}{2} \geqslant 4$), products of forms may be rejected.

There remain transvectants of each form of A_1 taken simply with

$$H_2 = (ab)^4 a_x^{n-4} b_x^{n-4}.$$

For the future we shall only write down the determinantal factors of a covariant.

Transvectants with f give rise to

$$(ab)^4(bc),\ (ab)^4(bc)^2.$$

Those with H give

$$(ab)^4(bc)(cd)^2,\ (ab)^4(bc)^2(cd)^2,$$

and finally those with t give

$$(ab)^4(bc)(cd)^2(dc),\ (ab)^4(bc)^2(cd)^2(dc).$$

Now by § 76

$$(ab)^4(bc)^2(cd)^2 \equiv (ab)^4(cd)^4,\ \bmod (ab)^6;$$

hence $$(ab)^4(bc)^2(cd)^2(de)$$

being a term of $$\{(ab)^4(bc)^2(cd)^2, e_x^n\}$$

we have

$$(ab)^4 (bc)^2 (cd)^2 (de) \equiv \{(ab)^4 (cd)^4, e_x^n\}, \text{ mod } (ab)^6,$$

for all expressions derived by convolution from

$$(ab)^4 (bc)^2 (cd)^2$$

are $$\equiv 0 \text{ mod } (ab)^6. \quad (\S\ 73.)$$

Now a term of the last transvectant is

$$(cd)^4 . (ab)^4 (ae),$$

$$\therefore\ (ab)^4 (bc)^2 (cd)^2 (de) \equiv (cd)^4 . (ab)^4 (ae), \text{ mod } (ab)^6$$

and accordingly may be rejected.

Finally $$(ab)^4 (bc) (cd)^2 (de)$$

is reducible as being the Jacobian of a Jacobian, and the system A_2 consists of

$$f,\ (ab)^2,\ (ab)^2 (bc),$$

$$(ab)^4,\ (ab)^4 (bc),\ (ab)^4 (bc)^2,\ (ab)^4 (bc) (cd)^2.$$

113. Before proceeding further we shall develope the results of § 76 by shewing that a symbolical product Γ containing the factor

$$(ab)^\lambda (bc)^\mu (cd)^\nu,$$

in which λ is even and equal to $\mu + \nu$, can in general be expressed in terms of covariants that are either reducible or of grade greater than λ.

The above reduction of

$$(ab)^4 (bc)^2 (cd)^2 (de)$$

is a case in point.

In fact Γ is a term of

$$\{(ab)^\lambda (bc)^\mu (cd)^\nu, \phi\}^\rho$$

which we write $$(T, \phi)^\rho.$$

Hence $$\Gamma = (T, \phi)^\rho + \Sigma (\overline{T}, \bar{\phi})^{\rho'}, \quad \rho' < \rho$$

$$\equiv (T, \phi)^\rho + \Sigma (T, \bar{\phi})^{\rho'}, \text{ mod } (ab)^{\lambda+1},$$

since $\overline{T}$ derived by convolution from $(ab)^\lambda (bc)^\mu (ca)^\nu$ is of grade greater than λ, § 73.

Again

$$T = (ab)^\lambda . (cd)^\lambda + C_{\lambda+1} \quad (\S\ 76),$$

therefore

$$\Gamma \equiv \{(ab)^\lambda . (cd)^\lambda, \phi\}^\rho + \{(ab)^\lambda . (cd)^\lambda, \bar{\phi}\}^{\rho'} + C_{\lambda+1}.$$

Now if $2n-2\lambda \geqslant \rho$ each of these transvectants contains terms having $(cd)^\lambda c_x^{n-\lambda} d_x^{n-\lambda}$ as a factor.

Hence

$$\{(ab)^\lambda . (cd)^\lambda, \phi\}^\rho$$
$$\equiv (cd)^\lambda \{(ab)^\lambda, \phi\}^\rho + \Sigma \{(ab)^\lambda . (cd)^\lambda, \bar{\phi}\}^\sigma, \text{ mod } (ab)^{\lambda+1}, \quad \sigma < \rho,$$

and by continuation of this process we can express Γ entirely in terms of reducible covariants and covariants of grade greater than λ; it suffices to remark that the index σ diminishes at every step.

It is quite easy to see that the condition

$$2n - 2\lambda \geqslant \rho$$

is satisfied in all our cases—at any rate it will be in the course of the subsequent work.

114. Returning now to the general form, B_2 consists of

$$H_3 = (ab)^6 a_x^{n-6} b_x^{n-6},$$

and A_3 is derived by transvection from A_2 and B_2.

The argument used in evolving A_1 and A_2 enables us to see

(i) that transvectants with index > 3 may be rejected,

(ii) thence that transvectants of products or powers of forms may be likewise rejected.

We are therefore left with transvectants of the forms of A_2 taken simply with H_2, the index being $\not> 3$.

Omitting Jacobians of Jacobians and forms having a factor

$$(ab)^6 (bc)^\mu (cd)^\nu, \text{ where } \mu + \nu \geqslant 6,$$

we have

from f, $\quad (ab)^6 (bc),\ (ab)^6 (bc)^2,\ (ab)^6 (bc)^3$;

„ H, $\quad (ab)^6 (bc) (cd)^2,\ (ab)^6 (bc)^2 (cd)^2,\ (ab)^6 (bc)^3 (cd)^2$;

„ $(ab)^2 (bc)$, $\quad (ab)^6 (bc)^2 (cd)^2 (de),\ (ab)^6 (bc)^3 (cd)^2 (de)$;

„ $(ab)^4$, $\quad (ab)^6 (bc) (cd)^4$;

„ $(ab)^4 (bc)$, $\quad$ none;

„ $(ab)^4 (bc)^2$, $\quad (ab)^6 (bc) (cd)^4 (de)^2$;

„ $(ab)^4 (bc) (cd)^2$, $\quad$ none.

Hence we have found for A_3 the above ten forms in addition to those of A_2. Putting aside the question as to whether any of these ten new forms are reducible, a continued repetition of the above process establishes the fact that all the forms of the system A_k, relatively complete for the modulus

$$(ab)^{2k+2}, \quad \left(k < \frac{n}{2}\right),$$

are included in the set

$$(ab)^{\lambda}(bc)^{\mu}(cd)^{\lambda'}(de)^{\mu'}(ef)^{\lambda''}(fg)^{\mu''}\ldots^{*}$$

where the exponents satisfy the following conditions:

(i) $\lambda \not> 2k$,

(ii) $\lambda, \lambda', \lambda'', \ldots$ are all even,

(iii) $\lambda > \lambda' + \mu,\ \lambda' > \lambda'' + \mu', \ldots,$

(iv) no two of the exponents $\mu, \mu', \ldots$ are equal to unity.

In fact

(ii) follows immediately from the way in which the covariants are formed.

(iii) results from the application of §§ 73, 76.

(iv) is the expression of the fact that the Jacobian of a Jacobian is reducible.

Ex. (i). If the orders m, n, p of the forms f, ϕ, ψ, be each greater than two, then

$$(f\phi, \psi)^2 = \frac{m}{m+n}(f, \psi)^2\phi + \frac{n}{m+n}(\phi, \psi)^2 f - \frac{mn}{(m+n)(m+n-1)}(f, \phi)^2\psi.$$

Ex. (ii). For a form whose order is greater than four the covariants

$$(ab)^3(bc)(cd), \quad (ab)(bc)^3(cd), \quad (ab)^2(bc)^2(cd)$$

all vanish identically.

Ex. (iii). If $n > 5$, then

$$(bc)^2(ca)^2(ab)^2 a_x^{n-4} b_x^{n-4} c_x^{n-4}$$
$$= (ab)^4(ac)^2 a_x^{n-6} b_x^{n-4} c_x^{n-2} - \tfrac{1}{2}(ab)^6 a_x^{n-6} b_x^{n-6} . c_x^{n}.$$

Ex. (iv). If $n > 4$, then

$$(H, f)^3 = \frac{n-4}{4n-10}(ab)^4(bc)\, a_x^{n-4} b_x^{n-5} c_x^{n-1} = \frac{n-4}{4n-10}\{(f, f)^4, f\}$$

and if $n > 5$,

$$(H, f)^4 = -\frac{n-1}{2n-5}\{(f, f)^4, f\}^2 + \frac{3n^2-25n+50}{4(2n-7)(n-4)}(f, f)^6 . f.$$

* Cf. Jordan, *Liouville's Journal*, 1876, 1879.

Ex. (v). For a form whose order is greater than three prove that

$$(H, H)^2 + \frac{(n-2)}{2(2n-5)} H(f, f)^4$$
$$= \{(f, H)^2, f\}^2 + \frac{2(2n-6)}{3n-8} \{(f, H)^3, f\} + \frac{2(2n-7)}{3(3n-8)} (f, H)^4 . f.$$

Hence replacing $(f, f)^4$ by i express $(H, H)^2$ as a linear combination of

$$Hi,\ f(i, f)^2,\ f^2 (f, f)^6$$

and finally express t^2 in terms of the irreducible forms of the system.

Ex. (vi). Prove that all irreducible covariants of degree four and rank not greater than $\frac{n}{2}$ are included in

$$(ab)^{2\lambda} (bc)^{\mu} (cd)^{\nu}$$

where $$2\lambda \ngtr \frac{n}{2} \text{ and } \lambda > \mu > \nu.$$

Ex. (vii). In § 103 if no A be of order greater than m and no B be of order greater than n, then no form of the system C is of order greater than $m+n-2$.

CHAPTER VII.

THE QUINTIC.

115. To obtain the complete irreducible system of covariants of the quintic, we follow step by step Gordan's proof of the finiteness. Let us briefly recapitulate.

The complete system of forms, which are not expressible in terms of covariants having a symbolical factor $(ab)^2$, is first found; this is called A_0, it is the complete system mod $(ab)^2$. Generally A_k is used to denote the complete system mod $(ab)^{2k+2}$. To obtain the system A_{k+1} from the system A_k, a subsidiary system of forms B_k is used. This system is a system of forms having $\phi = (ab)^{2k+2} a_x^{n-2k-4} b_x^{n-2k-4}$ for ground-form.

When the order of this form is less than n, B_k consists of its complete irreducible system. Otherwise if the order of ϕ is greater than n we may take for B_k the single form ϕ; while when the order of ϕ is equal to n, the system B_k consists of ϕ and the invariant $(ab)^{\frac{n}{2}} (bc)^{\frac{n}{2}} (ca)^{\frac{n}{2}}$.

Then it has been proved that the system A_{k+1} may be obtained by taking transvectants of products and powers of forms from A_k with products and powers of forms from B_k.

116. The quintic will be written

$$f = a_x^5 = b_x^5 = \dots\dots$$

The system A_0 contains f only.

The system B_0 contains only

$$(ab)^2 a_x^3 b_x^3 = H.$$

The system A_1 is then obtained from the transvectants

$$(f^\alpha, H^\beta)^\gamma.$$

If $\gamma > 2$, this transvectant contains a term having a factor $(bc)^3$; such a term can be expressed as a sum of symbolical products each containing a factor $(ab)^4$, and is therefore $\equiv 0$, mod $(ab)^4$.

Hence we may reject these transvectants when $\gamma > 2$, for all transvectants which contain reducible terms may be rejected.

The transvectant $(H, f)^2$ contains the term

$$(ab)^2 (bc)^2 a_x^3 b_x c_x^3 = \tfrac{1}{2} f . (ab)^4 a_x b_x \quad (\S\ 51, \text{Ex. (vi)}).$$

The system A_1 then consists of

$$f,\ H,\ (f,\ H) = t.$$

The system B_1 is built up from the form

$$(ab)^4 a_x b_x = i,$$

this is of order < 5, hence we must take the complete irreducible system of the quadratic i.

The system B_1 then consists of

$$i,\ (i,\ i)^2 = A.$$

The system A_2 is now the complete system of forms for the quintic, it is made up of the transvectants

$$U = (f^\alpha H^\beta t^\gamma,\ i^\delta A^\epsilon)^\eta.$$

Since A is an invariant we may suppose that $\epsilon = 0$ (if at the same time we remember that A belongs to the complete system).

Since H is a form of even order, and i is a quadratic, all transvectants are reducible except those which have

$$\text{(i)}\quad \beta = 0,$$

$$\text{(ii)}\quad \alpha = 0,\ \gamma = 0,\ \beta = 1.$$

Again t is the Jacobian of f and H, therefore

$$t^2 = -\tfrac{1}{2}\{(f,\ f)^2 H^2 - 2(f,\ H)^2 f . H + (H,\ H)^2 . f^2\}$$
$$\equiv -\tfrac{1}{2} H^3 \bmod (ab)^4. \quad (\S\ 78.)$$

Hence any transvectant Γ, in which $\gamma > 1$, can be expressed in terms of transvectants Γ in which the degree of the product on the left has been decreased and that on the right has been increased together with reducible terms (§ 105, Cor.): for as we have just seen if $\beta > 1$ then U is reducible.

Accordingly we have the following cases to consider:

(i) $\alpha = 1$ or 2, $\beta = 0$, $\gamma = 0$,

(ii) $\alpha = 0$, $\beta = 0$, $\gamma = 1$,

(iii) $\alpha = 1$, $\beta = 0$, $\gamma = 1$,

(iv) $\alpha = 0$, $\beta = 1$, $\gamma = 0$.

All other transvectants are reducible or are expressible in terms of these.

(i) $\alpha = 1, 2$, $\beta = 0$, $\gamma = 0$.

The irreducible transvectants are

$$(f, i), (f, i)^2, (f, i^2)^3, (f, i^2)^4, (f, i^3)^5, (f^2, i^5)^{10}.$$

To see that the other possibilities contain reducible terms we shall take one example. The transvectant $(f^2, i^5)^9$ contains the term $(f, i^3)^5 (f, i^2)^4$.

(ii) $\alpha = 0$, $\beta = 0$, $\gamma = 1$.

Since $$t = -(ab)^2 (bc) a_x^3 b_x^2 c_x^4,$$

and $$(bc)(bi) c_x i_x = \tfrac{1}{2}\{(bc)^2 i_x^2 + (bi)^2 c_x^2 - (ci)^2 b_x^2\},$$

the term $$-(ab)^2 (bc)(bi) a_x^3 b_x c_x^4 i_x$$

of $$(t, i)$$

is reducible. Similarly the transvectants

$$(t, i^2)^3, (t, i^3)^5, (t, i^4)^7$$

contain reducible terms.

Thus $(t, i^4)^7$ contains the term

$$-(ab)^2 (bc)(ai')^2 (ai'')(bi'')(ci''')^2 (bi) c_x^2 i_x$$

which is at once reduced by means of the above identity.

We are left with the forms

$$(t, i)^2, (t, i^2)^4, (t, i^3)^6, (t, i^4)^8, (t, i^5)^9.$$

(iii) $\alpha = 1$, $\beta = 0$, $\gamma = 1$.

The only irreducible transvectant here, is

$$(f.t, i^7)^{14}.$$

To see that the other transvectants are reducible it is sufficient to remark that for example $(f.t, i^7)^{13}$ contains the term

$$(f, i^3)^5 . (t, i^4)^8.$$

(iv) $\alpha = 0, \beta = 1, \gamma = 0.$

The transvectants

$$(H, i), (H, i)^2, (H, i^2)^3, (H, i^2)^4, (H, i^3)^5, (H, i^3)^6$$

prove to be all irreducible.

We are left with 23 forms which are as follows*:

Degree	Order 0	1	2	3	4	5	6	7	9
1						f			
2			i				H		
3				$(i, f)^2$		(i, f)			t
4	A				$(i, H)^2$		(i, H)		
5		$(i^2, f)^4$		$(i^2, f)^3$				$(i, t)^2$	
6			$(i^2, H)^4$		$(i^2, H)^3$				
7		$(i^3, f)^5$				$(i^2, t)^4$			
8	$(i^3, H)^6$		$(i^3, H)^5$						
9				$(i^3, t)^6$					
11		$(i^4, t)^8$							
12	$(i^5, f^2)^{10}$								
13		$(i^5, t)^9$							
18	$(i^7, ft)^{14}$								

* One very obvious remark is to be made regarding this, and all other complete systems obtained by the present methods. We are assured that every covariant can be expressed rationally and integrally in terms of those retained in the complete system, but there is nothing in the process to shew that the latter are all irreducible, except in so far as failure to reduce them may be taken as evidence in this direction. Theoretically then Gordan's process gives an upper limit to the irreducible system.

The enumerative method, depending on the generating function, introduced by Cayley and finally developed by Sylvester and Franklin (*Am. Jour.* vol. VII.) gives a lower limit to the system and when the two methods give the same result the irreducible set has been obtained. The results even when identical have to be received with some caution on account of the enormous labour involved.

117. It is found that for discussing the properties of covariants, it is convenient to have the indices of the transvectants which express them as low as possible. On this account it is usual to replace some of the forms in the irreducible system just given by others which differ from them by reducible terms only.

In the first place the covariant

$$j = -(f, i)^2$$

is of fundamental importance in the quintic system. It is a cubic, and the system of forms for which it is a ground-form are irreducible when considered as forms belonging to the quintic.

Now $(H, i^2)^4$ contains a term

$$(ab)^2 (ai)^2 (bi')^2 a_x b_x = (j, j)^2,$$

accordingly we shall take the irreducible form of degree 6 and order 2 to be

$$(j, j)^2 = \tau.$$

Similarly the form $(t, i^3)^6$ may be replaced by (j, τ), for $(t, i^3)^6$ contains a term

$$(ab)^2 (bc) (ai)^2 (bi')^2 (ci'')^2 a_x c_x^2$$
$$= ((ab)^2 (ai)^2 (bi')^2 a_x b_x, \ (ci'')^2 c_x^3) = (j, \tau).$$

And $(f^2, i^5)^{10}$ may be replaced by the invariant of j,

$$(\tau, \tau)^2.$$

This invariant will be denoted by C, the proof that it may be included in the system instead of $(f^2, i^5)^{10}$ will be given later (§ 121).

It will be found useful to denote the term

$$(ai)^2 (ai')^2 (bi'')^2 (bi''')^2 (ai^{\text{iv}}) (bi^{\text{iv}})$$

of this transvectant by M. Then M may be taken as the invariant of degree 12.

It may be recalled in fact in connection with the simultaneous system of a cubic and quartic (Gundelfinger, *Math. Ann.* Bd. IV.) that the two results originally agreed, but a revision of the generating function led to a reduction of the lower limit which it theoretically gives, and afterwards two forms included in the irreducible system as derived by the methods of Gordan and Clebsch were found to be reducible. The complete systems for the binary forms up to the octavic may be considered as accurately determined by the two methods combined.

Besides j and τ, there is one more quadratic covariant, given in the list as $(H, i^3)^5$. This is equal to $((H, i^2)^4, i)$; we may substitute τ for $(H, i^2)^4$, and hence take as the remaining quadratic covariant

$$(\tau, i) = -\vartheta.$$

118. *The linear covariants.*

$$(f, i^2)^4 = (ai)^2 (ai')^2 a_x = -(j, i)^2 = \alpha,$$

$$(f, i^3)^5 = (ai)^2 (ai')^2 (ai'') i_x'' = (\alpha, i) = -\beta.$$

$(t, i^4)^8$ contains the term

$$\begin{aligned}&(ab)^2 (bc) (ai)^2 (bi')^2 (ci'')^2 (ci''')^2 a_x \\ &= ((ab)^2 (ai)^2 (bi')^2 a_x b_x, \alpha) \\ &= (\tau, \alpha) = \gamma.\end{aligned}$$

$(t, i^5)^9$ contains the term

$$\begin{aligned}&(ab)^2 (bc) (ai)^2 (bi')^2 (ai'') (ci''')^2 (ci^{iv})^2 i_x'' \\ &= ((ab)^2 (ai)^2 (bi')^2 (ai'') i_x'' b_x, \alpha) - \tfrac{1}{2} (ab)^2 (ai)^2 (bi')^2 (ai'') (bi'') . \alpha,\end{aligned}$$

of which the second term is reducible and the first

$$= ((\tau, i), \alpha) = -(\vartheta, \alpha) = -\delta^*.$$

119. *The invariants.*

$$(i, i)^2 = A,$$

$$(H, i^3)^6 = ((H, i^2)^4, i)^2.$$

The latter may be replaced by

$$(\tau, i)^2 = B.$$

$(f^2, i^5)^{10}$ has been replaced already by M.

$(ft, i^7)^{14}$ contains a term

$$((f, i^2)^4, (t, i^5)^9),$$

and hence may be replaced by $(\alpha, \delta) = -R$.

Taking the Jacobians of the 6 linear forms two and two, we obtain the 6 invariants

$$(\alpha\beta), (\beta\gamma), (\gamma\alpha), (\alpha\delta), (\beta\delta), (\gamma\delta).$$

* This is the definition of the linear covariant δ of degree 13 given by Clebsch. In Gordan's book $\delta = (\tau, \beta) = -(\vartheta, \alpha) - \frac{1}{2}(i, \tau)^2 . \alpha$. In other respects the letters common to the two books are identical in meaning.

The values of these invariants are*

$$(\alpha\beta) = -(ai)^2(ai')^2(ai'')(bi''')^2(bi^{\text{iv}})^2(bi'') = -M = -(i\alpha)^2,$$
$$(\beta\gamma) = (i\alpha)(\tau\alpha)(i\tau) = (\vartheta\alpha)^2 = (\delta\alpha) = R,$$
$$(\gamma\alpha) = (\tau\alpha)^2 = N,$$
$$(\alpha\delta) = -R.$$

Now $\delta = ((i, \tau), \alpha) = (i\tau)(\tau\alpha)i_x + \frac{1}{2}(i\tau)^2 . \alpha_x,$

hence

$$(\beta\delta) = (i\alpha)(ii')(i'\tau)(\tau\alpha) + \tfrac{1}{2}B . (i\alpha)^2$$
$$= \frac{-1}{2}AN + \frac{1}{2}BM,$$

$$(\gamma\delta) = ((\tau\alpha)\tau_x, (i\tau')(i\alpha)\tau_x' - \tfrac{1}{2}B . \alpha)$$
$$= \tfrac{1}{2}(CM - BN).$$

Also $N = (\gamma\alpha) = (\tau\alpha)^2 = (ji)^2(j'i')^2(\tau j)(\tau j')$
$$= (ji)(j'i')^2(\tau j)\{(\tau i)(jj') + (\tau j)(j'i)\}.$$

Now from the theory of the cubic we know that any symbolical product which contains a factor $(\tau j)^2$ is zero.

Hence

$$N = (jj')(ji)(j'i')^2(\tau j)(\tau i)$$
$$= \tfrac{1}{2}(jj')(\tau i)\{(ji)(j'i')^2(\tau j) - (j'i)(ji')^2(\tau j')\}$$
$$= \tfrac{1}{2}(jj')(\tau i)\{(jj')(ii')(j'i')(\tau j) + (ji')(j'i)(j'j)(\tau i')\}$$
$$= \tfrac{1}{4}(jj')^2(\tau j')(\tau j) . (ii')^2 - \tfrac{1}{2}((jj')^2 j_x j_x', (\tau i)(\tau i') i_x i_x')^2$$
$$= \tfrac{1}{4}AC - \tfrac{1}{2}(\tau, \tfrac{1}{2}[(\tau i)^2 . i_x'^2 + (\tau i')^2 . i_x^2 - (ii')^2 \tau_x^2])^2$$
$$= \tfrac{1}{2}(AC - B^2).$$

120. The third transvectant of f with j is identically zero. For

$$(f, j)^3 = -(a_x^5, (bi)^2 b_x^3)^3 = -(ab)^3(bi)^2 a_x^2$$
$$= +\tfrac{1}{2}(ab)^3\{(ai)^2 b_x^2 - (bi)^2 a_x^2\} = \tfrac{1}{2}(ab)^4 i_x . \{(ai) b_x + (bi) a_x\}$$
$$= (i, i) = 0.$$

This property is sufficient to define j. For if ψ be an arbitrary cubic then $(f, \psi)^3$ is a quadratic. And in order that $(f, \psi)^3$ may

* In future when no confusion can arise the comma between the two forms in a transvectant will be omitted. This is the usual practice, cf. Stroh.

vanish identically the coefficients of x_1^2, x_1x_2, x_2^2 must be separately zero; giving three equations to determine the ratios of the four coefficients of ψ,—see Chap. XII.

Again

$$\begin{aligned}(f, \tau)^2 &= (aj)(aj')(jj')^2 a_x^3 \\ &= (aj)(aj') a_x \{(aj') j_x - (aj) j_x'\}^2 \\ &= -2(aj)^2 (aj')^2 a_x j_x j_x', \text{ since } (aj)^3 a_x^2 = 0 \\ &= -2(ab)^2 (ac)^2 (bi)^2 (ci')^2 a_x b_x c_x.\end{aligned}$$

Now
$$\begin{vmatrix} (ab)^2 & 0 & (i'b)^2 & (cb)^2 \\ (ai)^2 & (bi)^2 & A & (ci)^2 \\ (ac)^2 & (bc)^2 & (i'c)^2 & 0 \\ 0 & (ba)^2 & (i'a)^2 & (ca)^2 \end{vmatrix} = 0 \qquad (\S\ 77).$$

Hence, if Σ include all possible expressions obtained by interchanging a, b, c,

$$\begin{aligned}&\Sigma (ab)^2 (bi)^2 (i'c)^2 (ca)^2 \\ &\quad = (ab)^4 (ci)^2 (ci')^2 + (bc)^4 (ai)^2 (ai')^2 + (ca)^4 (bi)^2 (bi')^2 \\ &\qquad\qquad + 2A.(ab)^2 (bc)^2 (ca)^2.\end{aligned}$$

And therefore

$$\begin{aligned}(f, \tau)^2 &= -2(ab)^2 (ac)^2 (bi)^2 (ci')^2 a_x b_x c_x \\ &= -(ab)^4 a_x b_x . (ci)^2 (ci')^2 c_x - \tfrac{2}{3} A.(ab)^2 (bc)^2 (ca)^2 a_x b_x c_x.\end{aligned}$$

But
$$\begin{aligned}&(ab)^2 (bc)^2 (ca)^2 a_x b_x c_x \\ &= (ab)^2 (bc)^2 (ca) [-(bc) a_x - (ab) c_x] a_x c_x \\ &= -(ab)^2 (bc)^3 (ca) a_x^2 c_x - (ab)^3 (bc)^2 (ca) a_x c_x^2 \\ &= \tfrac{1}{2} (bc)^4 (ca)(ab) a_x^3 + \tfrac{1}{2} (ab)^4 (bc)(ca) c_x^3 \\ &= -(i, f)^2 = j.\end{aligned}$$

Therefore
$$(f, \tau)^2 = -i.\alpha - \tfrac{2}{3} A.j.$$

121. To obtain the relation between the invariants C and M, we take the expression for M and introduce in it so far as possible symbols referring to the cubic j and its Hessian τ, for

$$C = (\tau\tau)^2.$$

Now

$$\begin{aligned}
M &= (i\alpha)^2 \\
&= (ji)^2 (j'i')^2 (ji'') (j'i'') \\
&= (ji)^2 (j'i')^2 (ja) (j'b) (ab)^4 \\
&= (ja)(j'b)[(ja)(ib) - (jb)(ia)]^2 [(j'a)(i'b) - (j'b)(i'a)]^2 \\
&= (ja)(j'b)[-2(ja)(jb)(ia)(ib) + (jb)^2 (ia)^2][(j'a)^2 (i'b)^2 \\
&\qquad\qquad - 2(j'a)(j'b)(i'a)(i'b)],
\end{aligned}$$

since $(ja)^3 a_x^2 = 0$, and $(j'b)^3 b_x^2 = 0$.

In this expression we will introduce, as far as possible, symbols referring to j; for

$$(ia)^2 a_x^3 = j = (i'b)^2 b_x^3.$$

M is then seen to be the sum of four terms, viz.

$$\begin{aligned}
&(ja)(j'b)(jb)^2(ia)^2(j'a)^2(i'b)^2 \\
&= (jj'')(j'j''')(jj''')^2(j'j'')^2 = -(\tau\tau)^2 = -C: \\
&4(ja)^2(j'b)^2(jb)(j'a)(ia)(i'b)(i'a)(ib) \\
&= 2(ja)^2(j'b)^2(jb)(j'a)\{(ia)^2(i'b)^2 + (ib)^2(i'a)^2 - (ii')^2(ab)^2\} \\
&= 4(jj'')^2(j'j''')^2(jj''')(j'j'') \\
&\qquad - 2A\,.\,(ja)^2(j'b)^2(ab)^2\{(ja)(j'b) - (jj')(ab)\} \\
&= -4C + A\,.\,(jj')^2(ab)^4\{(ja)(j'b) + (jb)(j'a)\},
\end{aligned}$$

since $(ja)^3 a_x^2 = 0$,

$$\begin{aligned}
&= -4C + 2A\,.\,(\tau i)^2 = -4C + 2AB: \\
&-2(ja)^2(j'a)^2(jb)(j'b)(i'b)^2(ia)(ib) \\
&= 2(ja)^2(j'a)^2(jj'')(j'j'')(ia)(ij'') \\
&= (ja)(j'a)(ia)(ij'')\{(ja)^2(j'j'')^2 + (j'a)^2(jj'')^2 - (j''a)^2(jj')^2\} \\
&= -(\tau a)^2(j''a)^2(ia)(ij'') = -(\tau a)(\tau j'')(j''a)^2(ia)^2 = C:
\end{aligned}$$

the last term of M

$$-2(jb)^2(j'b)^2(ja)(j'a)(ia)^2(i'a)(i'b)$$

is obtained from the one just considered by interchanging i and i', a and b, its value is therefore C.

Hence
$$\begin{aligned}
M &= -C - 4C + 2AB + 2C \\
&= -3C + 2AB.
\end{aligned}$$

122. The covariants of orders 0, 1, 2, with the exception of i, have been replaced by transvectants, of index not greater than 2, of simpler forms. We may simplify the expressions of the others in the same way, in fact this has been done already for two of the covariants degree 3, viz. j and (j, τ). The remaining cubic covariant $(f, i^2)^3 = ((f, i)^2, i)$

$$= -(j, i).$$

The transvectant $(H, i)^2$ may be replaced by its term

$$(ab)^2 (bi)^2 a_x^3 b_x = -(aj)^2 a_x^3 j_x;$$

it will be convenient to write this covariant

$$(f, j)^2 = p_x^4.$$

The remaining quartic covariant is

$$(H, i^2)^3 = ((H, i)^2, i),$$

and may then be replaced by (p, i).

Now since $(aj)^3 a_x^2 = 0$,

$$\begin{aligned} (aj)^2 a_x^3 j_y &= (aj)^2 a_x^2 j_x a_y \\ &= \tfrac{1}{4} \{(aj)^2 a_x^3 j_y + 3 (aj)^2 a_x^2 j_x a_y\} \\ &= p_x^3 p_y. \end{aligned}$$

Hence

$$(p, i) = (aj)^2 (ji) a_x^3 i_x = (aj)^2 (ai) a_x^2 j_x i_x.$$

Now

$$\begin{aligned} (f, \alpha) &= (f, -(ji)^2 j_x) \\ &= -(aj)(ji)^2 a_x^4 \\ &= -(aj)\{(ai) j_x - (aj) i_x\}^2 a_x^2 \\ &= 2 (aj)^2 (ai) i_x j_x a_x^2 \\ &= 2 (p, i), \end{aligned}$$

which gives another expression for the same covariant.

For order 5 we have only to consider the transvectant $(t, i^2)^4$; this has a term

$$(ab)^2 (bc)(ci)^2 (ai')^2 a_x b_x^2 c_x^2,$$

hence this transvectant may be replaced by $((H, i)^2, (f, i)^2)$, and therefore by

$$(p, j).$$

Lastly the covariant order 7 may be replaced by the Jacobian (H, j).

123. To express any transvectant of two covariants of the quintic in terms of members of the irreducible system, it is in general advisable to use as far as possible symbolical letters referring to covariants such as j, i, τ, ϑ etc. instead of those belonging to the quintic itself. We give here the values of some of the transvectants, partly for the sake of reference, and partly as examples; we would recommend the student to verify a few of them*.

The following table gives the 2nd transvectants and Jacobians of the quadratic and cubic covariants.

2nd Trans-vectants	Jacobians					
	i	τ	ϑ	j	(j, i)	(j, τ)
i	A	ϑ	$\frac{1}{2}(Bi-A\tau)$	$-(j, i)$	$\frac{2}{3}ai+\frac{1}{2}Aj$	$\frac{1}{2}a\tau+\frac{1}{2}Bj$
τ	B	C	$\frac{1}{2}(Ci-B\tau)$	$-(j, \tau)$	$\frac{1}{6}a\tau+\frac{1}{2}Bj$	$\frac{1}{2}Cj$
ϑ	0	0	N	$-\frac{1}{2}a\tau$	$\frac{1}{6}a\vartheta+\frac{1}{2}\gamma i$	$\frac{1}{2}\gamma\tau$
j	$-a$	0	$-\gamma$	τ	$-\frac{1}{2}i\tau-\frac{1}{3}ja$	$-\frac{1}{2}\tau^2$
(j, i)	$\frac{1}{3}\beta$	$-\frac{2}{3}\gamma$	$\frac{1}{3}\delta-\frac{1}{2}Ba$	$-\frac{1}{3}\vartheta$	$-\frac{1}{9}\{2a^2-\frac{3}{2}A\tau-3Bi\}$	$-\frac{1}{6}\{3\gamma j+a(j, \tau)\}$
(j, τ)	γ	0	$-\frac{1}{2}Ca$	0	$\frac{1}{3}B\tau+\frac{1}{6}Ci$	$\frac{1}{2}C\tau$

The third transvectants of the cubic covariants are

$$(j, (j, i))^3 = B, \quad (j, (j, \tau))^3 = C, \quad ((j, i), (j, \tau))^3 = 0.$$

In obtaining the values of Jacobians it is well to remember the formula

$$((f, \phi), \psi) = \frac{m-n}{2(m+n-2)}(f, \phi)^2 . \psi + \tfrac{1}{2}\{(f, \psi)^2 \phi - (\phi, \psi)^2 . f\}$$

proved, Chap. IV. § 77; where f, ϕ are binary forms of orders m, n respectively; and the order of each of the forms f, ϕ, ψ is not less than 2. This will be of frequent assistance, since of the 15

* Should he experience any great difficulty he will find some of them worked out in Gordan's *Invariantentheorie* and Clebsch's *Binären Formen*.

	α	β	γ	δ
f	$2(p, i)$	$-Ap+\alpha j-\frac{3}{2}i\tau$	$\frac{3}{2}\tau^2+Bp$	$2B(p, i)+j\gamma-\frac{3}{2}\tau\vartheta$
i	β	$-\frac{1}{2}A\alpha$	$\delta-\frac{1}{2}B\alpha$	$\frac{1}{2}(B\beta-A\gamma)$
τ	γ	$-\delta-\frac{1}{2}B\alpha$	$-\frac{1}{2}C\alpha$	$\frac{1}{2}(C\beta-B\gamma)$
ϑ	δ	$\frac{1}{2}(A\gamma-B\beta)$	$\frac{1}{2}(B\gamma-C\beta)$	$-\frac{1}{2}N\alpha$
j	$-\vartheta$	$\alpha^2-\frac{1}{2}(A\tau-Bi)$	$\frac{1}{2}(Ci-B\tau)$	$\alpha\gamma$
(j, i)	$\frac{1}{3}\alpha^2-\frac{1}{2}(A\tau-Bi)$	$-\frac{2}{3}\alpha\beta+\frac{1}{2}A\vartheta$	$\frac{1}{3}\alpha\gamma+\frac{1}{2}B\vartheta$	$-\frac{1}{3}(\alpha\delta+\beta\gamma)+\frac{1}{3}B\alpha^2-\frac{1}{6}Ni$
(j, τ)	$\frac{1}{2}(Ci-B\tau)$	$\frac{1}{2}B\vartheta-\alpha\gamma$	$\frac{1}{2}C\vartheta$	$\frac{1}{2}(C\alpha^2-N\tau)$
p	$-(j, \tau)+\frac{2}{3}A(j, i)-\frac{1}{3}\beta i$	$(\frac{1}{2}B-\frac{1}{3}A^2)j-\frac{1}{2}Ai\alpha+\frac{3}{2}\tau\alpha$	$\frac{1}{2}Bi\alpha+\frac{1}{6}jM$	$\tau\gamma+\frac{1}{6}M(ji)-\frac{1}{6}Bi\beta+\frac{1}{2}B(p\alpha)$
(p, i)	$\frac{3}{4}\alpha\tau-\frac{1}{2}Ai\alpha+(\frac{1}{2}B-\frac{1}{3}A^2)j$	$\frac{1}{2}A(j\tau)-\frac{1}{3}A^2(ji)+\frac{1}{6}Ai\beta-\frac{3}{4}\tau\beta$	$\frac{1}{4}\tau\gamma+\frac{1}{6}M(ji)-\frac{1}{6}Bi\beta$	$(\frac{1}{4}B^2-\frac{1}{3}A^2B+\frac{1}{4}AC)j$ $-\frac{1}{2}ABi\alpha+\frac{3}{4}B\tau\alpha-\frac{3}{4}\tau\delta$

irreducible covariants of the quintic for which the order is not less than 2, 9 have been expressed as Jacobians.

It is useful to know the values of the following transvectants

$$(p, f)^2 = \tfrac{1}{6} ij ; \ (p, i)^2 = -\tau ; \ (p, j)^2 = \tfrac{1}{2} i\alpha + \tfrac{1}{3} Aj.$$

124. For purposes to be presently explained the transvectants $(f, \alpha^5)^5$, $(f, \alpha^4\beta)^5$, ... will be required. Such transvectants are calculated step by step, first (f, α), then $(f, \alpha^2)^2 = ((f, \alpha), \alpha)$, and so on.

It will be useful then to know the values of the transvectants of certain of the covariants with α, β, γ, δ; when these are known the values of such transvectants as $(f, \alpha^5)^5$ may be obtained with great ease.

125. Syzygies. It has been proved (§ 77) that the product of two Jacobians, or the square of a Jacobian can be expressed as a sum of terms, each term being the product of at least three forms. Now nine of the covariants of the quintic are Jacobians,—we must exclude the forms $(i\alpha)$, $(\tau\alpha)$, $(\vartheta\alpha)$, $(\alpha\beta)$, $(\alpha\delta)$ for one at least of the quantics in each of these Jacobians is of order less than 2: hence we have 45 syzygies.

A more general method of obtaining syzygies is given by Stroh. He considers four different forms, and seeks to obtain the syzygies which are of unit degree in each of these forms. First for three forms there is evidently only one such relation, that for weight unity

$$f_1 (f_2, f_3) + f_2 (f_3, f_1) + f_3 (f_1, f_2) = 0,$$

this is written for short $(f_1 f_2 f_3) = 0$.

The other syzygies obtained by Stroh arise from the symbolical relation

$$(ab)\, c_x d_x + (cd)\, a_x b_x = (ad)\, b_x c_x + (cb)\, a_x d_x,$$

which may easily be verified. Raise both sides of this identity to the power i and expand by the binomial theorem: hence

$$\Sigma \binom{i}{\lambda} (ab)^\lambda (cd)^{i-\lambda} a_x^{i-\lambda} b_x^{i-\lambda} c_x^\lambda d_x^\lambda$$

$$= \Sigma \binom{i}{\lambda} (ad)^\lambda (cb)^{i-\lambda} a_x^{i-\lambda} d_x^{i-\lambda} c_x^\lambda b_x^\lambda.$$

Hence

$$\Sigma \binom{i}{\lambda} (f_1, f_2)^\lambda (f_3, f_4)^{i-\lambda} - \Sigma \binom{i}{\lambda} (f_1, f_4)^\lambda (f_3, f_2)^{i-\lambda} = 0.$$

This is written

$$(f_1 f_2 f_3 f_4)_i = 0.$$

Other syzygies may be obtained from it by interchange of the various quantics concerned. But it will be noticed that $(f_1 f_2 f_3 f_4)_i$ is unaltered if one pair of letters is interchanged and at the same time the other pair. Also this expression is only changed in sign if f_2 and f_4 are interchanged. Hence only three distinct syzygies of weight i are obtained in this manner.

If $i = 2$

$$(f_1 f_2 f_3 f_4)_2 \equiv (f_1 f_2)^2 f_3 f_4 + (f_3 f_4)^2 f_1 f_2 - (f_1 f_4)^2 f_2 f_3 - (f_3 f_2)^2 f_1 f_4$$
$$+ 2 (f_1 f_2)(f_3 f_4) - 2 (f_1 f_4)(f_3 f_2) = 0.$$

Whence

$$\{f_1 f_2 f_3 f_4\}_2 \equiv \tfrac{1}{4} [(f_1 f_2 f_3 f_4)_2 + (f_1 f_3 f_4 f_2)_2 + (f_1 f_4 f_2 f_3)_2]$$
$$= (f_1 f_2)(f_3 f_4) + (f_1 f_3)(f_4 f_2) + (f_1 f_4)(f_2 f_3) = 0,$$

a result already well known.

Also

$$[f_1 f_2 f_3 f_4]_2 \equiv \tfrac{1}{2} [(f_1 f_2 f_3 f_4)_2 - (f_1 f_3 f_4 f_2)_2 - (f_1 f_4 f_2 f_3)_2]$$
$$= (f_1 f_2)^2 f_3 f_4 + (f_3 f_4)^2 f_1 f_2 - (f_1 f_4)^2 f_2 f_3$$
$$- (f_3 f_2)^2 f_1 f_4 + 2 (f_1 f_3)(f_2 f_4) = 0.$$

This is the same relation as that given (§ 78) for the product of two Jacobians. It will be seen at once that the other syzygy of weight 2 is deducible from this.

The syzygies of higher weight may often be simplified, in the same way.

The syzygy

$$\{f_1 f_2 f_3 f_4\}_2 = 0$$

is remarkable from the fact that the forms in it need not be of order higher than unity; while in the syzygy

$$(f_1 f_2 f_3 f_4)_2 = 0$$

from which it was deduced, each of the forms is of necessity of order 2 at least.

In general in the syzygy

$$(f_1 f_2 f_3 f_4)_i = 0$$

each of the forms f_1, f_2, f_3, f_4 must be of order i at least; but from these syzygies others may be deduced which are true for forms of lower order. Stroh, in his papers on syzygies*, deduces many such.

We will obtain such a syzygy from that of weight 3,

$$(f_1 f_2 f_3 f_4)_3 = 0.$$

If λ, μ, ν be three quantities whose sum is zero, we know that

$$\lambda^3 + \mu^3 + \nu^3 - 3\lambda\mu\nu = 0,$$

hence

$$3\,(ab)\,(bc)\,(ca)\,a_x b_x c_x = (bc)^3\,a_x^3 + (ca)^3\,b_x^3 + (ab)^3\,c_x^3.$$

We thus obtain

$$\begin{aligned}&\tfrac{1}{6}\left[(a_x^3 b_x^3 c_x^3 d_x^3)_3 - (a_x^3 c_x^3 d_x^3 b_x^3)_3 + (a_x^3 d_x^3 b_x^3 c_x^3)_3\right]\\ &= (ab)\,(bc)\,(ca)\,a_x b_x c_x d_x^3\\ &+ (b_x^3, c_x^3)\,(a_x^3, d_x^3)^2 + (c_x^3, a_x^3)\,(b_x^3, d_x^3)^2 + (a_x^3, b_x^3)\,(c_x^3, d_x^3)^2 = 0.\end{aligned}$$

It will be seen at once that a_x, b_x, c_x, d_x are factors of this syzygy, we may then divide by any one or all of these factors; or else we may multiply the syzygy by a power of any one of them. The syzygy is then true whenever the order of each form is greater than unity. It is not difficult to see that this derived syzygy is just as general as the original one

$$(f_1 f_2 f_3 f_4)_3 = 0,$$

in other words, the original syzygy might be derived from it.

Stroh writes the syzygy just obtained in the form

$$\{f\phi q q'\}_3 = (fq)^2\,(\phi q') - (fq')^2\,(\phi q) - (f\phi)^2\,(qq') + f((qq')\,\phi)^2 = 0,$$

where q, q' are quadratics, and f, ϕ are forms of order 2 at least.

Again from the syzygy

$$(f_1 f_2 f_3 f_4)_4 = 0$$

* *Math. Ann.* Bd. 33, pp. 61–107 (§§ 18–22); Bd. 34, pp. 306–320, 354–370; Bd. 36, pp. 262–288; in § 3 of this latter paper he gives a list of syzygies of the kind just mentioned which he has deduced.

we obtain

$$(ff\phi\phi)_4$$

$$= f^2(\phi\phi)^4 + 6(ff)^2(\phi\phi)^2 + \phi^2(ff)^4$$
$$- 2f\phi(f\phi)^4 + 8(f\phi)^3(f\phi) - 6[(f\phi)^2]^2 = 0.$$

In this write i^2 for ϕ, where i is a quadratic.

Now

$$(i^2, i^2)^4 = \tfrac{2}{3}[(ii)^2]^2,$$

$$(i^2, i^2)^2 = \tfrac{1}{2}(ii)^2 . i^2,$$

$$2(f, i^2)^3(f, i^2) = 2i((f, i)^2, i) . (f, i)$$

$$= i^2 f(f, i^2)^4 + i^2[(f, i)^2]^2 - if(i, i)^2(f, i)^2 - i^3((f, i)^2, f)^2$$

—see § 78.

Hence substituting these values in the syzygy obtained by writing i^2 for ϕ, and then dividing the result by $2i^2$, we obtain

$$(ffii)_4$$

$$= f(fi^2)^4 - 2i((fi)^2 f)^2 - [(fi)^2]^2 + (ii)^2(ff)^2 + \tfrac{1}{2}i^2(ff)^4 = 0.$$

From this may be obtained a syzygy

$$(f\psi ii)_4 = 0$$

by means of the operator $\left(\psi \frac{\partial}{\partial f}\right)$. This operator requires that ψ and f should be of the same order, but when the syzygy is written symbolically it will be seen at once that factors of the form a_x and b_x may be introduced so that the relation is true whatever be the orders of f and ψ, provided that neither is less than 4.

In the same way we may obtain a syzygy

$$[f\psi i\tau]_4$$

where both i and τ are quadratics.

126. Application to the quintic. The forms of the quintic consist of $f, H, i, j, \tau, p, \alpha, \beta, \gamma, \delta$ and some of the Jacobians of these forms.

A large number of syzygies may be at once obtained by writing for the forms in the general syzygies of the last paragraph covariants of f.

We shall content ourselves with a few examples.

$$(fiH) = f.(iH) + H(fi) - it = 0.$$

$$(fjH) = f.(jH) - H(jf) - jt = 0.$$

$$[ffii]_2 = ((fi))^2 + Hi^2 + 2fij + f^2A = 0.$$

$$\{ij\tau i\}_3 = B(ji) - A(j\tau) - \alpha\vartheta + i\gamma = 0.$$

$$(ppii)_4 = -Bp + \tfrac{2}{3}i\alpha^2 - \tfrac{1}{6}Ai\tau - \tfrac{1}{6}Bi^2 - \tau^2 + \tfrac{1}{3}Aj\alpha = 0.$$

$$\{\alpha\beta\gamma\delta\}_2 = \tfrac{1}{2}M(BN - CM) - \tfrac{1}{2}N(AN - BM) - R^2 = 0.$$

From this last we deduce the relation connecting the irreducible invariants A, B, C, R.

It is easy to write down a great number of syzygies in this way. Stroh (*Math. Ann.* Bd. 34, pp. 354—370) has given a list of 168 syzygies of the quintic. The notation for the elementary syzygies given here is not quite the same as that used in Stroh's paper, but the notation is there explained; the notation here has been mainly taken from a later paper by the same author (Stroh, *Math. Ann.* Bd. 36, pp. 262—288).

127. Reducibility of syzygies. If

$$S_1 = 0, \quad S_2 = 0, \quad \ldots$$

be any syzygies, and if P_1, P_2, ... be any products of forms such that P_1S_1, P_2S_2, ... are expressions all of the same degree and order, then

$$P_1S_1 + P_2S_2 + \ldots = 0$$

is a syzygy. In this way it will be seen that an infinite number of syzygies may be built up.

A syzygy $$S = 0$$

is said to be *reducible*, when

$$S \equiv P_1S_1 + P_2S_2 + \ldots,$$

where $S_1 = 0$, $S_2 = 0$, ... are syzygies whose degree is less than the degree of $S = 0$ and the P's are covariants. Otherwise it is said to be *irreducible*.

It will be seen at once that any syzygy which contains a product of only two irreducible forms must be irreducible. All the irreducible syzygies which have yet been found for the quintic are of this nature.

128. It may happen that certain syzygies

$$S_1 = 0, \quad S_2 = 0, \ldots S_i = 0$$

are such that certain products of forms P_1, P_2, ... P_i may be found, for which

$$P_1 S_1 + P_2 S_2 + \ldots + P_i S_i \equiv 0,$$

the expressions P, S being regarded as functions of the concomitants—which for the moment are treated as independent variables.

Such a relation is called a syzygy of the second kind.

The following is an example,

$$(f_1 f_2 f_3 f_4) \equiv f_1 (f_2 f_3 f_4) - f_2 (f_1 f_3 f_4) + f_3 (f_1 f_2 f_4) - f_4 (f_1 f_2 f_3) \equiv 0.$$

Thus for the quintic

$$(fHij) \equiv f(Hij) - H(fij) + i(fHj) - j(fHi) \equiv 0$$

is a syzygy of the second kind.

Syzygies of the second kind may clearly be reducible or irreducible. Between them may arise syzygies of the third kind, and so on.

The following questions at once present themselves. 'Is the number of syzygies finite when the system of forms is finite?' 'When the syzygies of the first kind are finite in number, are also those of the second and of higher kinds finite?' 'Is there any limit to the number of *kinds* of syzygies which arise from a finite system of forms?'

All these questions have been answered in the affirmative by Hilbert (*Math. Ann.* Bd. 36, pp. 473—534). They are partly considered in Chapter IX.

129. The typical representation of the binary quintic. For special purposes, some particular linear transformation of a binary quantic may have peculiar advantages. Thus any particular term of the quantic may by a special transformation be made to vanish. If the quantic has two linear covariants, such that the determinant formed by their coefficients does not vanish, these may be taken for the variables: the transformed quantic will then possess the property that every one of its coefficients is an invariant. We proceed to prove this. Let α_x, β_x be two linear covariants of the quantic a_x^n, which are such that $(\alpha\beta)$ is not zero. Then raising the identity

$$(\alpha\beta)\, a_x = (a\beta)\, \alpha_x - (a\alpha)\, \beta_x$$

to the nth power, we obtain

$$(\alpha\beta)^n . a_x^n = (a\beta)^n . \alpha_x^n - n\,(a\beta)^{n-1}\,(a\alpha)\,\alpha_x^{n-1}\beta_x + \ldots$$

The expression on the right is the transformed quantic, and from the symbolical form of the coefficients, it follows that they are all invariants.

For the general quintic any pair of linear covariants may be chosen; for example those which we have written α and β.

The coefficients may be easily calculated with the help of the table given on p. 139; they are as follows:

$$(f, \alpha^5)^5 = -(B - \tfrac{2}{3}A^2)\,R = -\lambda R.$$

$$(f, \alpha^4\beta)^5 = -\left(\tfrac{1}{2}N - \tfrac{1}{3}AM + \lambda\frac{B}{2}\right)M + \lambda\tfrac{1}{2}\,(2BM - AN) = \mu.$$

$$(f, \alpha^3\beta^2)^5 = R\left(M + \frac{\lambda}{2}A\right).$$

$$(f, \alpha^2\beta^3)^5 = \tfrac{1}{2}M\,(AN - BM) - \mu\,\frac{A}{2}.$$

$$(f, \alpha\beta^4)^5 = -AR\left(M + \frac{\lambda A}{4}\right).$$

$$(f, \beta^5)^5 = -M^3 + \tfrac{1}{2}AM\,(AN - BM) + \mu\,\frac{A^2}{4}.$$

Further the invariant $(\alpha\beta) = -M = 3C - 2AB$ must not be zero.

To be more accurate the coefficients given above should be divided by $(-M)^5$. In the expression for any covariant in terms of the actual coefficients, the above transformed coefficients may be substituted, the covariant multiplied by a power of the determinant of transformation is then equal to the expression thus obtained. In this way any covariant may be expressed in terms of the invariants and two of the linear covariants.

To illustrate a different method of expressing any covariant in terms of the invariants and two linear covariants we shall obtain j in terms of the invariants and the covariants α and δ. Raising the identity

$$(\alpha\delta) j_x = (j\delta)\alpha_x - (j\alpha)\delta_x$$

to the third power

$$-R^3 j = (j, \delta^3)^3 \alpha^3 - 3(j, \delta^2\alpha)^3 \alpha^2\delta + 3(j, \delta\alpha^2)^3 \alpha\delta^2 - (j, \alpha^3)^3 \delta^3.$$

But by the method of § 124

$$(j, \alpha^3)^3 = -R, \quad (j, \alpha^2\delta)^3 = 0,$$

$$(j, \alpha\delta^3) = -\tfrac{1}{2}NR, \quad (j, \delta^3)^3 = -\tfrac{1}{2}R(CM - BN).$$

Therefore

$$R^2 j = \tfrac{1}{2}(CM - BN)\alpha^3 - \tfrac{3}{2}N\alpha^2\delta - \delta^3.$$

130. Given two binary forms of the same order, in particular two quintics, can one be linearly transformed into the other, and if so how?

The reply (in part) to the first question is that if the absolute invariants of the two quantics are equal to one another, and if they each possess a corresponding pair of linear covariants of which the determinants do not vanish, then the quantics are transformable into each other. The question will be found discussed in Clebsch, *Binären Formen*, § 92; and for the case where there are no linear covariants in § 105.

When two quintics have equal absolute invariants and one of the 6 invariants $(\alpha\beta)$, $(\alpha\gamma)$ ... is other than zero, say $(\alpha\beta)$, we may transform one quintic into the other thus:—

Let unaccented letters refer to one quintic, and accented letters to the other; we transform each quintic, so that the variables in the first are α, β, in the second are α', β'.

Thus $$f = A_0\alpha^5 + 5A_1\alpha^4\beta + \dots\dots$$

$$f' = A_0'\alpha'^5 + 5A_1'\alpha'^4\beta' + \dots\dots$$

Let the ratio $\dfrac{A}{A'} = r$, then since the absolute invariants for the two quintics are equal it follows that

$$\frac{B}{A^2} = \frac{B'}{A'^2},$$

and hence $$\frac{B}{B'} = r^2.$$

Similarly

$$\frac{C}{C'} = r^3, \text{ and } \frac{R}{R'} = r^{\frac{9}{2}}.$$

Hence also

$$\frac{A_0}{A_0'} = r^{-6}, \quad \frac{A_1}{A_1'} = r^{-\frac{13}{2}}, \quad \frac{A_2}{A_2'} = r^{-7},$$

$$\frac{A_3}{A_3'} = r^{-\frac{15}{2}}, \quad \frac{A_4}{A_4'} = r^{-8}, \quad \frac{A_5}{A_5'} = r^{-\frac{17}{2}}.$$

The quintic f' may now be transformed into f by means of the transformation

$$\alpha = r^{\frac{6}{5}}\alpha',$$

$$\beta = r^{\frac{6}{5}+\frac{1}{2}}\beta'.$$

131. Associated forms. If y_1, y_2 is a pair of variables cogredient with x_1, x_2; then the two forms

$$\eta = (xy), \quad \xi = a_x^{n-1} a_y$$

are invariantive. Now regard x for the moment as a constant, and the two equations just written down as equations of linear transformation to transform from the variables y_1, y_2 to new variables ξ, η. The variables of the transformed form are covariants, hence its coefficients are invariants—or to be more accurate covariants, for they contain x but not y. Let us proceed exactly as in § 6.

The determinant of transformation is

$$(\xi\eta) = a_x^n = f.$$

From the identity

$$(\xi\eta)\, b_y = (b\eta)\, \xi - (b\xi)\, \eta$$

we obtain the transformed quantic

$$(\xi\eta)^n\, b_y^n = (b\eta)^n\, \xi^n - n\,(b\eta)^{n-1}(b\xi)\, \xi^{n-1}\eta + \ldots \quad \ldots\ldots(\text{I}).$$

Let us calculate the coefficients of the transformed form for the case of the quintic.

$$(b\eta)^5 = b_x^5 = f,$$

$$(b\eta)^4 (b\xi) = b_x^4 (ba)\, a_x^4 = 0,$$

$$(b\eta)^3 (b\xi)^2 = b_x^3 (ba)\, a_x^4 (ba')\, a'^4_x = \tfrac{1}{2} H . f,$$

$$(b\eta)^2 (b\xi)^3 = b_x^2 (ba)(ba')(ba'')\, a_x^4 a'^4_x a''^4_x$$

$$= \tfrac{1}{2} b_x^2 (ba)\, a_x^4 a'^3_x a''^3_x \{(ba')^2 a''^2_x + (ba'')^2 a'^2_x - (a'a'')^2 b_x^2\}$$

$$= -t.f - \tfrac{1}{2} H . (ba)\, b_x^4 a_x^4 = -t.f,$$

$$(b\eta)(b\xi)^4 = b_x(ba)(ba')(ba'')(ba''')\,a_x^4 a'^4_x a''^4_x a'''^4_x$$
$$= (ba')^2(ba)(ba''')\,b_x a'^3_x a_x^4 a'''^4_x \,.\, a''^5_x$$
$$-\tfrac{1}{2}H\,.\,(ba)(ba''')\,b_x^3 a_x^4 a'''^4_x$$
$$= (ba')^2(ba)^2 b_x a'^3_x a_x^3\,.\,f^2 - \tfrac{1}{2}H^2 f$$
$$-\tfrac{1}{4}H^2 f$$
$$= \tfrac{1}{2}if^3 - \tfrac{3}{4}H^2 f,$$
$$(b\xi)^5 = (ba)(ba')(ba'')(ba''')(ba^{\text{iv}})\,a_x^4 a'^4_x a''^4_x a'''^4_x a^{\text{iv}\,4}_x$$
$$= (ba')^2(ba)^2(ba^{\text{iv}})\,a'^3_x a_x^3 a^{\text{iv}\,4}_x\,.\,f^2 + Htf.$$

Now $$((ba)^2(ba')^2 b_x a_x^3 a'^3_x, f)$$
$$= \tfrac{1}{7}(ba)^2(ba')^2\{(bc)\,a_x^3 a'^3_x + 3(ac)\,b_x a_x^2 a'^3_x + 3(a'c)\,b_x a_x^3 a'^2_x\}\,c_x^4$$
$$= (ba)^2(ba')^2(bc)\,a_x^3 a'^3_x c_x^4$$
$$+\tfrac{6}{7}(ab)^3(ba')^2 a_x^2 a'^3_x\,.\,f$$
$$= (ba)^2(ba')^2(bc)\,a_x^3 a'^3_x c_x^4 + \tfrac{3}{7}(ab)^4\{(a'b)\,a_x + (a'a)\,b_x\}\,a'^4_x\,.\,f$$
$$= (ba)^2(ba')^2(bc)\,a_x^3 a'^3_x c_x^4 + \tfrac{6}{7}(f, i)\,.\,f.$$

But the transvectant

$$((ba)^2(ba')^2 b_x a_x^3 a'^3_x, f) = (\tfrac{1}{2}if, f) = \frac{-1}{7}(f, i)\,.\,f.$$

Hence $$(ba)^2(ba')^2(ba^{\text{iv}})\,a_x^3 a'^3_x c_x^4$$
$$= -(f, i)\,.\,f,$$
and therefore
$$(b\xi)^5 = -(f, i)\,.\,f^3 + Htf.$$

The transformation is then

$$f^4\,.\,f(y) = \xi^5 + 5H\xi^3\eta^2 + 10t\xi^2\eta^3$$
$$+\,5\,.\left(\frac{i}{2}f^2 - \frac{3}{4}H^2\right)\xi\eta^4 + ((f, i)\,f^2 - Ht)\,\eta^5.$$

Now let $\Phi(y)$ be any covariant of $f(y)$, then when the above transformation is made, the coefficients of $f(y)$ are replaced by the corresponding coefficients of the powers and products of ξ, η in the expression on the left. Let Φ thus transformed become $\Phi'(\xi, \eta)$, then Φ is equal to Φ' divided by a power of f the determinant of transformation; thus

$$\Phi(y) = \frac{\Phi'(\xi, \eta)}{f^\lambda}.$$

This equation is an identity. We may replace in it y by x; when this is done, ξ becomes f and η becomes zero, hence

$$\Phi(x) = \frac{\Phi'(f, 0)}{f^\lambda}.$$

Hence any covariant of the quintic is equal to its leading coefficient, when the original coefficients of the quintic are replaced by the corresponding coefficients in the form (I), divided by some power of f.

From this we see that all covariants of the quintic may be expressed rationally in terms of the covariants f, H, i, t, (f, i), in such a way that f alone occurs in the denominator. Such a system of covariants in terms of which all covariants of a system may be algebraically expressed is called a system of associated forms. We have confined ourselves to the case of the quintic, the results obtained are however true in general. The coefficients of the transformed quantic may always be expressed as rational integral functions of f, the covariants of degree 2, and the Jacobians of these latter with f. And this is in fact the simplest system of associated forms.

The matter will be found fully discussed in Clebsch, *Binären Formen*, ch. VII. The student who requires further information on the subject of typical representation will find it in the chapter just quoted and the two succeeding chapters of Clebsch's book.

The reduction of the quintic to a sum of three fifth powers will be discussed in Ch. XI., and so nothing need be said on the subject here, especially since it concerns the non-symbolical treatment of the subject rather than the symbolical treatment. The special canonical forms to which the quintic may be reduced, when one or other of its invariants vanishes, will be found in Prof. Elliott's *Algebra of Quantics*.

For a symbolical treatment of the subject the student is referred to Gordan's *Invariantentheorie*, or Clebsch, *Binären Formen*, §§ 93—96.

132. The Sextic. The difficulty in obtaining the complete irreducible system of concomitants of a binary form increases very much with the order. The system for the sextic is obtained here;

it affords examples of a method of reduction applicable to forms of a higher order, but not required when dealing with the quintic.

The arrangement in systems of forms whose grade does not exceed a certain number is followed as before.

The system A_0 contains only f; B_0 contains only

$$(f, f)^2 = H.$$

The system A_1 consists of

$$f, H, (f, H) = t.$$

For B_1 we must take the complete system of

$$i = (f, f)^4.$$

Now i is a quartic, and its complete system is

$$i, (i, i)^2 = \Delta, \quad (i, \Delta) = v, \quad (i, i)^4, \quad (i\Delta)^4.$$

To find the system A_2 we must take the transvectants of powers and products of forms of A_1 with powers and products of forms of B_1.

Now the form $(if)^3$ can be shewn to vanish, for

$$\begin{aligned}(if)^3 &= (ab)^4 (bc)^2 (ac)\, a_x c_x^3 \\ &= -\tfrac{1}{3}(ab)(bc)(ca)\,[(ab)^3 (bc)\, a_x c_x^3 + (bc)^3 (ca)\, b_x a_x^3 \\ &\qquad\qquad\qquad + (ca)^3 (ab)\, c_x b_x^3] \\ &= \tfrac{1}{6}(ab)(bc)(ca)\,[(ab)^4 c_x^4 + (bc)^4 a_x^4 + (ca)^4 b_x^4],\end{aligned}$$

on using Stroh's series

$$\begin{pmatrix} f & f & f \\ 1 & 1 & 1 \end{pmatrix}_7.$$

But $$(ab)^5 (bc)(ca)\, c_x^4 = 0,$$

since it changes sign when a and b are interchanged.

Hence $$(if)^3 = 0.$$

The quadratic covariant $(if)^4$ is of great importance; it is usually denoted by the symbol l.

If any covariant can be expressed as a symbolical product in which the factor $(ia)^3$ appears, it can be expressed as a sum of transvectants of l with other forms. For such a covariant

$$\begin{aligned}&= ((ia)^3 i_x a_x^3, \Phi)^\rho + \Sigma\, ((ia)^4 a_x^2, \Phi')^\rho \\ &= \Sigma\, (l, \Phi')^{\rho'}.\end{aligned}$$

Again

$$\Delta = (i, i)^2 = (ia)^2 (ab)^4 i_x^2 b_x^2 + \lambda (ab)^6 . i.$$

And

$$((ia)^3 i_x a_x^3,\ b_x^6)^3 = \tfrac{1}{4} (ia)^3 b_x^3 \{(ab)^3 i_x + 3 (ab)^2 (ib) a_x\}$$
$$= (ia)^3 (ab)^3 i_x b_x^3 + \tfrac{3}{4} (ia)^4 (ab)^2 b_x^4,$$

but

$$(ia)^3 i_x a_x^3 = 0,$$

hence

$$(ia)^3 (ab)^3 i_x b_x^3 = - \tfrac{3}{4} (ia)^4 (ab)^2 b_x^4 = - \tfrac{3}{4} (lf)^2.$$

Now since a and b are equivalent symbols

$$(ia)^3 (ab)^3 i_x b_x^3 = - \tfrac{1}{2} (ab)^4 i_x^2 \{(ia)^2 b_x^2 + (ia)(ib) a_x b_x + (ib)^2 a_x^2\}$$
$$= - \tfrac{3}{2} (ia)^2 (ab)^4 i_x^2 b_x^2 + \tfrac{1}{4} (ab)^6 . i,$$

and therefore

$$\Delta \equiv \tfrac{1}{2} (lf)^2 \text{ mod. } (ab)^6.$$

Thus every form of B_1 except i and the invariant $(i, i)^4$

$$\equiv 0 \text{ modd. } (ia)^4, (ab)^6.$$

133. We shall first find the system which is relatively complete with respect to the moduli $(ai)^4$ and $(ab)^6$ (§ 107, Cor. II.). This is obtained by taking the transvectants of powers of i, with powers and products of forms of the system A_1. Let us call this the system C.

First consider the forms

$$(i^\alpha, f^\beta)^\gamma.$$

We have (i, f), $(i, f)^2$. Every other one of these forms

$$\equiv 0 \text{ mod. } (ia)^4.$$

Next consider forms

$$(i^\alpha, H^\beta)^\gamma.$$

We retain only (i, H), for $(i, H)^2$ contains the term

$$(ia)^2 (ab)^2 i_x^2 a_x^2 b_x^4;$$

this can (§ 63) be linearly expressed in terms of covariants

$$(ia)^4 a_x^2 b_x^6,\ \ (ib)^4 b_x^2 a_x^6,\ \ (ia)^3 (ab) i_x a_x^2 b_x^5,$$
$$(ib)^3 (ab) i_x b_x^2 a_x^5,\ \ (ab)^4 a_x^2 b_x^2 . i_x^4.$$

Hence

$$(i, H)^2 \equiv \lambda i^2 + \mu l . f.$$

The form $(i, H)^3$ contains the term

$$(ia)^3 (ab)^2 i_x a_x b_x^4;$$

and the form $(i, H)^4$ contains the term

$$(ia)^4 (ab)^2 b_x^4;$$

hence these may both be rejected.

All other forms $(i^\alpha, H^\beta)^\gamma$ contain a term having a factor $(iH)^4$.

No one of the forms $(i^\alpha, t^\beta)^\gamma$ need be retained, for if $\gamma = 1$, the transvectant is the Jacobian of a Jacobian and another form and therefore reducible (§ 77).

If $\gamma > 2$, the transvectant always contains a term which involves a factor $(ia)^3$ or a factor $(ia)^4$, and therefore which

$$\equiv 0 \text{ mod. } (ai)^4.$$

If $\gamma = 2$, the transvectant $(i, t)^2$ contains the term

$$\begin{aligned} &(iH)^2 (Hf)\, i_x^2 H_x^5 f_x^5 \\ &= ((i, H)^2, f) \text{ mod. } (iH)^3 \\ &= \lambda\, (i^2, f) \text{ mod. } (ai)^4 \\ &= \lambda i\, (i, f) \text{ mod. } (ai)^4. \end{aligned}$$

The general transvectant

$$(i^\alpha, f^\beta H^\gamma t^\delta)^\epsilon$$

may be treated in the same way. If $\epsilon > 2$, the transvectant contains a term which

$$\equiv 0 \text{ mod. } (ai)^4.$$

And if $\epsilon \not> 2$, it is certainly reducible, except for the cases already discussed.

The system C then contains the forms

$$f, H, t, i, (i, i)^4, (f, i), (H, i), (f, i)^2.$$

134. To find the system A_2 we must now take all possible transvectants of powers and products of the system C, with powers and products of the complete system of l.

Now l is a quadratic, its complete system must then consist of

$$l, (l, l)^2.$$

The invariant $(l, l)^2$ is the same as an invariant already found, viz.:

$$\begin{aligned} (i, \Delta)^4 &= \tfrac{1}{2} (i, (fl)^2)^4 \text{ mod. } (ab)^6 \\ &= \tfrac{1}{2} (ll)^2 \text{ mod. } (ab)^6. \end{aligned}$$

Since l is a quadratic and all the covariants of C are of even degree, we need only consider the transvectants of powers of l with each form separately.

The forms

$$(l, f),\ (l, f)^2 = 2\Delta,\ (l^2, f)^3,\ (l^2, f)^4,\ (l^3, f)^5,\ (l^3, f)^6$$

are all irreducible; so also is

$$(l, H).$$

The covariant $(l, H)^2$ contains the term

$$\begin{aligned}(la)^2 (ab)^2 a_x^2 b_x^4 &= ((l, f)^2, f)^2 \\ &= 2\,((i, i)^2, f)^2 \\ &= \lambda\,(ii')^2 (i'a)^2 i_x^2 a_x^4 + \mu\,(i, i)^4 f.\end{aligned}$$

But the term

$$(ii')^2 (i'a)^2 i_x^2 a_x^4$$

is linearly expressible in terms of the covariants (§ 63)

$$(ii')^4 a_x^6,\ (ii')^3 (i'a)\, i_x a_x^5,\ (ia)^4 a_x^2 i_x'^4,$$

$$(ia)^3 (ii')\, i_x'^3 a_x^3,\ (i'a)^4 a_x^2 i_x^4,$$

each of which is reducible.

Hence $\qquad (l, H)^2 = \lambda_1 (ii)^4 f + \lambda_2 li.$

Now, if $\beta > 2$

$$\begin{aligned}&(l^\alpha, H)^\beta \\ &= ((l, H)^2, l^{\alpha-1})^{\beta-2} = (ii)^4\,\Phi + \lambda_2\,(li, l^{\alpha-1})^{\beta-2},\end{aligned}$$

hence these transvectants are all reducible.

The covariant (l, t) is reducible, § 77.

The covariant $(l, t)^2$ contains the term

$$((l, H)^2, f),$$

and is therefore reducible.

The covariant, $\beta > 2$,

$$(l^\alpha, t)^\beta = ((l, t)^2, l^{\alpha-1})^{\beta-2}$$

and is reducible.

Hence all the covariants

$$(l^\alpha, t)^\beta$$

are reducible.

The covariants $(l, i),\ (l, i)^2,\ (l^2, i)^3$

are irreducible; but

$$\begin{aligned}(l^2, i)^4 &= (l^2, (ab)^4 a_x^2 b_x^2)^4 \\ &\equiv ((l, f)^2, (l, f)^2)^4 \text{ mod. } (ab)^6 \\ &\equiv 4\,((i, i)^2, (i, i)^2)^4 \text{ mod. } (ab)^6,\end{aligned}$$

which is reducible when considered as an invariant of the quartic i.

The form $(l, (f, i))$ is reducible, § 77;

$(l, (f, i))^2$ contains the term

$$((l, f)^2, i) = (\Delta, i) = -v;$$

$(l^2, (f, i))^3$ contains the reducible term

$$(l, ((l, f)^2, i)).$$

Similarly $(l^3, (f, i))^5,\ (l^4, (f, i))^7$

may be reduced.

The forms

$$(l^2, (f, i))^4,\ (l^3, (f, i))^6,\ (l^4, (f, i))^8$$

are however irreducible.

The form $(l, (f, i)^2)$ is not reducible.

Now $(l, (f, i)^2)^2$ contains the term

$$((l, f)^2, i)^2 = 2\,(\Delta, i)^2 = \tfrac{1}{3}\,i(ii)^4 \ldots$$

(see § 89).

Hence $(l^\alpha, (f, i)^2)^\beta$ is reducible when $\beta \geqslant 2$, for this

$$\begin{aligned}&= (l^{\alpha-1}, (l, (f, i)^2)^2)^{\beta-2} \\ &= \tfrac{1}{3}\,(ii)^4\,(l^{\alpha-1}, i)^{\beta-2} + \lambda\,(l^{\alpha-1}, l^2)^{\beta-2}.\end{aligned}$$

Lastly $(l, (H, i))$ is reducible by § 77.

$(l, (H, i))^2$ contains a term

$$((l, H)^2, i) = \lambda_1\,(ii)^4\,(f, i) + \lambda_2\,(li, i)$$

and $(l^\alpha, (H, i))^\beta,\quad \beta > 2,$

$$= (l^{\alpha-1}, (l, (H, i))^2)^\beta$$

which is reducible.

Thus the system A_2 contains the forms of the system C together with l, $(l, l)^2$ and

$$(l, f), (l, f)^2, (l^2, f)^3, (l^2, f)^4, (l^3, f)^5, (l^3, f)^6,$$
$$(l, H), (l, i), (l, i)^2, (l^2, i)^3, (l, (f, i))^2, (l^2, (f, i))^4,$$
$$(l^3, (f, i))^6, (l^4, (f, i))^8, (l, (f, i)^2).$$

The system B_2 contains only the invariant $(f, f)^6$; we merely add this to the system A_2, and the result is the complete system for f.

We append the following table giving the 26 irreducible concomitants of the sextic.

Degree	Order						
	0	2	4	6	8	10	12
1				f			
2	$(f, f)^6$		$(f, f)^4 = i$		$(f, f)^2 = H$		
3		$(f, i)^4 = l$		$(f, i)^2$	(f, i)		$(f, H) = t$
4	$(i, i)^4$		$(f, l)^2$	(f, l)		(H, i)	
5		$(i, l)^2$	(i, l)		(H, l)		
6	$(l, l)^2$			$((f, i)^2, l)$ $((f, i), l)^2$			
7		$(f, l^2)^4$	$(f, l^2)^3$				
8		$(i, l^2)^3$					
9			$((f, i), l^2)^4$				
10	$(f, l^3)^6$	$(f, l^3)^5$					
12		$((f, i), l^3)^6$					
15	$((f, i), l^4)^8$						

Ex. (i). Prove that if the covariant α, of a quintic f, is identically equal to zero, then also

$$\beta \equiv 0, \quad \gamma \equiv 0, \quad \delta \equiv 0, \quad \vartheta \equiv 0, \quad (pi) \equiv 0, \quad M \equiv 0, \quad N \equiv 0, \quad R \equiv 0,$$
$$Bi \equiv A\tau, \quad (B - \tfrac{2}{3}A^2) j \equiv 0, \quad Bf + \tfrac{5}{2} j\tau \equiv 0.$$

Ex. (ii). In the last example either

$$j \equiv 0 \text{ or } B \equiv \tfrac{2}{3}A^2.$$

Prove that in the former case every covariant vanishes with the exception of

$$f,\ H,\ i,\ t,\ (fi),\ A\ ;$$

and that these are connected by the relations

$$AH+\tfrac{1}{2}i^3=0,$$
$$At+\tfrac{1}{2}i^2(fi)=0,$$
$$2[(f,i)]^2+Hi^2+f^2A=0.$$

Ex. (iii). Prove that if a vanishes identically and j is not zero, then

$$B\equiv\tfrac{2}{3}A^2,\quad t\equiv\tfrac{2}{3}Ai,\quad A\,(Af+\tfrac{15}{4}ij)\equiv 0.$$

The latter result gives an alternative, but if

$$Af=-\tfrac{15}{4}ij,$$

then

$$Aj=\tfrac{15}{4}(ij,i)^2=\tfrac{3}{2}Aj.$$

Hence in either case since j is other than zero,

$$A\equiv 0,\quad B\equiv 0,\quad C\equiv 0.$$

Shew further that j is a perfect cube: that it is a factor of f: that i is a factor of j: that p is a perfect fourth power, having j for a factor: and that p is a factor of H.

Ex. (iv). If all the invariants of a quintic vanish shew that a must vanish and that j must be a perfect cube and a factor of f.

Ex. (v). If $\qquad \eta=(xy),\quad \xi=a_x^{n-1}a_y,\quad f=a_x^n,$

then

(i) $n=2$,

$$f\cdot f(y)=\xi^2+\tfrac{1}{2}(f,f)^2\cdot\eta.$$

(ii) $n=3$,

$$f^2\cdot(y)=\xi^3+\tfrac{3}{2}H\xi\eta^2+t\eta^3.$$

(iii) $n=4$,

$$f^3\cdot f(y)=\xi^4+3H\xi^2\eta^2+4t\xi\eta^3+\left(\frac{if^2}{2}-\frac{3}{4}H^2\right)\eta^4.$$

(iv) $n=6$,

$$f^5\cdot f(y)=\xi^6+\frac{15}{2}H\xi^4\eta^2+20t\xi^3\eta^3+15\left(\frac{i}{2}f^2-\frac{3}{4}H^2\right)\xi^2\eta^4$$
$$+6\left((fi)f^2-Ht\right)\xi\eta^5+\left(\frac{A}{2}f^4-\frac{15}{4}iHf^2+\frac{45}{8}H^3+10t^2\right)\eta^6$$

(*Clebsch*).

[The student, who wishes for information concerning special quintics—when some of the invariants vanish—is referred to Clebsch, *Binären Formen*, ch. VIII., and to Elliott, *Algebra of Quantics*, ch. XIII.]

CHAPTER VIII.

SIMULTANEOUS SYSTEMS.

135. It was proved in Chap. VI. § 103, that if S_1, S_2 be any two finite and complete systems of forms, then the system S formed by taking transvectants of powers and products of powers of forms of S_1 with powers and products of powers of forms of S_2 is both finite and complete. If S_1, S_2 be the complete systems for any two binary forms f_1, f_2; then S is the complete system of concomitants for the forms f_1, f_2 taken simultaneously; for S is complete and contains both f_1 and f_2. Hence the complete irreducible system of concomitants of a pair of binary forms is finite.

136. To make the matter clearer, let us briefly recapitulate the argument.

(i) Any concomitant of the simultaneous system can be expressed as a sum of symbolical products; the factors in which are all of the following types

$$(ab),\ (\alpha\beta),\ (a\alpha),\ a_x,\ \alpha_x:$$

where letters of the Roman alphabet refer to the quantic f_1, and letters of the Greek alphabet to the quantic f_2.

(ii) Any concomitant of the simultaneous system can be expressed linearly in terms of transvectants of products of forms belonging to the complete system for f_1, with products of forms belonging to the complete system for f_2.

For any symbolical product in which the letters are partly Roman and partly Greek is a term of a transvectant $(U, V)^\rho$, where U is a product containing only Roman letters and V a product containing only Greek letters. But by § 51 any term of the transvectant $(U, V)^\rho$ is equal to

$$(U, V)^\rho + \Sigma\lambda(\overline{U}, \overline{V})^{\rho'},$$

where $\overline{U}$, $\overline{V}$ are obtained by convolution from U, V respectively; λ is numerical and ρ' is less than ρ.

Now U, $\overline{U}$ are covariants of f_1 and hence may be expressed as a sum of products of the irreducible forms of f_1; similarly V, $\overline{V}$ may be expressed as a sum of products of the irreducible forms of f_2.

Hence the theorem is true for any symbolical product, the letters of which refer some to f_1 and some to f_2: and therefore it is true for any concomitant of the simultaneous system.

(iii) The system of transvectants $(U, V)^\rho$, where U is a product of concomitants of f_1 and V a product of the concomitants of f_2, is both finite and complete. This was proved in § 103.

137. *The complete irreducible system of concomitants of a finite number of quantics is finite.*

The proof of this theorem is inductive. Let us suppose that it has been proved that the complete system of concomitants of any n quantics is finite.

Consider a set of $n + 1$ quantics,

$$f_1, f_2, \ldots f_{n+1}.$$

The n quantics $f_1, f_2, \ldots f_n$ possess, by hypothesis, a finite system of concomitants which may be called S_1. The single form f_{n+1} also possesses a finite system of concomitants, which may be called S_2. The complete system, S, of concomitants of the $n + 1$ quantics is obtained by combining S_1 with S_2. And since the systems S_1 and S_2 are both finite and complete, it follows that the complete system S is finite. Hence if the complete system of concomitants of any n quantics is finite then that for any $n + 1$ quantics is also finite. But the complete system for any one or any two quantics is finite. Hence the complete system of concomitants for any finite number of quantics is finite.

We proceed to find the complete systems, in a few of the simpler cases.

138. Linear form and quadratic. The complete system of concomitants of a quadratic f consists simply of the quadratic itself and the invariant $(f, f)^2$.

Thus the system S_1 is

$$f, \ (f, f)^2 = D.$$

The system S_2—of the linear form l—is simply l.

The combined system S is obtained by taking all possible transvectants,

$$(f^\alpha D^\beta, l^\gamma)^\delta.$$

But unless $\beta = 0$, this is equal to

$$D^\beta (f^\alpha, l^\gamma)^\delta,$$

and is certainly reducible.

Again $$(f^\alpha, l^\gamma)^\delta = (f^\alpha, l^\delta)^\delta \,.\, l^{\gamma-\delta},$$

which is reducible unless $\gamma = \delta$.

Further $(f^\alpha, l^\delta)^\delta$ contains a term

$$(f, l)^2 \,.\, (f^{\alpha-1}, l^{\delta-2})^{\delta-2},$$

and is reducible if $\delta > 2$.

The system S then consists of

$$f, \Delta, l, (f, l), (f, l)^2.$$

138 A. Linear form and any finite system. Let the finite system referred to be denoted by S_1. The system S_2 consists simply of the linear form l.

Let the system S_1 consist of the forms $C_1, C_2, \ldots C_\lambda$ which are of orders $s_1, s_2, \ldots s_\lambda$.

Then we have to consider all possible transvectants

$$(C_1^{a_1} C_2^{a_2} \ldots C_\lambda^{a_\lambda}, l^\beta)^\gamma.$$

If $\beta > \gamma$ this transvectant contains a factor l, and is therefore reducible. It may then be supposed that

$$\beta = \gamma.$$

Let us suppose that $a_r \neq 0$; then if $\gamma > s_r$ the transvectant contains the terms

$$(C_r, l^{s_r})^{s_r} (C_1^{a_1} \ldots C_r^{a_r-1} \ldots C_\lambda^{a_\lambda}, l^{\gamma-s_r})^{\gamma-s_r},$$

and is reducible. If $\gamma \not> s_r$ it contains the term

$$(C_r, l^\gamma)^\gamma . C_1^{a_1} C_2^{a_2} \ldots C_r^{a_r-1} \ldots C_\lambda^{a_\lambda},$$

and is therefore reducible.

Hence the system S contains the forms

$$C_1, C_2, \ldots C_\lambda; \; l; \; (C_r, l^\gamma)^\gamma,$$

$$\gamma = 1, 2, \ldots s_r,$$

$$r = 1, 2, \ldots \lambda;$$

and these forms only.

Ex. Prove that the complete system for a linear form l, and a given finite system of forms, consists of the linear form, the given system, and the forms obtained by operating with powers of

$$\left(l_2 \frac{\partial}{\partial x_1} - l_1 \frac{\partial}{\partial x_2}\right)$$

on the members of the given system.

139. Two quadratics. Let f_1, f_2 be the two quadratics. We have to combine the two systems S_1, S_2, where S_1 consists of

$$f_1, \; (f_1, f_1)^2 = D_1,$$

and S_2 consists of

$$f_2, \; (f_2, f_2)^2 = D_2.$$

Since D_1 and D_2 are invariants, they give rise to no new forms. Hence we have only to consider transvectants

$$(f_1^\alpha, f_2^\beta)^\gamma.$$

The only irreducible transvectants which can be obtained are

$$J_{12} = (f_1, f_2),$$

and

$$D_{12} = (f_1, f_2)^2.$$

The required system is then

$$f_1, f_2, J_{12}, D_1, D_2, D_{12}.$$

139 A. Any number of quadratics. Consider first three quadratics f_1, f_2, f_3. To obtain their simultaneous system we combine the system S_1 for f_1, f_2 with the system S_2 for f_3.

Leaving invariants out of account we must consider all transvectants

$$(f_1^\alpha f_2^\beta J^\gamma_{1,2}, f_3^\delta)^\epsilon.$$

Since all the forms are quadratics the only irreducible transvectants obtainable are

$$(f_1, f_3),\ (f_2, f_3),\ (J_{12}, f_3);$$
$$(f_1, f_3)^2,\ (f_2, f_3)^2,\ (J_{12}, f_3)^2.$$

Of these (J_{12}, f_3) is reducible, for it is the Jacobian of a Jacobian and another form.

The rest are

$$J_{13},\ J_{23},\ D_{13},\ D_{23},$$

and another invariant which may be called

$$E_{123}.$$

The complete system for the three quadratics is then seen to be

$$f_1, f_2, f_3, J_{12}, J_{13}, J_{23}, D_1, D_2, D_3, D_{12}, D_{23}, D_{31}, E_{123}.$$

There is only one form of a new kind, and this is an invariant. Hence in forming the system for four quadratics, we shall not meet with any new kind of concomitant. And in fact it is easy to see that every irreducible concomitant in the system for any number of quadratics belongs to one or other of the *types*

$$f,\ J,\ D,\ E.$$

139 B. It is easy to obtain the syzygies between the forms of the last paragraph. First J_{12} is a Jacobian, and therefore, § 78

$$2J^2_{12} = -D_1 f_2^2 - D_2 f_1^2 + 2D_{12} f_1 f_2 \ \ldots\ldots\ldots\ldots (1).$$

It will be convenient to use the notation

$$f_1 = a_x^2,\ f_2 = b_x^2,\ f_3 = c_x^2,\ \ldots.$$

Then as in § 77 we obtain

$$2J_{12}J_{34} = 2(ab)a_x b_x . (cd) c_x d_x$$
$$= \begin{vmatrix} (ac)^2 & (ad)^2 & a_x^2 \\ (bc)^2 & (bd)^2 & b_x^2 \\ c_x^2 & d_x^2 & 0 \end{vmatrix}$$
$$= -D_{13} f_2 f_4 - D_{24} f_1 f_3 + D_{14} f_2 f_3 + D_{23} f_1 f_4 \ \ldots\ldots (2).$$

By replacing f_4 by f_1, a syzygy for $2J_{12}J_{31}$ is obtained.

Again $$E_{123} = -(ab)(bc)(ca),$$

hence, as in § 77,

$$2E_{123} \, . \, E_{456} = \begin{vmatrix} (ad)^2 & (ae)^2 & (af)^2 \\ (bd)^2 & (be)^2 & (bf)^2 \\ (cd)^2 & (ce)^2 & (cf)^2 \end{vmatrix}$$

$$= \begin{vmatrix} D_{14} & D_{15} & D_{16} \\ D_{24} & D_{25} & D_{26} \\ D_{34} & D_{35} & D_{36} \end{vmatrix} \quad \text{............} (3).$$

From this may be obtained syzygies for

$$2E^2_{123}, \; 2E_{123} \, . \, E_{124}, \; 2E_{123} \, . \, E_{145}.$$

Similarly, the syzygy

$$2E_{123}J_{45} = \begin{vmatrix} D_{14} & D_{15} & f_1 \\ D_{24} & D_{25} & f_2 \\ D_{34} & D_{35} & f_3 \end{vmatrix} \quad \text{...............} (4)$$

may be obtained, and other particular cases may be deduced.

Again $$\begin{vmatrix} a_1^2 & a_1a_2 & a_2^2 & a_x^2 \\ b_1^2 & b_1b_2 & b_2^2 & b_x^2 \\ c_1^2 & c_1c_2 & c_2^2 & c_x^2 \\ d_1^2 & d_1d_2 & d_2^2 & d_x^2 \end{vmatrix} = 0,$$

for the last column of this determinant is a sum of multiples of the first three columns.

But it has been shewn, § 77, that

$$E_{123} = -(ab)(bc)(ca) = \begin{vmatrix} a_1^2 & a_1a_2 & a_2^2 \\ b_1^2 & b_1b_2 & b_2^2 \\ c_1^2 & c_1c_2 & c_2^2 \end{vmatrix}.$$

Hence

$$f_1E_{234} - f_2E_{134} + f_3E_{124} - f_4E_{123} = 0 \text{...........} (5).$$

If in the above determinant the elements of the last column are replaced by $(ae)^2$, $(be)^2$, $(ce)^2$, $(de)^2$ respectively, another syzygy is obtained,

$$D_{15}E_{234} - D_{25}E_{134} + D_{35}E_{124} - D_{45}E_{123} = 0 \text{ .... } (6).$$

In § 77, it was proved that

$$\begin{vmatrix} (ae)^2 & (af)^2 & (ag)^2 & (ah)^2 \\ (be)^2 & (bf)^2 & (bg)^2 & (bh)^2 \\ (ce)^2 & (cf)^2 & (cg)^2 & (ch)^2 \\ (de)^2 & (df)^2 & (dg)^2 & (dh)^2 \end{vmatrix} = 0 \quad\ldots\ldots\ldots\ldots (7)$$

Similarly we obtain the syzygies

$$\begin{vmatrix} D_{15} & D_{16} & D_{17} & f_1 \\ D_{25} & D_{26} & D_{27} & f_2 \\ D_{35} & D_{36} & D_{37} & f_3 \\ D_{45} & D_{46} & D_{47} & f_4 \end{vmatrix} = 0 \quad\ldots\ldots\ldots\ldots (8),$$

and

$$\begin{vmatrix} D_{14} & D_{15} & D_{16} & f_1 \\ D_{24} & D_{25} & D_{26} & f_2 \\ D_{34} & D_{35} & D_{36} & f_3 \\ f_4 & f_5 & f_6 & 0 \end{vmatrix} = 0 \quad\ldots\ldots\ldots\ldots (9).$$

Every kind of syzygy which occurs in the irreducible system of concomitants for any number of quadratics has now been obtained.

Ex. (i). Shew that the last three syzygies just written down are not independent of those which come before, but may be obtained from them on multiplying by forms of the types E and J.

Ex. (ii). Obtain the syzygies (1), (2), (5) by means of Stroh's method.

Ex. (iii). Obtain (4) from (3), and (6) from (5) by transvection.

140. Quadratic and cubic. Let ϕ be the quadratic and f the cubic. Then we have to combine the systems of forms S_1 and S_2; where S_1 contains

$$\phi, \quad (\phi, \phi)^2 = D,$$

and S_2 contains

$$f, \quad (f, f)^2 = H, \quad (f, H) = T, \quad (H, H)^2 = \Delta.$$

All transvectants

$$(\phi^\alpha D^\beta, f^\gamma H^\delta T^\epsilon \Delta^\eta)^\zeta$$

must be considered. Any transvectant for which either β or η is other than zero is obviously reducible, it may then be supposed that

$$\beta = 0, \quad \eta = 0.$$

Again both ϕ and H are quadratics, hence if δ is not zero and $\zeta > 2$ the transvectant contains the reducible term

$$(\phi,\ H)^2\,(\phi^{\alpha-1},\ f^{\gamma}H^{\delta-1}T^{\epsilon})^{\zeta-2}.$$

We have then only to discuss the transvectants

$$(\phi,\ H),\quad (\phi,\ H)^2,\quad (\phi^{\alpha},\ f^{\gamma}T^{\epsilon})^{\zeta}.$$

Of the transvectants

$$(\phi^{\alpha},\ f^{\gamma})^{\zeta},$$

all contain reducible terms except

$$(\phi,\ f),\quad (\phi,\ f)^2,\quad (\phi^2,\ f)^3,\quad (\phi^3,\ f)^6.$$

Now since, by § 91,

$$T^2 = -\tfrac{1}{2}(H^3 + \Delta f^2),$$

those transvectants for which $\epsilon > 1$ are all reducible. Hence of the transvectants

$$(\phi^{\alpha},\ T^{\epsilon})^{\zeta}$$

all are reducible except

$$(\phi,\ T)^2,\quad (\phi^2,\ T)^3,$$

for $(\phi,\ T)$ is reducible by § 77.

The only other irreducible transvectant is readily seen to be

$$(\phi^3,\ fT)^6.$$

The simultaneous system for the quadratic ϕ and the cubic f then consists of:

five invariants

$$D,\quad \Delta,\quad (\phi,\ H)^2,\quad (\phi^3,\ f^2)^6,\quad (\phi^3,\ fT)^6;$$

four linear covariants

$$(\phi,\ f)^2,\quad (\phi^2,\ f)^3,\quad (\phi,\ T)^2,\quad (\phi^2,\ T)^3;$$

three quadratic covariants

$$\phi,\quad H,\quad (\phi,\ H);$$

three cubic covariants

$$f,\quad T,\quad (\phi,\ f).$$

141. Quadratic and any system of forms. Let the system of forms referred to be denoted by S_1, the system S_2 for the quadratic f consists of

$$f,\quad D = (f,\ f)^2.$$

The invariants of both systems may be left out of account as they produce no new forms.

All possible transvectants

$$U = (P, f^r)^\rho$$

must be discussed, where P is a product of the forms

$$C_1, C_2, \ldots C_\lambda$$

of the system S_1.

If $\rho < 2r - 1$, U contains the term

$$f \cdot (P, f^{r-1})^\rho,$$

and is therefore reducible. Since ρ cannot be greater than $2r$ we may confine ourselves to the cases

$$\rho = 2r, \quad 2r - 1.$$

If P be a product of two factors one of which is of even order, then U is reducible.

For let $P = P_1 P_2$, where P_1 is of order $2t$, then

$$(P_1 P_2, f^r)^\rho$$

contains the term

$$(P_1, f^t)^{2t} \cdot (P_2, f^{r-t})^{\rho - 2t},$$

and is in consequence reducible.

Also if P is of order $> \rho$, and is a product of two factors, U is reducible. By what we have just proved, if one of the factors of P is of even order U is reducible; let then, $P = P_1 P_2$ where P_1 is of order $2t_1 + 1$ and P_2 of order $2t_2 + 1$, then

$$(P_1 P_2, f^r)^\rho$$

contains the term

$$(P_1, f^{t_1})^{2t_1} \cdot (P_2, f^{r-t_1})^{\rho - 2t_1},$$

since $\rho < 2t_1 + 2t_2 + 2$, and therefore $\rho - 2t_1 < 2t_2 + 2$.

Thus U must always be reducible except when P consists of a single term C; or when P consists of a product of two terms C_i, C_j each of which is of odd order, their total order being $\rho = 2r$, so that

$$U = (C_i C_j, f^r)^{2r}.$$

Thus the irreducible forms belong to three classes:

$$\text{(i)}\quad (C, f^r)^{2r-1},$$
$$\text{(ii)}\quad (C, f^r)^{2r},$$
$$\text{(iii)}\quad (C_i C_j, f^r)^{2r},$$

where C_i is of order $2t+1$, and C_j of order $2r-2t-1$; this latter class furnishes invariants only.

It has not been proved that all transvectants belonging to these three classes are irreducible; on the contrary we proceed to examine a case in which certain of the transvectants thus retained are reducible.

142. Let C be a Jacobian $=(C_l,\ C_m)$.

(i) Let $$C_l = \phi_x^{2\sigma},\quad C_m = \psi_x^{2\tau},$$
$$f = a_x^2 = b_x^2 = \ldots,\quad C = (\phi\psi)\,\phi_x^{2\sigma-1}\psi_x^{2\tau-1}.$$

Then the form $$(C, f^r)^{2r-1}$$

is reducible. For if $2r > 2\sigma + 2\tau - 1$, this transvectant vanishes; and if $2r < 2\sigma + 2\tau - 1$ it contains the term

$$(\phi\psi)(\phi a)\,a_x\psi_x(\phi b^{(1)})^2(\phi b^{(2)})^2\ldots(\phi b^{(\lambda)})^2$$
$$(\psi c^{(1)})^2(\psi c^{(2)})^2\ldots(\psi c^{(r-\lambda-1)})^2\,\phi_x^{2\sigma-2\lambda-2}\,\psi_x^{2\tau+2\lambda-2r}.$$

But

$$(\phi\psi)(\phi a)\,a_x\psi_x = \tfrac{1}{2}\left[-(a\psi)^2\phi_x^2 + (\phi\psi)^2 a_x^2 + (\phi a)^2\psi_x^2\right].$$

And hence the term written down is

$$-\tfrac{1}{2}(\phi, f^{\lambda})^{2\lambda}.(\psi, f^{\mu+1})^{2\mu+2} + \tfrac{1}{2}f.T + \tfrac{1}{2}(\phi, f^{\lambda+1})^{2\lambda+2}.(\psi, f^{\mu})^{2\mu},$$

where T is a term of $((C_\lambda, C_\mu)^2, f^{r-1})^{2r-2}$.

(ii) If $$C_l = \phi_x^{2\sigma},\quad C_m = \psi_x^{2\tau+1},$$

then the transvectant
$$(C, f^r)^{2r-1}$$

vanishes if $2r > 2\sigma + 2\tau$, and if $2r < 2\sigma + 2\tau$ it contains the reducible term

$$(\phi\psi)(\phi a)\,a_x\psi_x(\phi b^{(1)})^2\ldots(\phi b^{(\lambda)})^2$$
$$(\psi c^{(1)})^2\ldots(\psi c^{(r-\lambda-1)})^2\,\phi_x^{2\sigma-2\lambda-2}\,\psi_x^{2\tau+2\lambda-2r+1}.$$

We are left with the case

$$2r = 2\sigma + 2\tau.$$

(iii) If $C_l = \phi_x^{2\sigma+1}, \quad C_m = \psi_x^{2\tau+1}$,

then the transvectant

$$(C, f^r)^{2r-1}$$

vanishes if $2r > 2\sigma + 2\tau + 1$, and contains a reducible term if $2r < 2\sigma + 2\tau$; but not if $2r = 2\sigma + 2\tau$.

Hence: "*The transvectant* $(C, f^r)^{2r-1}$ *is reducible, if C is a Jacobian, except when one at least of the forms of which C is composed is of odd order, and the order of C itself is equal to* $2r$ *or* $2r-1$."

143. Quadratic and Quartic. The simultaneous system of irreducible concomitants when the ground-forms are a quadratic and a quartic may now be written down.

The complete system for the quartic ϕ is known to be

$$\phi, \quad H = (\phi, \phi)^2, \quad T = (\phi, H), \quad i = (\phi, \phi)^4, \quad j = (H, \phi)^4.$$

Since there are no forms here of odd order, there can arise no irreducible concomitants belonging to the third of the three classes mentioned above. The simultaneous forms are

$$(\phi, f), \quad (\phi, f)^2, \quad (\phi, f^2)^3, \quad (\phi, f^2)^4,$$
$$(H, f), \quad (H, f)^2, \quad (H, f^2)^3, \quad (H, f^2)^4,$$
$$(T, f)^2, \quad (T, f^2)^4, \quad (T, f^3)^6.$$

It follows from the theorem of § 142 that the forms (T, f), $(T, f^2)^3$, $(T, f^3)^5$ are reducible, T being a Jacobian.

To complete the simultaneous system we must take into account the forms which belong to the quartic and quadratic separately; thus in all we have 18 concomitants.

Ex. Prove that all the forms of the complete system for the two covariants j and i of a binary quintic f, considered as separate quantics, are irreducible when considered as concomitants of the quintic; with the single exception of one invariant of degree 18 in the coefficients of f.

CHAPTER IX.

HILBERT'S THEOREM.

Hilbert's Proof of Gordan's Theorem.

144. WE shall now give another proof of Gordan's theorem that the irreducible system of invariants and covariants of any number of binary forms is finite. The method, which is due to Hilbert*, is of more general application than that of Gordan, inasmuch as with slight and non-essential modifications it applies to forms with any number of variables; on the other hand, unlike Gordan's process it gives practically no information as to the actual determination of the finite system whose existence it establishes, in other words it proves that the problem always has a solution, while the other method, although only proving this for binary forms, gives much information as to the nature of the solution.

In the exposition of Hilbert's proof we shall confine ourselves to binary forms, and to save trouble we shall deal with pure invariants only; inasmuch as the complete system of invariants and covariants of any number of forms is really equivalent to the system of invariants of the set of forms obtained by adjoining an arbitrary linear form to the original set, the proof for invariants is sufficient for the most general case. Cf. § 139.

145. The proof may conveniently be divided into two parts of the following purport.

I. *Proof of the fact that any invariant I of the system may be written in the form*

$$I = A_1 I_1 + A_2 I_2 + \dots + A_n I_n,$$

* *Math. Ann.* XXXVI. Story, *Math. Ann.* XLII.

where I_1, $I_2 \ldots I_n$ are a finite number of fixed invariants of the system, and the A's while not necessarily invariants are integral functions of the coefficients.

II. *The application to both sides of the equation just given of a differential operator which leaves an invariant unaltered except for a numerical multiplier, and changes a term*

$$A_r I_r$$

into one of the form $J_r I_r$, where J_r is an invariant.

As a result of I. and II. any invariant may be obtained in the form

$$I_1 J_1 + I_2 J_2 + \ldots + I_n J_n.$$

Then by applying the same argument to the J's and so on it follows at once that the I's form the complete system.

146. As a matter of fact the result I. is a particular case of a much more general proposition which we shall first enunciate, then illustrate, and finally prove.

THEOREM. *If a homogeneous function of any number of variables be formed according to any definite laws, then, although there may be an infinite number of functions F satisfying the conditions laid down, nevertheless a finite number F_1, F_2, ... F_r can always be chosen so that any other F can be written in the form*

$$F = A_1 F_1 + A_2 F_2 + \ldots + A_r F_r,$$

where the A's are homogeneous integral functions of the variables but do not necessarily satisfy the conditions for the F's.

Suppose for example that we have three variables x, y, z which we take to represent coordinates and that $F = 0$ represents a curve through the point $y = 0$, $z = 0$ (this being the law according to which F is formed), then F may be written in the form

$$yP + zQ,$$

as follows at once since the highest power of x must be wanting in the equation; $y = 0$, $z = 0$ being two curves of the system, this is the application of the theorem to this case.

As another example, if the curve pass through all the vertices of the fundamental triangle its equation may be written

$$yzP + zxQ + xyR = 0,$$

where P, Q, R are integral functions of the coordinates, and here $yz=0$ etc. are curves of the system.

Again, we have the famous theorem that the equation of any curve through all the points common to $\phi=0$ and $\psi=0$ may be written

$$A\phi+B\psi=0.$$

In each of these cases it will be noted that the system of forms $F_1, F_2, \dots F_r$ is determined; in the general case it is not actually determined, the essential point being that it is finite and that the A's are integral functions.

To establish the theorem in its general form we first remark that it is manifestly true when there is only one variable x, because in this case each F consists of a power of x and therefore all the F's are divisible by that which is of lowest degree; thus there is only one form in the system $F_1, F_2, \dots F_r$.

We now assume that the theorem is true when there are $n-1$ variables and deduce that it is true when there are n variables.

Let $x_1, x_2, \dots x_n$ be the variables and first suppose that the system contains a form H of order r in which the coefficient of x_n^r does not vanish. Then we can divide any form in which x_n occurs to a power equal to or greater than r by H without introducing coefficients fractional in the x's, and we can continue the process until the remainder contains no power of x_n higher than the $(r-1)$th.

Hence we can write any form of the system thus

$$F=HP+Mx_n^{r-1}+N,$$

where P is the quotient, M is a function of $x_1, x_2, \dots x_{n-1}$, and N is a function of the variables but of degree $r-2$ at the most in x_n.

Now the functions M are formed according to definite laws if the F's are, because each M is deduced from the corresponding F by a definite process, and as they only contain $n-1$ variables the theorem is true by hypothesis for them.

Accordingly we can choose a finite number of M's, say $M_1, M_2, \dots M_k$, such that any other may be written in the form

$$M=B_1M_1+B_2M_2+\dots+B_kM_k,$$

where the B's are integral functions of $x_1, x_2, \dots x_{n-1}$.

But since

$$x_n^{r-1} M = F - HP - N, \quad x_n^{r-1} M_1 = F_1 - HP_1 - N_1, \text{ etc.,}$$

we have

$$F - HP - N = B_1(F_1 - HP_1 - N_1) + B_2(F_2 - HP_2 - N_2) + \dots + B_k(F_k - HP_k - N_k),$$

or

$$F = H(P - B_1P_1 - B_2P_2 - \dots - B_kP_k) + B_1F_1 + \dots + B_kF_k + N - B_1N_1 - \dots - B_kN_k.$$

Now the part of the right-hand side which does not contain one of the forms as a factor consists of B's and N's and therefore only contains x_n to the power $r-2$ at most. Hence we may write

$$F = HQ_1 + B_1F_1 + B_2F_2 + \dots + B_kF_k + M^{(1)}x_n^{r-2} + N^{(1)},$$

and now $M^{(1)}$ is a function of $x_1, x_2, \dots x_{n-1}$ formed according to definite laws and $N^{(1)}$ is of order $r-3$ at most in x_n.

Thus we can write F as the sum of a finite number of terms each containing a form of the system for factor together with expressions of order $r-2$ at most in the last variable.

Then applying precisely the same argument to the $M^{(1)}$'s as we applied to the M's we see that by adding a finite number of F's to

$$H, F_1, F_2, \dots F_k$$

we can reduce the order of the remaining portion in x_n to $r-3$.

Proceeding in this way and adding only a finite number of F's at each step we can finally write F in the form

$$HQ_r + C_1F_1 + C_2F_2 + \dots + C_mF_m + M^{(r)},$$

where $M^{(r)}$ only involves $x_1, x_2, \dots x_{n-1}$ and in the nature of things is formed according to definite laws. Hence applying the same process to the $M^{(r)}$'s as we applied to the M's we finally have F in the form

$$A_1F_1 + A_2F_2 + \dots + A_sF_s,$$

where the F's include H and the number s is finite.

Consequently if the theorem be true for $n-1$ variables it is true for n, but it is true for one variable, therefore by induction it is true universally.

We have now to remove the limitation imposed above, viz., that there exists a form of the system in which the coefficient of the highest power of x_n is not zero.

If there is no such form among the F's let F_t be one of the forms and apply to all a linear substitution

$$x_r = a_{r1} y_1 + a_{r2} y_2 + \dots + a_{rn} y_n; \quad r = 1, 2, \dots n.$$

Suppose that $F_t(x)$ becomes $G_t(y)$, then the coefficient of the highest power of y_n in G_t is $F_t(a_{1n}, a_{2n}, \dots a_{nn})$, and therefore unless F_t is identically zero we can choose the linear substitution so that this coefficient is not zero*. Hence the theorem is true for the forms G in the variables y, and therefore by changing the variables back again from y to x we see that it is true for the F's.

Q. E. D.

147. Returning now to the consideration of invariants it is clear that such an expression regarded as a homogeneous function of the coefficients of the forms is formed according to definite laws; hence, if I be any invariant of the system, we have

$$I = A_1 I_1 + A_2 I_2 + \dots + A_n I_n,$$

where $I_1, I_2, \dots I_n$ are n fixed invariants and the A's are homogeneous integral functions of the coefficients but not necessarily invariants. As to the functions A a simple remark may be added. All that is asserted in the general statement of the foregoing theorem is that they are homogeneous in all the coefficients, but an invariant is homogeneous in each set of coefficients involved taken separately, and consequently since I and I_m are homogeneous in each set of coefficients, A_m is also homogeneous in each set. That this is so could of course be seen in the proof of the general theorem because at no point of the investigation is the homogeneity disturbed.

148. We now come to the second part of the proof, but before proceeding with it we must prove a necessary lemma on the properties of the operator Ω so often used in the course of this work.

If P be a function of $\xi_1, \xi_2, \eta_1, \eta_2$ which is homogeneous and of order λ in ξ_1, ξ_2 and homogeneous and of order μ in η_1, η_2, then

$$\Omega^m (D^n P) = C_0 D^{n-m} P + C_1 D^{n-m+1} \Omega (P) + \dots + C_m D^n \Omega^m (P),$$

* We assume here that unless a form vanishes identically values of the variables can be found for which it is not zero. It is easy to give a formal proof of this theorem. Cf. Weber's *Algebra*, Vol. I. p. 457.

where $D=\xi_1\eta_2-\xi_2\eta_1$, m and n are positive integers and the C's are either zero or constant.

The result can be readily proved by induction, for we have

$$\Omega(DP)$$
$$=P+\eta_2\frac{\partial P}{\partial\eta_2}+\xi_1\frac{\partial P}{\partial\xi_1}+D\frac{\partial^2 P}{\partial\xi_1\partial\eta_2}-\left(-P-\xi_2\frac{\partial P}{\partial\xi_2}-\eta_1\frac{\partial P}{\partial\eta_1}+D\frac{\partial^2 P}{\partial\xi_2\partial\eta_1}\right),$$

and by Euler's theorem for homogeneous functions the right-hand side becomes

$$(\lambda+\mu+2)P+D\Omega P.$$

Now in this result change P into $D^{n-1}P$ so that λ and μ are increased by $n-1$, and we have

$$\Omega(D^nP)=(\lambda+\mu+2n)D^{n-1}P+D\Omega(D^{n-1}P).$$

Hence

$$D\Omega(D^{n-1}P)=(\lambda+\mu+2n-2)D^{n-2}P+D^2\Omega(D^{n-2}P),$$
$$D^2\Omega(D^{n-2}P)=(\lambda+\mu+2n-4)D^{n-1}P+D^3\Omega(D^{n-3}P),$$
$$\cdots\cdots\cdots\cdots\cdots\cdots\cdots\cdots\cdots\cdots\cdots\cdots$$
$$D^{n-2}\Omega(D^2P)=(\lambda+\mu+4)D^{n-1}P+D^{n-1}\Omega(DP),$$
$$D^{n-1}\Omega(DP)=(\lambda+\mu+2)D^{n-1}P+D^n\Omega(P).$$

By adding these results together we obtain

$$\Omega(D^nP)=\{n(\lambda+\mu)+n(n+1)\}D^{n-1}P+D^n\Omega(P),$$

which establishes the result when $m=1$ for all values of n.

Assume that the result is true for any value of m so that

$$\Omega^m(D^nP)=C_0D^{n-m}P+C_1D^{n-m+1}\Omega(P)+\ldots+C_mD^m\Omega^m(P),$$

then operating again with Ω we have

$$\Omega^{m+1}(D^nP)=\sum_{r=1}^{r=m}C_r\Omega\{D^{n-m+r}\Omega^r(P)\}.$$

But

$$\Omega(D^{n-m+r}P)$$
$$=(n-m+r)(\lambda+\mu+n-m+r+1)D^{n-m+r-1}P+D^{n-m+r}\Omega(P),$$

and changing P into $\Omega^r(P)$ so that λ, μ are each diminished by r, we deduce

$$\begin{aligned}&\Omega\{D^{n-m+r}\Omega^r(P)\}\\&=(n-m+r)(\lambda+\mu-r-m+n+1)\,D^{n-m+r-1}\Omega^r(P)+D^{n-m+r}\Omega^{r+1}(P)\\&=\alpha_r D^{n-m+r-1}\Omega^r(P)+D^{n-m+r}\Omega^{r+1}P\end{aligned}$$

when α_r is numerical.

Thus we have

$$\Omega^{m+1}(D^nP)=\sum_{r=1}^{r=m}(C_r\alpha_r+C_{r-1})\,D^{n-m+r-1}\Omega^r(P),$$

in other words, if the result be true for m it is true for $m+1$, for the right-hand side is of the stipulated form. Hence by induction the theorem is true universally.

Ex. Prove that

$$\Omega^n(D^mP)=\sum_r\binom{m}{r}\frac{n!\,(\lambda+\mu+n+1-r)!}{(\lambda+\mu+n-m+1)!\,(n-m+r)!}D^{n-m+r}\Omega^rP.$$

Cor. It clearly follows that if in the formal statement any exponent of D on the right-hand side be negative the corresponding coefficient C is zero because only integral functions can appear in the process.

149. With the aid of the above lemma the proof of Gordan's theorem may be easily completed.

For the sake of convenience we shall regard x_1, x_2 as the variables in the fundamental binary forms, despite the fact that in the general theorem proved above they play the rôle that the coefficients do in the remaining portion of the investigation.

Suppose the variables are changed by the linear transformation

$$\left.\begin{aligned}x_1&=\xi_1x_1+\eta_1x_2\\x_2&=\xi_2x_1+\eta_2x_2\end{aligned}\right\},$$

then an invariant I of the forms becomes $(\xi\eta)^\mu I$.

Further we have

$$I=A_1I_1+A_2I_2+\ldots+A_nI_n$$

and an invariant I_m on the right becomes

$$(\xi\eta)^{\mu_m}I_m.$$

We now write down the identity which is the transformation of

$$I = A_1 I_1 + A_2 I_2 + \ldots + A_n I_n,$$

i.e. the same identity for the transformed quantics; we suppose a coefficient A_m written in symbolical letters entirely, so that it is the sum of a number of terms each of which contains only factors of the types a_ξ and a_η, where a is a symbol belonging to one of the quantics.

If after transformation A_m become B_m we have

$$I(\xi\eta)^\mu = \sum_1^n B_m (\xi\eta)^{\mu_m} I_m$$

and the equation shews at once that B_m is of order $\mu - \mu_m$ in both ξ and η.

Now operate on both sides of this identity with

$$\Omega^\mu = \left(\frac{\partial^2}{\partial\xi_1\partial\eta_2} - \frac{\partial^2}{\partial\xi_2\partial\eta_1}\right)^\mu.$$

The left-hand side becomes a numerical multiple of I, viz. $(\mu+1)(\mu\,!)^2 I$, and on the right-hand side we have

$$\Omega^\mu \{(\xi\eta)^{\mu_m} B_m\} I_m$$

$$= I_m \{C_0(\xi\eta)^{\mu_m - \mu} B_m + C_1(\xi\eta)^{\mu_m - \mu + 1}\Omega B_m + \ldots + C_\mu(\xi\eta)^{\mu_m}\Omega^\mu B_m\}$$

by the lemma, since I_m does not involve ξ or η.

But if $\mu - \mu_m = \nu$, then

$$\mu_m - \mu,\ \mu_m - \mu + 1, \ldots\ \mu_m - \mu + (\nu - 1)$$

are all negative.

Consequently

$$C_0,\ C_1, \ldots\ C_{\nu-1}$$

are all zero.

Again B_m is of order $\mu - \mu_m = \nu$ in both ξ and η,

hence $$\Omega^{\nu+1}(B_m),\ \Omega^{\nu+2}(B_m), \ldots\ \Omega^\mu(B_m)$$

are all zero, and the effect of the operator on

$$(\xi\eta)^{\mu_m} B_m$$

therefore reduces to a single term, namely

$$C_\nu \Omega^\nu (B_m).$$

Now B_m is the sum of a number of terms each containing ν factors of the type a_ξ and r factors of the type a_η, hence by a fundamental theorem,

$$\Omega^\nu (B_m)$$

is an invariant of the system.

Therefore after operating with Ω^μ on both sides of the equation we are left with

$$I = \Sigma J_m I_m, \text{ where } J_m = \frac{C_r}{(\mu+1)(\mu!)^2} \Omega^\nu (B_m)$$

and is an invariant.

Since J_m is an invariant we can express it also as the sum of a number of terms each containing an I_m as a factor, hence by continual reduction we can ultimately express I as a rational integral function of $I_1, I_2, \ldots I_m$, that is to say, these invariants constitute a complete system and, as we have seen, their number is finite; Gordan's theorem is thus completely established.

150. Syzygies between the irreducible invariants.

Examples of relations between the members of an irreducible system of invariants or covariants have already been given, and in fact a very large number were obtained for the quintic.

It can be deduced from Hilbert's Lemma that the system of syzygies is finite, that is to say if $S = 0$ be any syzygy we can find a finite number of syzygies

$$S_1 = 0, \quad S_2 = 0, \ldots S_r = 0,$$

such that $$S = C_1 S_1 + C_2 S_2 + \ldots + C_r S_r,$$

where $C_1, \ldots C_r$ are invariants.

If there be such a relation, then of course all other syzygies are necessary consequences of

$$S_1 = 0, \quad S_2 = 0, \ldots S_r = 0,$$

and these constitute the finite system.

The proof is very simple. Let $I_1, I_2, \ldots I_m$ be the members of the irreducible system of invariants, then S is a function of $I_1, I_2, \ldots I_m$ formed according to the law that it must vanish when for the I's we substitute their actual values in terms of the coefficients.

Hence we have

$$S = C_1S_1 + C_2S_2 + \dots + C_rS_r,$$

where $S_1 = 0$, etc. are a finite number of syzygies and the C's, being functions of the I's, are invariants. (Cf. § 127.)

151. Gordan's Proof of Hilbert's Lemma.

Many versions have been given of the fundamental lemma of Hilbert on functions formed according to given laws, but the majority of them do not differ materially from the original proof due to Hilbert. Nevertheless Gordan has recently given a demonstration* which is so interesting and depends on such simple principles that we cannot refrain from giving an account of it here. We shall state it in the form of two theorems.

Theorem I. *If a simple product of positive integral powers of n letters*

$$x_1^{k_1}x_2^{k_2}\dots x_n^{k_n},$$

be formed in such a way that the exponents $k_1, k_2, \dots k_n$ *satisfy certain prescribed conditions, then, although the number of products satisfying the conditions may be infinite, yet a finite number of them can be chosen so that every other is divisible by one at least of this finite number.*

To illustrate the scope of this theorem take the case of products of three letters and suppose the conditions are

$$k_1 \equiv 0 \text{ (mod. 3)},$$

$$k_2^2 - k_3 = 7.$$

The simple products satisfying the conditions are

$$x_2^3x_3^2, \quad x_2^4x_3^9, \dots$$

$$x_1^3x_2^3x_3^2, \quad x_1^3x_2^4x_3^9 \dots,$$

and it is evident that all such products are divisible by $x_2^3x_3^2$.

Again, suppose the sole condition is

$$k_1 - k_2 + k_3 > 0;$$

the products are

$$x_1,\ x_3;\quad x_1^2,\ x_3^2,\ x_1x_3;\quad x_1^2x_2,\ x_1x_2x_3,\ x_2x_3^2,\ x_1^3, \dots$$

and all the products are divisible by x_1 or x_3.

* *Liouville's Journal*, 1900.

Other examples could be given, but the above will suffice to shew the nature of the theorem which we now proceed to prove.

If $n=1$ the truth of the theorem is evident because all the products are powers of a single letter and are therefore divisible by that having the least exponent.

We shall now assume that the result is true for $n-1$ letters and prove that it is true for n letters.

Let
$$x_1^{a_1} x_2^{a_2} \dots x_n^{a_n}$$
be a definite product P satisfying the given conditions and let
$$x_1^{k_1} x_2^{k_2} \dots x_n^{k_n}$$
be a typical product K of the system.

If K be not divisible by P one of the k's must be less than the corresponding a.

Suppose that $k_r < a_r$, then, consistently with this, k_r must have one of the values
$$0, 1, 2, \dots a_r - 1.$$

Hence if K be not divisible by P one of a number

$$a_1 + a_2 + \dots + a_n = N \text{ contingencies arises, viz.}$$

either

k_1 has one of the values $0, 1, 2, \dots a_1 - 1$,

or k_2 has one of the values $0, 1, 2, \dots a_2 - 1$, etc.

Suppose that $k_r = m$, and that this is the pth of the possible cases; then the remaining exponents $k_1, k_2, \dots k_{r-1}, k_{r+1} \dots k_n$ satisfy definite conditions which are obtained by making $k_r = m$ in the original conditions.

Let
$$K_p = x_1^{k_1} x_2^{k_2} \dots x_r^{m} \dots x_n^{k_m}$$
be a product of the system for which $k_r = m$ and write
$$K_p = x_r^m K'_p.$$

Then K'_p contains only $n-1$ letters and the exponents satisfy definite conditions, and when these are satisfied the exponents of K_p satisfy the original conditions. Hence by hypothesis a finite number of products of the type K'_p can be found such that every other such product is divisible by one at least of these.

Denote this finite system by

$$L_1, L_2, \dots L_{a_p}$$

so that K_p is divisible by one at least of the L's.

Thus $K_p = x_r^m K'_p$ is divisible by one at least of the products

$$x_r^m L_1, x_r^m L_2, \dots x_r^m L_{a_p},$$

which all belong to the original system of products because every L belongs to the subsidiary system.

Denote these latter products by

$$M_p^{(1)}, M_p^{(2)}, \dots M_p^{(a_p)};$$

then in the pth of the N possible contingencies K is divisible by one of the products

$$M_p^{(1)}, M_p^{(2)}, \dots M_p^{(a_p)}.$$

Now one of these N contingencies certainly does arise when K is not divisible by P, and hence K must be divisible by one of the products

$$M_1^{(1)}, M_1^{(2)}, \dots M_1^{(a_1)}; \quad M_2^{(1)}, \dots M_2^{(a_2)}; \dots M_N^{(a_N)},$$

or else by P.

The exponents of the M's all satisfy the prescribed conditions and they are finite in number, hence if the theorem be true for $n-1$ letters it is true for n letters, but it is true for one letter and hence by induction it is true universally.

152. Theorem II. *If a system of homogeneous forms be constructed according to given laws, then a finite number of definite forms of the system can be chosen such that every other form of the system is an aggregate of terms each of which involves one of the finite number of forms as a factor, and the coefficients are integral in the variables.* (Hilbert's Lemma.)

Suppose in fact that $x_1, x_2, \dots x_n$ are the variables and that ϕ is a typical form of the system. Now construct an auxiliary system of functions η of the same variables according to the law that a function is an η function when it can be written in the form

$$\eta = \Sigma A\phi,$$

the A's being integral functions of the variables which make the right-hand side homogeneous, but otherwise unrestricted except that the number of terms on the right-hand side must be finite.

The class of functions η is infinitely more comprehensive than the class ϕ, and it possesses the important property that a function of the form $\Sigma B\eta$ which is homogeneous in the variables is also an η function.

Now in examining the functions η we arrange the terms of one of them of order r in such a way that x_1^r comes first and, generally, a term

$$S = x_1^{a_1} x_2^{a_2} \dots x_n^{a_n}$$

comes before a term

$$T = x_1^{b_1} x_2^{b_2} \dots x_n^{b_n},$$

when the first of the quantities

$$a_1 - b_1, \quad a_2 - b_2, \dots a_n - b_n$$

which does not vanish is positive.

In such a case we say that the term S is simpler than the term T and T is more complex than S, so that any function η is arranged with its terms in ascending order of complexity. Now the functions η being formed according to fixed laws, their first terms satisfy given conditions relating to the exponents, and hence by Theorem I. a finite number of η's, say $\eta_1, \eta_2, \dots \eta_p$, can be chosen such that the first term of any other η is divisible by the first term of one of these.

Take any function η of the auxiliary system, and suppose its first term is divisible by the first term of η_{m_1}, and that P_1 is the quotient.

Then $\eta - P_1\eta_{m_1}$ is an η function with a more complex first term than η because η and $P_1\eta_{m_1}$ have the same first term; if we denote this new function by $\eta^{(1)}$ we have

$$\eta = P_1\eta_{m_1} + \eta^{(1)}.$$

Next, if the first term of $\eta^{(1)}$ be divisible by that of η_{m_2} we have

$$\eta^{(1)} = P_2\eta_{m_2} + \eta^{(2)},$$

and the first term of $\eta^{(2)}$ is more complex than that of $\eta^{(1)}$.

Continuing this process of reduction we find

$$\eta^{(r-1)} = P_r\eta_{m_r} + \eta^{(r)} \text{ and so on.}$$

Now the first terms of

$$\eta,\ \eta^{(1)},\ \eta^{(2)}, \ldots \eta^{(r)} \ldots$$

are in ascending order of complexity, and hence the time must come when there is no η function of the same order as η with a more complex first term than $\eta^{(r)}$; in that case we have

$$\eta^{(r)} = P_{r+1}\eta_{m_{r+1}}.$$

Hence

$$\eta = P_1\eta_{m_1} + P_2\eta_{m_2} + \ldots + P_{r+1}\eta_{m_{r+1}}{}^*$$

where the η's on the right-hand side are all members of the finite system $\eta_1, \eta_2, \ldots \eta_p$.

Now the η system includes all the ϕ's, moreover each η contains only a finite number of ϕ's and hence every ϕ can be expressed in the form

$$A_1\phi_1 + A_2\phi_2 + \ldots + A_r\phi_r,$$

where $\phi_1, \phi_2, \ldots \phi_r$ are the ϕ's contained in the expression for $\eta_1, \eta_2, \ldots \eta_p$ and the A's are integral functions of the variables.

153. Remark. If all the conditions satisfied by $k_1, k_2, \ldots k_n$ in Theorem I. be linear homogeneous equations, then the theorem establishes the existence of a finite number of solutions by means of which any other solution can be reduced. The difference of two solutions being now a solution, it follows that by continual reduction we can express all solutions of the linear equations in terms of a finite number—this is the result otherwise proved in § 97.

* The η's on the right being members of a finite system are finite in number; hence even though the number of steps in the reduction be infinite, there can only be a finite number of terms on the right-hand side. The same η may of course occur in more than one term, but in that case we should add all such terms together.

CHAPTER X.

THE GEOMETRICAL INTERPRETATION OF BINARY FORMS.

154. GIVEN two points of reference A, B on any straight line, the position of any other point P may be determined by the value of the ratio $\frac{AP}{PB}$ of the distances of P from A and B. A convention as regards *sign* is necessary to complete the definition; it is convenient to regard the ratio as positive if P lie between A and B, otherwise as negative.

A P B

When a binary form of order n is equated to zero, the ratio $\frac{x_1}{x_2}$ may have any one of n values. These determine n points on the straight line AB, such that the ratio $\frac{AP}{PB}$ for each of these points is equal to one of the roots of the equation for $\frac{x_1}{x_2}$. The coordinates (x_1, x_2) then define the position of P on the straight line by means of the equation

$$\frac{AP}{PB} = \frac{x_1}{x_2}.$$

It is found advisable, as will be evident immediately, to define the position of the point P whose coordinates are (x_1, x_2) by means of the equation

$$\frac{AP}{PB} = \lambda \frac{x_1}{x_2},$$

where λ is a fixed numerical multiplier.

A further convention will be useful, viz. the positive direction of measurement from A is towards B and that from B is towards A.

To find the distance between (x_1, x_2) and (y_1, y_2), in terms of the length l of AB.

Denoting the two points by P and Q, we have

$$\frac{AP}{\lambda x_1} = \frac{PB}{x_2} = \frac{AB}{\lambda x_1 + x_2} = \frac{l}{\lambda x_1 + x_2},$$

and similarly

$$\frac{QB}{y_2} = \frac{l}{\lambda y_1 + y_2}.$$

Hence

$$PQ = (PB - QB) = l\frac{x_2(\lambda y_1 + y_2) - y_2(\lambda x_1 + x_2)}{(\lambda y_1 + y_2)(\lambda x_1 + x_2)}$$

$$= \frac{\lambda l(yx)}{(\lambda y_1 + y_2)(\lambda x_1 + x_2)}.$$

155. Let us consider the effect of a change in the points of reference. Let the new points of reference A', B' in terms of the original system of coordinates be (ξ_1, ξ_2), (η_1, η_2); if the new multiplier be μ and (X_1, X_2) be the new coordinates of P, then

$$\frac{X_1}{X_2} = \mu\frac{A'P}{PB'}.$$

Hence
$$\frac{X_1}{X_2} = \mu.\frac{(x\xi)}{\lambda\xi_1 + \xi_2}.\frac{\lambda\eta_1 + \eta_2}{(\eta x)}.$$

The change in coordinates is thus equivalent to the linear transformation

$$X_1 = \rho_1(x\xi),$$
$$X_2 = \rho_2(\eta x),$$

where
$$\frac{\rho_1}{\rho_2} = \mu\frac{\lambda\eta_1 + \eta_2}{\lambda\xi_1 + \xi_2}.$$

156. A linear transformation

$$x_1 = \xi_1 X_1 + \eta_1 X_2,$$
$$x_2 = \xi_2 X_1 + \eta_2 X_2,$$

may be regarded geometrically from two different points of view:

(i) As changing the points of reference and the constant multiplier, but leaving the other points on the straight line in their original position.

(ii) If the points of reference are regarded as fixed, the

transformation alters the positions of all the points defined by the algebraic forms under discussion.

Consider the first of these points of view. When $x_1 = 0$, the point $P(x_1, x_2)$ coincides with one of the points of reference A. Similarly, if $x_2 = 0$, P coincides with B. Hence to find the new points of reference in the original system of coordinates, it is only necessary to write $X_1 = 0$, and $X_2 = 0$. We obtain them at once as (η_1, η_2) and (ξ_1, ξ_2).

The distances of P from these new points of reference A', B' are

$$\lambda l \frac{(x\eta)}{(\lambda x_1 + x_2)(\lambda \eta_1 + \eta_2)} = \lambda l \frac{(\xi\eta) X_1}{(\lambda x_1 + x_2)(\lambda \eta_1 + \eta_2)},$$

and
$$\lambda l \frac{(\xi x)}{(\lambda x_1 + x_2)(\lambda \xi_1 + \xi_2)} = \lambda l \frac{(\xi\eta) X_2}{(\lambda x_1 + x_2)(\lambda \xi_1 + \xi_2)}.$$

The ratio of these distances is

$$\frac{A'P}{PB'} = \frac{\lambda \xi_1 + \xi_2}{\lambda \eta_1 + \eta_2} \cdot \frac{X_1}{X_2}.$$

Hence the new multiplier is $\dfrac{\lambda \xi_1 + \xi_2}{\lambda \eta_1 + \eta_2}$; and the coordinates (X_1, X_2) define the same point as that defined by the coordinates (x_1, x_2). It should be observed that the sign of the expression $\dfrac{\lambda \xi_1 + \xi_2}{\lambda \eta_1 + \eta_2} \cdot \dfrac{X_1}{X_2}$ is positive if X lie between A' and B', otherwise it is negative.

Ex. (i). Shew that by properly choosing quantities a_1, a_2 the distance between the points (x_1, x_2), (y_1, y_2) may be written $\dfrac{(yx)}{(ax)(ay)}$. And that in this case the constant multiplier after transformation becomes $\dfrac{(a\xi)}{(a\eta)}$.

Ans. $a_1 = -\sqrt{\dfrac{1}{l\lambda}}$, $a_2 = \sqrt{\dfrac{\lambda}{l}}$.

Ex. (ii). The point (a_1, a_2) of the last example is the point at infinity on the range.

When an invariant of a binary quantic is zero, there exists some relation between its roots which is unaffected by any linear transformation. Hence when the binary quantic is regarded as the analytical expression of n points on a range, the vanishing of an invariant is the condition that there may be some definite geometrical relation between the points, independent of the points of reference and of the constant multiplier.

For example if two of the points coincide an invariant—the discriminant—vanishes. Again, as will be shewn later, if four points on a straight line which form a harmonic range are represented analytically by a quartic, then the invariant j of that quartic is zero.

157. In the second point of view stated in the last paragraph, the points of reference and multiplier are regarded as fixed, the point P takes a new position P' given by the coordinates (X_1, X_2).

Let the points Q, R, S, viz. (y_1, y_2), (z_1, z_2), (w_1, w_2) become Q', R', S'.

Then
$$PQ = \lambda l \frac{(yx)}{(\lambda x_1 + x_2)(\lambda y_1 + y_2)}$$
and
$$(yx) = (\xi\eta)(YX).$$
Hence
$$\frac{PQ \,.\, RS}{PS \,.\, RQ} = \frac{(yx)(wz)}{(wx)(yz)}$$
$$= \frac{(YX)(WZ)}{(WX)(YZ)} = \frac{P'Q' \,.\, R'S'}{P'S' \,.\, R'Q'}.$$

The expression $\dfrac{PQ \,.\, RS}{PS \,.\, RQ}$ is called the cross (or anharmonic) ratio of the four points P, Q, R, S; it is usually denoted by $\{PQRS\}$*. The result just proved may be written
$$\{PQRS\} = \{P'Q'R'S'\}.$$

* By rearranging the letters P, Q, R, S we obtain 24 such cross-ratios. It is easy to see, however, that only six of these are different. Then if λ, μ, ν are written for the three products $PQ \,.\, RS$, $PR \,.\, SQ$, $PS \,.\, QR$, the six different cross-ratios are
$$-\frac{\lambda}{\mu},\ -\frac{\mu}{\nu},\ -\frac{\nu}{\lambda},\ -\frac{\mu}{\lambda},\ -\frac{\nu}{\mu},\ -\frac{\lambda}{\nu}.$$
Since the four points are collinear,
$$PQ \,.\, RS + PR \,.\, SQ + PS \,.\, QR = 0$$
or
$$\lambda + \mu + \nu = 0;$$
by means of this relation all six cross-ratios of four points may be expressed in terms of one of them. This mode of presenting the subject is due to Mr R. R. Webb.

That is, the cross-ratio of four points on the range is unaltered by any linear transformation. Hence the transformed range is homographic with the original range.

Further any range homographic with the original range may be obtained from it by a linear transformation. To prove this, it is only necessary to prove that the coefficients of transformation may be so chosen that three non-coincident points P, Q, R of the original range are changed to any three non-coincident points P', Q', R' chosen at random on the straight line. For when P', Q', R' are known, the point S' of the transformed range corresponding to S is given by

$$\{P'Q'R'S'\} = \{PQRS\}*.$$

Let us suppose then that the values of the ratios $\frac{X_1}{X_2}, \frac{Y_1}{Y_2}, \frac{Z_1}{Z_2}$ are given.

Then $$x_1 = \left(\xi_1 \frac{X_1}{X_2} + \eta_1\right) X_2, \quad x_2 = \left(\xi_2 \frac{X_1}{X_2} + \eta_2\right) X_2$$

and $$\frac{x_1}{x_2} = \frac{\xi_1 \frac{X_1}{X_2} + \eta_1}{\xi_2 \frac{X_1}{X_2} + \eta_2}$$

or $$\xi_2 \frac{X_1}{X_2} \cdot \frac{x_1}{x_2} - \xi_1 \frac{X_1}{X_2} + \eta_2 \frac{x_1}{x_2} - \eta_1 = 0.$$

Similarly $$\xi_2 \frac{Y_1}{Y_2} \cdot \frac{y_1}{y_2} - \xi_1 \frac{Y_1}{Y_2} + \eta_2 \frac{y_1}{y_2} - \eta_1 = 0$$

and $$\xi_2 \frac{Z_1}{Z_2} \cdot \frac{z_1}{z_2} - \xi_1 \frac{Z_1}{Z_2} + \eta_2 \frac{z_1}{z_2} - \eta_1 = 0.$$

We have three equations to determine the ratios of the coefficients $\xi_1, \xi_2, \eta_1, \eta_2$.

These ratios are thus determined uniquely.

Hence a range of points on a straight line may be transformed into any other range homographic with itself by a linear transformation.

If an invariant of a binary quantic representing a range of n points is zero, these points must possess some special property,

* This must give a unique position for S' since it is equivalent to a linear relation between its coordinates.

which is also a property of all ranges homographic with the original range. Such a property is said to be projective, and thus the vanishing of an invariant must be the condition for the existence of some projective property of the points which the quantic represents.

Conversely, if a system of n points on a straight line possesses some projective property, there will exist a corresponding analytical relation between the coefficients of the quantic represented by these points, which is unaltered by any linear transformation. It does not necessarily follow that the condition is represented by the vanishing of an invariant; it sometimes happens that a projective property necessitates the vanishing of all the coefficients of a covariant.

Again a covariant of a quantic will define a certain number of points on the straight line. These points are related to the original points of the range in the same way as their homologues are related to the homologues of the original points on a homographic range. It is usual to denote this by saying that the points are projectively related to the points of the original range.

158. A binary form is homogeneous in two variables, we are then—in such forms—only concerned with the ratio of the variables. Let $f(x_1, x_2)$ be any binary form of order n, then the equation

$$f(x_1, x_2) = 0$$

defines n values of the ratio $\frac{x_1}{x_2}$. Hence in any geometrical figure in which the geometric element is completely defined in position by a single parameter, the form $f(x_1, x_2)$ may be considered as defining n of these elements. For example a point which lies on a *unicursal* curve is such an element. If x, y, z are its Cartesian coordinates, it is well known that we may express x, y, z as rational algebraic functions of a single parameter. Again the tangent to a fixed unicursal curve may be taken to be the element. Or else the element might be the osculating plane of a twisted unicursal curve.

Now the two simplest figures of the kind, are a range of points on a fixed straight line, and a *pencil* of straight lines passing through a fixed point and lying in a fixed plane. We may deduce the properties of the latter from the former. For if any straight line be drawn to cut the pencil there is a one-to-one correspondence between the points on the range thus formed and the rays of the pencil. In fact any ray of the pencil may be defined by the coordinates (x_1, x_2) of the point in which it intersects the straight line. With this definition it appears that everything that has been said for the range applies equally well to the pencil.

159. A binary form may be expressed as a product of n linear factors. A covariant of the binary form is necessarily a covariant of the system of linear forms of which it is a product. Thus let

$$a_x^n = (xx^{(1)})(xx^{(2)}) \dots (xx^{(n)}) \dots\dots\dots\dots\dots\dots \text{(I)},$$

where $\frac{x_1^{(1)}}{x_2^{(1)}}, \frac{x_1^{(2)}}{x_2^{(2)}}, \dots$ are the roots of the equation $a_x^n = 0$.

Any invariant may be written in the form

$$I = \Sigma (x^{(1)}x^{(2)})^{a_{12}} (x^{(1)}x^{(3)})^{a_{13}} \dots (x^{(2)}x^{(3)})^{a_{23}} \dots\dots\dots \text{(II)},$$

since it is an invariant of the linear forms $(xx^{(1)})$, $(xx^{(2)})$ ....

The coefficients of the quantic are given in terms of the roots by equating the different powers of x in (I). Two things are at once apparent.

(i) The coefficients are symmetric functions of the quantities $x^{(1)}$, $x^{(2)}$ ....

(ii) The coefficients are functions homogeneous and linear in each of the n sets of variables $x_1^{(1)}, x_2^{(1)}$; $x_1^{(2)}, x_2^{(2)}$; ... $x_1^{(n)}, x_2^{(n)}$.

It follows that any function of the coefficients must, when expressed in terms of the quantities $x^{(1)}, x^{(2)}$ ..., be symmetrical in them. And further such a function must be homogeneous and of the same order in each of the sets of variables $x_1^{(1)}, x_2^{(1)}$; ... $x_1^{(n)}, x_2^{(n)}$.

Hence in the expression for an invariant (II) it is necessary that

$$a_{12} + a_{13} + \dots + a_{1n} = p,$$
$$a_{21} + a_{23} + \dots + a_{2n} = p,$$

where p is a quantity which is the same for each term of the sum representing the invariant.

Let T be any one term of this sum, then let

$$\frac{I}{T} = \Sigma\, (x^{(1)}x^{(2)})^{\beta_{12}}\, (x^{(1)}x^{(3)})^{\beta_{13}} \ldots (x^{(2)}x^{(3)})^{\beta_{23}} \ldots.$$

Then
$$\beta_{12} + \beta_{13} + \ldots + \beta_{1n} = 0,$$
$$\beta_{21} + \beta_{23} + \ldots + \beta_{2n} = 0.$$

We are going to prove that $\frac{I}{T}$ is a function of anharmonic ratios of the roots. It will be assumed that when the number of quantities $x^{(1)}, x^{(2)} \ldots$ is less than n, then the term

$$(x^{(1)}x^{(2)})^{\beta_{12}}\, (x^{(1)}x^{(3)})^{\beta_{13}} \ldots (x^{(2)}x^{(3)})^{\beta_{23}} \ldots,$$

where
$$\sum_r \beta_{1r} = 0,\ \sum_r \beta_{2r} = 0,$$

may be expressed as a function of the anharmonic ratios.

Now the ratio

$$\{x^{(1)}x^{(2)}x^{(4)}x^{(t)}\} = \frac{(x^{(1)}x^{(2)})\,(x^{(4)}x^{(t)})}{(x^{(1)}x^{(t)})\,(x^{(4)}x^{(2)})}.$$

Hence

$$(x^{(1)}x^{(t)}) = (x^{(1)}x^{(2)}) \cdot \frac{(x^{(4)}x^{(t)})}{(x^{(4)}x^{(2)})} \cdot \frac{1}{\{x^{(1)}x^{(2)}x^{(4)}x^{(t)}\}}.$$

On replacing $(x^{(1)}x^{(t)})$ wherever possible by the value just found

$$(x^{(1)}x^{(2)})^{\beta_{12}}\, (x^{(1)}x^{(3)})^{\beta_{13}} \ldots (x^{(2)}x^{(3)})^{\beta_{23}} \ldots$$

becomes

$$(x^{(1)}x^{(2)})^{\beta_{12} + \beta'_{13} + \ldots}\, PQ = PQ,$$

where P is a function of anharmonic ratios, and Q is of the form

$$(x^{(2)}x^{(3)})^{\beta_{23}}\, (x^{(2)}x^{(4)})^{\beta_{24}} \ldots,$$

where
$$\sum_r \beta_{2r} = 0,\ \sum_r \beta_{3r} = 0.$$

The theorem has been assumed true for Q, hence with this assumption it is true when there are n quantities $x^{(1)}, x^{(2)}, \ldots$. If there are only three quantities, then

$$\beta_{12} + \beta_{13} = 0,$$
$$\beta_{21} + \beta_{23} = 0,$$
$$\beta_{31} + \beta_{32} = 0,$$

and
$$\beta_{12} = 0 = \beta_{23} = \beta_{31}.$$

Hence it is true when there are 4 quantities x, and therefore also when there are 5, and so on universally. Thus, *any invariant of a binary form is the numerator of a rational function of the anharmonic ratios of the roots.* If the invariant contains only one term, there is an apparent exception. The invariant equated to zero then represents the condition for the equality of a pair of roots, it can only be the discriminant.

160. So far our remarks have been confined to the case of a single quantic; a slight alteration in the wording of the previous paragraphs is all that is necessary to make them applicable to any system of binary forms. Each binary form of a system is geometrically represented by a set of points on a range, or of rays of a pencil. Points belonging to the same quantic must be regarded as indistinguishable from one another. Thus if we have a set of n points on a straight line, we may regard them as given by a single quantic of order n; by two quantics of orders r and $n-r$ respectively, or even by n separate linear forms.

Now let two of these points coincide; then, if the n points are regarded as a single quantic, the discriminant is zero; but there is nothing to tell us which roots coincide. We may regard the n points as two quantics, in this case either the discriminant of one of the quantics or else the resultant of the two is zero.

161. We shall now discuss the geometrical representation of the invariants and covariants of the binary forms of lowest order.

A quadratic has only one invariant, this vanishes when the points representing the quadratic are coincident; it is the discriminant.

A pair of quadratics a_x^2, b_x^2 have a simultaneous invariant $(ab)^2$.

Then if $x^{(1)}$, $x^{(2)}$ are the roots of a_x^2 and $y^{(1)}$, $y^{(2)}$ those of b_x^2,

$$(ab)^2 = (ay^{(1)})(ay^{(2)}) = \tfrac{1}{2}[(x^{(1)}y^{(1)})(x^{(2)}y^{(2)}) + (x^{(1)}y^{(2)})(x^{(2)}y^{(1)})].$$

If $(ab)^2 = 0$, it follows that

$$\frac{(x^{(1)}y^{(1)})(x^{(2)}y^{(2)})}{(x^{(1)}y^{(2)})(x^{(2)}y^{(1)})} = -1.$$

Hence the pair of points $y^{(1)}$, $y^{(2)}$ is harmonic with the pair $x^{(1)}$, $x^{(2)}$.

The quadratics have a covariant, their Jacobian, $\vartheta = (ab)\, a_x b_x$.

Now $$(a\vartheta)^2 = 0,\ (b\vartheta)^2 = 0,$$
hence ϑ represents the pair of points which is at the same time harmonic with a_x^2 and with b_x^2. In other words ϑ represents the pair of double points of the involution defined by the two pairs a_x^2, b_x^2.

The discriminant of ϑ is
$$(\vartheta\vartheta')^2 = \tfrac{1}{2}\{(aa')^2(bb')^2 - [(ab)^2]^2\};$$
if this is zero, ϑ is a perfect square; the double points of the involution coincide. Hence, as may be verified either geometrically or algebraically, one of each of the pairs a_x^2, b_x^2 coincides with the point represented by ϑ, and the other two points may be anywhere on the range. Thus a_x^2, b_x^2 have in this case a common point; hence $(\vartheta\vartheta')^2$ may be taken to be the resultant of the two quadratics. If there are more than two quadratics, there is only one more type of concomitant to be discussed; viz. the invariant
$$(ab)(bc)(ca).$$
This is equal to $-(\vartheta, c_x^2)^2$.

If this is zero then the pair of points ϑ is harmonic with the pair c_x^2. Hence
$$(ab)(bc)(ca) = 0$$
represents the condition that the three pairs of points a_x^2, b_x^2, c_x^2 should be pairs in involution.

To find the anharmonic ratio of the four points defined by a_x^2, b_x^2, we have
$$(x^{(1)}y^{(1)})(x^{(2)}y^{(2)}) + (x^{(1)}y^{(2)})(x^{(2)}y^{(1)}) = (ab)^2,$$
$$(x^{(1)}y^{(1)})(x^{(2)}y^{(2)}) - (x^{(1)}y^{(2)})(x^{(2)}y^{(1)}) = (x^{(1)}x^{(2)})(y^{(1)}y^{(2)})$$
$$= \sqrt{(aa')^2} \,.\, \sqrt{(bb')^2}.$$

Hence
$$2\,(x^{(1)}y^{(1)})(x^{(2)}y^{(2)}) = (ab)^2 + \sqrt{(aa')^2 \,.\, (bb')^2},$$
$$2\,(x^{(1)}y^{(2)})(x^{(2)}y^{(1)}) = (ab)^2 - \sqrt{(aa')^2 \,.\, (bb')^2},$$
and therefore
$$\{x^{(1)}y^{(1)}x^{(2)}y^{(2)}\} = \frac{(x^{(1)}y^{(1)})(x^{(2)}y^{(2)})}{(x^{(1)}y^{(2)})(x^{(2)}y^{(1)})}$$
$$= \frac{(ab)^2 + \sqrt{(aa')^2 \,.\, (bb')^2}}{(ab)^2 - \sqrt{(aa')^2 \,.\, (bb')^2}}.$$

Denoting the anharmonic ratio by ρ, and squaring, this equation becomes

$$D'^2(\rho-1)^2 = D \, . \, D''(\rho+1)^2,$$

or
$$\rho^2 - 2\rho\frac{D'^2+DD''}{D'^2-DD''}+1=0,$$

where
$$D=(aa')^2,\; D'=(ab)^2,\; D''=(bb')^2.$$

The two values of ρ correspond to the two anharmonic ratios $\{x^{(1)}y^{(1)}x^{(2)}y^{(2)}\}$ and $\{x^{(1)}y^{(2)}x^{(2)}y^{(1)}\}$.

If $\vartheta \equiv 0$, then the two quadratics are such that one is a multiple of the other. This is merely a particular case of the general property of the Jacobian; it is not necessary to do more than mention it here.

162. When a range possesses geometrical peculiarities which are unaffected by projection, there exist analytical relations of an invariant nature among the coefficients of the corresponding binary form; but it must not be supposed that these relations can always be expressed in terms of the pure invariants of the form. If there is only one such relation

$$A=0,$$

which expresses the necessary and sufficient condition that the range may possess a certain projective property, then it will be found that A is an invariant, for it is unaltered by linear transformation. On the other hand, when the condition is expressed by a set of algebraical relations

$$A=0,\; B=0, \ldots$$

$A, B, \ldots$ will not, in general, be invariants. Thus the condition that a binary form of order n may be a perfect nth power is that *all* the coefficients of its Hessian vanish*.

163. The Cubic. Any three collinear non-coincident points can be projected into any other three collinear non-coincident points; it is not to be expected then, that the geometry of a binary cubic will be of much interest from a projective point of view. But in respect of the associated points furnished by covariants, the geometry of the binary cubic is highly interesting.

* See later, Chap. XI.

The single invariant Δ is its discriminant and

$$\Delta = 0$$

is the necessary and sufficient condition that two of the three points represented by the cubic should coincide.

If all three points coincide, then Δ, H and t are all identically zero, as may be easily verified; but

$$H \equiv 0$$

represents the necessary and *sufficient* condition*.

164. Let us consider the *pencil*

$$\kappa f + \lambda t \equiv f_{\kappa,\lambda},$$

where κ and λ are new constants which determine the particular members of the pencil of cubics. Then $f_{\kappa,\lambda}$ represents three points, which are called a *triad*; by varying κ and λ a pencil of triads is obtained. The covariants of $f_{\kappa,\lambda}$ will be denoted by the symbols $H_{\kappa,\lambda}$, $t_{\kappa,\lambda}$, $\Delta_{\kappa,\lambda}$. These may be at once calculated; the following table, most of the results of which are proved in Chap. V., will be found useful for this purpose.

Index	Transvectant					
	(f, f)	(f, H)	(f, t)	(H, H)	(H, t)	(t, t)
1	0	t	$-\frac{1}{2}H^2$	0	$\frac{1}{2}\Delta f$	0
2	H	0	0	Δ	0	$\frac{1}{2}\Delta H$
3	0		Δ			0

The fundamental forms, it will be remembered, are connected by one syzygy,

$$t^2 = -\tfrac{1}{2}\{H^3 + \Delta f^2\}.$$

To obtain $H_{\kappa,\lambda}$ we have

$$\begin{aligned} H_{\kappa,\lambda} &= (\kappa f + \lambda t,\ \kappa f + \lambda t)^2 \\ &= \kappa^2 (f, f)^2 + 2\kappa\lambda\,(f, t)^2 + \lambda^2 (t, t)^2 \\ &= (\kappa^2 + \tfrac{1}{2}\Delta\lambda^2)\, H. \end{aligned}$$

* Chap. XI.

Hence, if we use the notation

$$\Theta = (\kappa^2 + \tfrac{1}{2}\Delta\lambda^2),$$

$$H_{\kappa,\lambda} = \Theta H.$$

In the same way, we obtain

$$t_{\kappa,\lambda} = (f_{\kappa,\lambda}, H_{\kappa,\lambda}) = \Theta (f_{\kappa,\lambda}, H)$$
$$= \Theta (\kappa t - \tfrac{1}{2}\Delta\lambda f)$$
$$= \tfrac{1}{2}\Theta \left(t\frac{\partial\Theta}{\partial\kappa} - f\frac{\partial\Theta}{\partial\lambda}\right).$$

And lastly $$\Delta_{\kappa,\lambda} = \Theta^2\Delta.$$

It is worth noticing that if we introduce the arguments κ, λ of Θ as suffixes, thus

$$\Theta_{\kappa,\lambda} = \kappa^2 + \tfrac{1}{2}\Delta\lambda^2,$$

the syzygy may be written

$$H^3 = -2\Theta_{t,f}.$$

165. Consider the relation

$$H^3 = -2(t^2 + \tfrac{1}{2}\Delta f^2),$$

if any pair of the three forms

$$f, \quad H, \quad t$$

have a common factor, then all three must have this factor. Let us suppose that such a factor exists, and let us change the variables so that the common factor is x_2. Then f is of the form

$$(0, a_1, a_2, a_3 \between x_1, x_2)^3,$$

and the coefficient of x_1^2 in H is $-a_1^2$; hence if x_2 is a factor of H, we must have $a_1 = 0$. This means that f has a double factor, and therefore

$$\Delta = 0.$$

The syzygy then becomes

$$H^3 = -2t^2,$$

whence it is easy to deduce that H is a perfect square and t a perfect cube. In fact if

$$f = \zeta^2\theta,$$

then $$H = A\zeta^2,\ t = \sqrt{\frac{-A^3}{2}}\,\zeta^3,\ \Delta = 0,$$

where $$A = -\tfrac{2}{9}[(\zeta\theta)]^2.$$

166. It will now be supposed that f, H, t have no common factor. The syzygy may be written

$$H^3 = -2\left(t + f\sqrt{\frac{-\Delta}{2}}\right)\left(t - f\sqrt{\frac{-\Delta}{2}}\right);$$

hence if ξ, η are the factors of H, so determined that

$$H = -2\xi\eta,$$

we may take
$$2\xi^3 = \left(t + f\sqrt{\frac{-\Delta}{2}}\right),$$

and
$$2\eta^3 = \left(t - f\sqrt{\frac{-\Delta}{2}}\right),$$

and therefore
$$\sqrt{\frac{-\Delta}{2}}\, f = \xi^3 - \eta^3,$$

$$t = \xi^3 + \eta^3.$$

Hence also

$$\sqrt{-\frac{\Delta}{2}} \cdot f_{\kappa,\lambda} = \left(\kappa + \lambda\sqrt{-\frac{\Delta}{2}}\right)\xi^3 - \left(\kappa - \lambda\sqrt{-\frac{\Delta}{2}}\right)\eta^3,$$

$$t_{\kappa,\lambda} = \Theta\left[\left(\kappa - \lambda\sqrt{-\frac{\Delta}{2}}\right)\xi^3 + \left(\kappa + \lambda\sqrt{-\frac{\Delta}{2}}\right)\eta^3\right],$$

$$H_{\kappa,\lambda} = -2\Theta\xi\eta.$$

It is at once apparent that the only members of the pencil which possess double factors are ξ^3 and η^3.

Let P_1, P_2, P_3 be the three points determined by any one member of the pencil, and A, B the two points determined by the Hessian (which are the same for every member of the pencil). Then if ω be a cube root of unity, the points P_1, P_2, P_3 correspond to linear forms

$$\xi - a\eta,\ \xi - \omega a\eta,\ \xi - \omega^2 a\eta.$$

The ranges

$$A,\ P_1,\ P_2,\ P_3,\ B$$

$$A,\ P_2,\ P_3,\ P_1,\ B$$

$$A,\ P_3,\ P_1,\ P_2,\ B$$

are projective. A set of points such as P_1, P_2, P_3 are said to be cyclically projective.

The simplest way to find the anharmonic ratio $\{AP_1P_2P_3\}$ is to transform the variables to ξ, η. This transformation may be regarded as merely changing the points of reference, and the constant multiplier. Then

$$\{AP_1P_2P_3\} = \frac{(-a)(-\omega^2 a + \omega a)}{(-\omega^2 a)(-a + \omega a)} = -\omega^2 = \{AP_2P_3P_1\} = \{AP_3P_1P_2\}.$$

Hence the range formed by a triad and one of its Hessian points is equianharmonic. The six distinct cross ratios of such a range are each $-\omega$ or $-\omega^2$.

167. The Quartic. Just as in the case of the cubic we considered a pencil of cubics instead of the single one, so here we shall find it convenient to consider the pencil

$$f_{\kappa,\lambda} = \kappa f + \lambda H,$$

instead of the single quartic f. Each member of the pencil will define four points, one of these points may be chosen at will on the range considered, but when this is done the ratio $\kappa : \lambda$ is fixed, and the remaining three points are uniquely determined. The calculation of the covariants of $f_{\kappa,\lambda}$ presents no serious difficulty. For convenience a table of the transvectants of the quartic is appended; most of the calculations were effected in Chap. V.

Index	Transvectant					
	(f, f)	(f, H)	(H, H)	(f, t)	(H, t)	(t, t)
1	0	t	0	$\frac{1}{12}if^2 - \frac{1}{2}H^2$	$\frac{1}{6}f(jf - iH)$	0
2	H	$\frac{1}{6}if$	$\frac{1}{3}jf - \frac{1}{6}iH$	0	0	$\frac{1}{6}jHf - \frac{1}{12}iH^2 - \frac{1}{72}i^2f^2$
3	0	0	0	$\frac{1}{4}(jf - iH)$	$\frac{1}{24}i^2f - \frac{1}{4}jH$	0
4	i	j	$\frac{1}{6}i^2$	0	0	0

The only transvectants which are not contained in this table are

$$(t, t)^5 = 0, \ (t, t)^6 = \tfrac{1}{24} i^3 - \tfrac{1}{4} j^2.$$

The syzygy between the forms is

$$t^2 = -\tfrac{1}{2}(H^3 - \tfrac{1}{2}iHf^2 + \tfrac{1}{3}jf^3).$$

In connection with the system for $f_{\kappa,\lambda}$ the expression

$$\Omega_{\kappa,\lambda} = \kappa^3 - \frac{i}{2}\kappa\lambda^2 - \frac{j}{3}\lambda^3,$$

or more briefly Ω, will be found of great importance. We observe at once that the syzygy may be written

$$2t^2 = -\Omega_{H,-f}.$$

Now

$$\begin{aligned} H_{\kappa,\lambda} &= (\kappa f + \lambda H, \kappa f + \lambda H)^2 \\ &= \kappa^2 H + \tfrac{1}{3}i\kappa\lambda f + \lambda^2(\tfrac{1}{3}jf - \tfrac{1}{6}iH) \\ &= \tfrac{1}{3}\left(H\frac{\partial\Omega}{\partial\kappa} - f\frac{\partial\Omega}{\partial\lambda}\right), \end{aligned}$$

$$\begin{aligned} t_{\kappa,\lambda} &= \left(\kappa f + \lambda H,\ \tfrac{1}{3}H\frac{\partial\Omega}{\partial\kappa} - \tfrac{1}{3}f\frac{\partial\Omega}{\partial\lambda}\right) \\ &= \tfrac{1}{3}t\left(\kappa\frac{\partial\Omega}{\partial\kappa} + \lambda\frac{\partial\Omega}{\partial\lambda}\right) = \Omega t, \end{aligned}$$

$$\begin{aligned} i_{\kappa,\lambda} &= \kappa^2 i + 2\kappa\lambda j + \lambda^2\tfrac{1}{6}i^2 = -3(\Omega, \Omega)^2 \\ &= -3H_\Omega, \end{aligned}$$

where Ω is regarded as a binary cubic in κ and λ.

Lastly

$$\begin{aligned} j_{\kappa,\lambda} &= \tfrac{1}{3}\left\{-i\kappa\frac{\partial\Omega}{\partial\lambda} + j\left(\kappa\frac{\partial\Omega}{\partial\kappa} - \lambda\frac{\partial\Omega}{\partial\lambda}\right) + \tfrac{1}{6}i^2\lambda\frac{\partial\Omega}{\partial\lambda}\right\} \\ &= -3t_\Omega. \end{aligned}$$

168. The invariant of the cubic Ω is

$$\Delta_\Omega = \tfrac{1}{27}(i^3 - 6j^2).$$

This, as we proceed to shew, may be taken to be the discriminant of the quartic. The discriminant is the condition that the equation

$$f = 0$$

may have a pair of equal roots.

It is an invariant; for if the range represented by f be projected, the pair of coincident points project into a pair of coincident points. Further it is well known to be of degree

$2(n-1)$ for the quantic of order n; hence it is of degree 6 for the quartic. It is then a linear function of i^3 and j^2.

Let us suppose that f has a double linear factor α_x, the remaining quadratic factor being p_x^2, and then find what relation exists between i and j for the quartic

$$a_x^4 = f = \alpha_x^2 . p_x^2.$$

In the first place

$$H = (\alpha_x^2 . p_x^2, \ a_x^4)^2$$

$$= \tfrac{1}{6}\{(\alpha a)^2 p_x^2 + (pa)^2 \alpha_x^2 + 4(\alpha a)(pa)\alpha_x p_x\} a_x^2$$

$$= \tfrac{1}{6}\{3(\alpha a)^2 p_x^2 + 3(pa)^2 \alpha_x^2 - 2(\alpha p)^2 a_x^2\} a_x^2 \ldots\ldots\ldots\ldots(\text{I}).$$

But

$$(\alpha a)^2 a_x^2 = (f, \ \alpha_x^2)^2 = \tfrac{1}{6}(\alpha p)^2 . \alpha_x^2,$$

$$(pa)^2 a_x^2 = (f, \ p_x^2)^2 = \tfrac{1}{6}\{(\alpha p)^2 p_x^2 + 3(p, \ p)^2 \alpha_x^2\}.$$

Therefore

$$H = -\tfrac{1}{6}(\alpha p)^2 f + \tfrac{1}{4}(p, \ p)^2 \alpha^4.$$

Similarly

$$i = (a\alpha)^2 (ap)^2$$

$$= ((a\alpha)^2 a_x^2, \ p_x^2)^2 = \tfrac{1}{6}[(\alpha p)^2]^2,$$

$$j = (f, \ H)^4 = -\tfrac{1}{6}(\alpha p)^2 . i$$

$$= -\tfrac{1}{36}[(\alpha p)^2]^3.$$

Therefore

$$i^3 - 6j^2 = 0.$$

Hence Δ_Ω may be taken to be the discriminant of the quartic f.

From the above form for H, it is evident that H contains the factor α_x twice over; hence if f contains a repeated linear factor then every form of the pencil $\kappa f + \lambda H$ contains the same repeated factor. This leads us to expect that the discriminant of $\kappa f + \lambda H$ is a multiple of the discriminant of f. This is so; for from the syzygy for the cubic we obtain

$$-H_\Omega^3 - 2t_\Omega^2 = \Delta_\Omega . \Omega^2,$$

or

$$(i_{\kappa,\lambda}^3 - 6j_{\kappa,\lambda}^2) = \Omega^2 (i^3 - 6j^2).$$

It is easy to obtain in the same way the condition that f may have two pairs of repeated factors. For writing as before

$$f = \alpha_x^2 . p_x^2,$$

we obtain

$$t = (f, \ H) = \tfrac{1}{4}(p, \ p)^2 (f, \ \alpha) \alpha_x^3.$$

Now if f contains two pairs of repeated factors, $(p, p)^2$ must vanish, for p_x^2 is a perfect square; hence in this case

$$t \equiv 0.$$

This is the necessary and sufficient condition, for if it is satisfied

$$(t, t)^6 = \tfrac{1}{4}(\tfrac{1}{6}i^3 - j^2) = 0,$$

and we are at liberty to assume that f has one repeated factor; then using the relation

$$(p, p)^2 (f, \alpha) \alpha_x^3 \equiv 0$$

we see that either

$$(p, p)^2 = 0,$$

in which case p is a perfect square; or

$$(f, \alpha) \equiv 0,$$

in which case

$$f = \alpha_x^4.$$

This furnishes another illustration of the remarks in § 162.

Ex. (i). Shew that the necessary and sufficient condition that f may have a three times repeated factor is

$$i=0, \quad j=0.$$

Ex. (ii). There are in general three different members of the pencil $f_{\kappa,\lambda}$ which are perfect squares.

They are given by solving the equation

$$\Omega = 0.$$

169. As has been already pointed out, the syzygy for the quartic may be written

$$2t^2 = -\Omega_{H,-f}.$$

If m_1, m_2, m_3 be the roots of the cubic

$$\Omega = 0,$$

then

$$\Omega_{\kappa,\lambda} = (\kappa - m_1\lambda)(\kappa - m_2\lambda)(\kappa - m_3\lambda).$$

Hence also

$$2t^2 = -(H + m_1 f)(H + m_2 f)(H + m_3 f).$$

If H and f have a common factor, by transformation this may be made x_2. Then f is of the form

$$(0, a_1, a_2, a_3, a_4 \between x_1, x_2)^4.$$

In order that x_2 may be a factor of H, we must have

$$-a_1^2=0.$$

Hence x_2^2 is a factor of f; and therefore

$$\Delta_\Omega=0.$$

Excluding this exceptional case, it is evident that no pair of the expressions

$$H+m_1 f,\ H+m_2 f,\ H+m_3 f$$

have a common factor (m_1, m_2, m_3 are distinct) for Δ_Ω is the discriminant of Ω. Hence the above relation shews that each of the expressions $H+mf$ must be a perfect square—since t^2 is a perfect square.

Let

$$H+m_1 f=-2\phi^2$$
$$H+m_2 f=-2\psi^2$$
$$H+m_3 f=-2\chi^2,$$

where ϕ, ψ, χ are binary quadratics.

Then

$$t=2\phi\psi\chi.$$

As an example it may be verified that for the quartic

$$x_1^4+6a\,x_1^2x_2^2+x_2^4,$$
$$H=2ax_1^4+2(1-3a^2)\,x_1^2x_2^2+2ax_2^4,$$
$$\Omega=\kappa^3-(3a^2+1)\,\kappa\lambda^2-2a(1-a^2)\,\lambda^3,$$

and that the roots of

$$\Omega=0$$

are $a-1$, $a+1$, $-2a$; which are identical with the values of m which make

$$H+mf$$

a perfect square.

Now

$$(H+m_1 f,\ H+m_2 f)=(m_1-m_2)\,t$$
$$=(-2\phi^2,\ -2\psi^2)=4(\phi,\ \psi)\,\phi\psi;$$

putting in the value of t we obtain

$$2(\phi,\ \psi)=(m_1-m_2)\,\chi.$$

Similarly

$$2(\psi,\ \chi)=(m_2-m_3)\,\phi,$$
$$2(\chi,\ \phi)=(m_3-m_1)\,\psi.$$

Now from the expression for the Jacobian of a Jacobian we obtain

$$((\phi, \psi), \chi) = -\tfrac{1}{2}\{\phi(\psi, \chi)^2 - \psi(\phi, \chi)^2\}.$$

Hence by repeated use of this formula

$$\begin{aligned}
0 &= (\phi\chi)^2\psi - (\phi\psi)^2\chi,\\
-\frac{(m_1-m_2)(m_1-m_3)}{2}\chi &= (\phi\phi)^2\chi - (\phi\chi)^2\phi,\\
-\frac{(m_1-m_2)(m_1-m_3)}{2}\psi &= (\phi\phi)^2\psi - (\phi\psi)^2\phi,\\
-\frac{(m_2-m_3)(m_2-m_1)}{2}\chi &= (\psi\psi)^2\chi - (\psi\chi)^2\psi,\\
0 &= (\psi\phi)^2\chi - (\psi\chi)^2\phi,\\
-\frac{(m_2-m_3)(m_2-m_1)}{2}\phi &= (\psi\psi)^2\phi - (\psi\phi)^2\psi,\\
-\frac{(m_3-m_1)(m_3-m_2)}{2}\psi &= (\chi\chi)^2\psi - (\chi\psi)^2\chi,\\
-\frac{(m_3-m_1)(m_3-m_2)}{2}\phi &= (\chi\chi)^2\phi - (\chi\phi)^2\chi,\\
0 &= (\chi\psi)^2\phi - (\chi\phi)^2\psi.
\end{aligned}$$

Since ϕ, ψ, χ have no common factors these equations give the following six relations

$$\begin{aligned}
(\phi\phi)^2 &= -\tfrac{1}{2}(m_1-m_2)(m_1-m_3), \quad (\psi\chi)^2 = 0,\\
(\psi\psi)^2 &= -\tfrac{1}{2}(m_2-m_3)(m_2-m_1), \quad (\chi\phi)^2 = 0,\\
(\chi\chi)^2 &= -\tfrac{1}{2}(m_3-m_1)(m_3-m_2), \quad (\phi\psi)^2 = 0.
\end{aligned}$$

The remaining invariant of these three quadratics is

$$\begin{aligned}
(\phi\psi)(\psi\chi)(\chi\phi) &= -((\phi\psi)\phi_x\psi_x, \chi_x^2)^2\\
&= -\tfrac{1}{2}(m_1-m_2).(\chi\chi)^2 = -\tfrac{1}{4}(m_1-m_2)(m_2-m_3)(m_3-m_1).
\end{aligned}$$

170. By means of the equations

$$\begin{aligned}
H + m_1 f &= -2\phi^2,\\
H + m_2 f &= -2\psi^2,\\
H + m_3 f &= -2\chi^2 \quad \ldots\ldots\ldots\ldots\ldots\ldots\ldots\text{(II)},
\end{aligned}$$

the quartic f, or more generally $\kappa f + \lambda H$, may be separated into quadratic factors.

Thus $$f = \frac{2}{m_2 - m_3}(\chi^2 - \psi^2),$$

$$H = \frac{2}{m_2 - m_3}(m_3 \psi^2 - m_2 \chi^2),$$

and $$\kappa f + \lambda H = \frac{2}{m_2 - m_3}\{(\kappa - \lambda m_2)\chi^2 - (\kappa - \lambda m_3)\psi^2\}$$

$$= \frac{2}{m_2 - m_3}(\sqrt{\kappa - \lambda m_2}\,\chi + \sqrt{\kappa - \lambda m_3}\,\psi)(\sqrt{\kappa - \lambda m_2}\,\chi - \sqrt{\kappa - \lambda m_3}\,\psi).$$

The second transvectant of either of these quadratic factors with ϕ is zero. Hence the two points determined by ϕ are the harmonic conjugates of the two pairs of points represented by the above quadratic factors of $\kappa f + \lambda H$. Now we have only used the last two of equations (II) to find the quadratic factors of $\kappa f + \lambda H$. Any pair of the three equations might be taken. The three results represent the three ways into which the quartic $\kappa f + \lambda H$ may be separated into quadratic factors. Then the three quadratic factors of t are the three pairs of points harmonically conjugate with respect to the four points $\kappa f + \lambda H$, when divided into two pairs of points.

Now $$(\phi\psi)^2 = 0,\ (\psi\chi)^2 = 0,\ (\chi\phi)^2 = 0,$$

hence the pair of points ϕ is harmonically conjugate with respect to each of the pairs ψ and χ. If the points $\kappa f + \lambda H$ are divided into two pairs in any way, these pairs determine an involution, one of the quadratic factors of t represents the pair of double points of the involution. The other two quadratic factors of t represent pairs of points belonging to the involution.

Now the points determined by t are independent of κ and λ, hence the pencil $\kappa f + \lambda H$ represents sets of four points such that when any set is separated into two pairs of points, these are pairs of one of three fixed involutions.

The quartic f is arbitrary, it may represent any four points. Hence the pairs of double points of the three involutions determined by four points on a line are harmonically conjugate two and two.

171. To determine the anharmonic ratio ρ of the four points f. We have obtained the quadratic factors of f, one pair is

$$\chi - \psi, \ \chi + \psi.$$

The anharmonic ratio of the four points determined by a pair of quadratics has been obtained in § 161, as a root of the equation

$$\rho^2 - 2\rho \frac{D'^2 + DD''}{D'^2 - DD''} + 1 = 0.$$

In our case

$$\begin{aligned} D &= (\chi - \psi, \ \chi - \psi)^2 \\ &= -\tfrac{1}{2}(m_3 - m_1)(m_3 - m_2) - \tfrac{1}{2}(m_2 - m_3)(m_2 - m_1) \\ &= -\tfrac{1}{2}(m_3 - m_2)^2 = D''. \\ D' &= -\tfrac{1}{2}(m_3 - m_1)(m_3 - m_2) + \tfrac{1}{2}(m_2 - m_3)(m_2 - m_1). \end{aligned}$$

Hence $$\rho^2 - \frac{(m_3 - m_1)^2 + (m_2 - m_1)^2}{(m_3 - m_1)(m_2 - m_1)}\rho + 1 = 0.$$

As there are six different values for the anharmonic ratio of four points, a sextic for ρ is to be expected. This will be obtained by multiplying together the three equations similar to the above.

It will be more convenient to write these equations in the form

$$\rho^2 - 2\rho + 1 - \frac{(m_3 - m_2)^2}{(m_3 - m_1)(m_2 - m_1)}\rho = 0.$$

Now m_1, m_2, m_3 are the roots of the cubic

$$\kappa^3 - \frac{i}{2}\kappa\lambda^2 - \frac{j}{3}\lambda^3 = 0.$$

The discriminant of this is

$$(m_1 - m_2)^2 (m_2 - m_3)^2 (m_3 - m_1)^2 = \tfrac{1}{2}(i^3 - 6j^2) = \Delta,$$

the exact expression is most quickly obtained by using the equation of the squared differences of the cubic.

The equation whose roots are $m_1 - m_2$, $m_2 - m_3$, $m_3 - m_1$ is obtained thus

$$\Sigma(m_1 - m_2) = 0,$$

$$\begin{aligned} \Sigma(m_3 - m_1)(m_1 - m_2) &= -\Sigma m_1^2 + \Sigma m_2 m_3 \\ &= 3\Sigma m_2 m_3 = -3\frac{i}{2}, \end{aligned}$$

$$(m_1 - m_2)(m_2 - m_3)(m_3 - m_1) = \sqrt{\Delta}.$$

The equation is $y^3 - 3\frac{i}{2}y - \sqrt{\Delta} = 0.$

The equation whose roots are $(m_1 - m_2)^3, \ldots$ is

$$z - 3\frac{i}{2}z^{\frac{1}{3}} - \sqrt{\Delta} = 0,$$

or

$$(z - \sqrt{\Delta})^3 = 27\frac{i^3}{8}z.$$

But

$$(m_2 - m_3)^3 = -\sqrt{\Delta}\,\frac{\rho^2 - 2\rho + 1}{\rho}.$$

Hence

$$\Delta^{\frac{3}{2}}\left\{-\frac{\rho^2 - 2\rho + 1}{\rho} - 1\right\}^3 = -27\frac{i^3}{8}\frac{\rho^2 - 2\rho + 1}{\rho}\Delta^{\frac{1}{2}},$$

or

$$\left(1 - 6\frac{j^2}{i^3}\right)\left\{1 + \frac{\rho^2 - 2\rho + 1}{\rho}\right\}^3 = \frac{\rho^2 - 2\rho + 1}{\rho} \cdot \frac{27}{4}.$$

Therefore

$$\frac{i^3}{j^2} = \frac{24(\rho^2 - \rho + 1)^3}{4(\rho^2 - \rho + 1)^3 - 27\rho^2(\rho - 1)^2}$$

$$= \frac{24(\rho^2 - \rho + 1)^3}{(\rho + 1)^2(\rho - 2)^2(2\rho - 1)^2}.*$$

* When the quartic is not treated by the symbolical method it is usual to define the invariants as follows:—the quartic itself is

$$f = (a,\ b,\ c,\ d,\ e)(x_1,\ x_2)^4,$$

$$I = ae - 4bd + 3c^2,$$

$$J = \begin{vmatrix} a & b & c \\ b & c & d \\ c & d & e \end{vmatrix}.$$

The invariants i, j in the text above differ from these by numerical factors only. Thus

$$i = (ab)^4 = 2I, \quad j = (ab)^2(bc)^2(ca)^2 = 6J.$$

In connection with the calculation of the values of invariants given symbolically in terms of the actual coefficients the reader may find it interesting to discover the fallacy in the following:—

$$j = (bc)^2(ca)^2(ab)^2$$

$$= \{(bc)(ca)(ab)\}^2$$

$$= \begin{vmatrix} a_1^2 & b_1^2 & c_1^2 \\ a_1a_2 & b_1b_2 & c_1c_2 \\ a_2^2 & b_2^2 & c_2^2 \end{vmatrix} \begin{vmatrix} a_1^2 & b_1^2 & c_1^2 \\ a_1a_2 & b_1b_2 & c_1c_2 \\ a_2^2 & b_2^2 & c_2^2 \end{vmatrix}$$

$$= \begin{vmatrix} a_1^4 + b_1^4 + c_1^4 & \ldots & \ldots & \ldots \\ a_1^3a_2 + b_1^3b_2 + c_1^3c_2 & \ldots & \ldots & \ldots \\ a_1^2a_2^2 + b_1^2b_2^2 + c_1^2c_2^2 \ldots & \ldots & \ldots \end{vmatrix}$$

$$= 27J.$$

Thus the anharmonic ratio is expressed by means of a sextic equation in terms of the absolute invariant $\frac{i^3}{j^2}$.

We see from this equation at a glance that if $i=0$, the points represented by the quartic form an equianharmonic range, for then,

$$\rho^2 - \rho + 1 = 0.$$

Similarly if $j = 0$, the four points form a harmonic range.

Again, if two of the points of the range are coincident, one value of ρ is unity; hence

$$j^2 = 6i^3,$$

as it should be.

172. The anharmonic ratio for the four points

$$\kappa f + \lambda H$$

may be obtained at once by writing $i_{\kappa,\lambda}$ for i, and $j_{\kappa,\lambda}$ for j.

To determine those values of $\kappa : \lambda$ for which the four points have any definite anharmonic ratio ρ; let

$$a = 24 \frac{(1-\rho+\rho^2)^3}{(\rho+1)^2(\rho-2)^2(2\rho-1)^2}.$$

Then

$$i^3_{\kappa,\lambda} - aj^2_{\kappa,\lambda} = 0,$$

or

$$3H_\Omega^3 + at_\Omega^2 = 0,$$

this is a sextic for $\kappa : \lambda$.

Now

$$H_\Omega^3 = -2t_\Omega^2 - \Omega^2\Delta_\Omega.$$

Hence

$$(a-6)\,t_\Omega^2 = \Omega^2\Delta_\Omega,$$

or

$$t_\Omega = \pm\,\Omega\sqrt{\frac{\Delta_\Omega}{a-6}}.$$

The sextic thus reduces to two cubics.

If $a=6$, it is easy to see that $\rho = 0, 1, \infty$, hence two points must coincide. In this case $\Omega = 0$, and $\frac{\kappa}{\lambda}$ has one of the three values

$$m_1,\ m_2,\ m_3.$$

Hence the three members of the pencil for which $a = 6$ are

$$H + m_1 f = \phi^2,$$
$$H + m_2 f = \psi^2,$$
$$H + m_3 f = \chi^2,$$

shewing that if one pair of points coincide, the other pair must also coincide.

If $a = \infty$, the four points form a harmonic range, and $t_\Omega = 0$. There are three members of the pencil for which the range is harmonic. If $a = 0$, then $H_\Omega = 0$; hence there are only two members of the pencil which form equianharmonic ranges.

In all other cases, there are six members of the pencil having a definite anharmonic ratio.

Ex. If $l_x l_x'$ is the Hessian of the cubic a_x^3, prove that the quartic $a_x^3 l_x$ is equianharmonic.

173. Case when $\Delta_\Omega = 0$. The discussion was limited in § 169 to the case when Δ_Ω is other than zero. Now Δ_Ω is the discriminant of the quartic f, and hence when Δ_Ω vanishes, two of the roots of $f = 0$ are the same. We may, as in § 168, write

$$f = \alpha_x^2 p_x^2,$$

where α_x^2 is the square of a linear form α_x, and p_x^2 is a quadratic. Then as before

$$H = \alpha_x^2 \{ -\tfrac{1}{6} (\alpha p)^2 p_x^2 + \tfrac{1}{4} (pp)^2 \alpha_x^2 \},$$
$$i = \tfrac{1}{6} \{(\alpha p)^2\}^2, \quad j = -\tfrac{1}{36} \{(\alpha p)^2\}^3.$$

The invariant Δ_Ω is also the discriminant of the cubic

$$\Omega = \kappa^3 - \frac{i}{2} \kappa \lambda^2 - \frac{j}{3} \lambda^3,$$

hence, in the present case, Ω has a repeated root. Let this be m_2, and let the other root be m_1. Then

$$2m_2 + m_1 = 0, \quad m_2^2 + 2m_1 m_2 = -\frac{i}{2}, \quad m_2^2 m_1 = \frac{j}{3}.$$

Hence

$$m_2 = -\frac{j}{i} = \tfrac{1}{6} (\alpha p)^2, \quad m_1 = \frac{2j}{i} = -\tfrac{1}{3} (\alpha p)^2,$$

and therefore

$$H + m_2 f = \tfrac{1}{4}(pp)^2 \,.\, \alpha_x^4,$$

$$H + m_1 f = \alpha_x^2 \{-\tfrac{1}{2}(\alpha p)^2 p_x^2 + \tfrac{1}{4}(pp)^2 \alpha_x^2\}.$$

Again, since

$$2t^2 = -(H + m_1 f)(H + m_2 f)^2,$$

we obtain as before

$$H + m_1 f = -2\phi^2,$$

where ϕ is a quadratic. One of the factors of ϕ must be α_x, and if the other is β_x, then

$$2\beta_x^2 = \tfrac{1}{2}(\alpha p)^2 p_x^2 - \tfrac{1}{4}(pp)^2 \alpha_x^2.$$

The value of t is then seen to be

$$t = \tfrac{3}{4}(pp)^2 \alpha_x^5 \beta_x.$$

174. We shall now briefly explain another interesting method of representing invariant properties of binary forms geometrically.

If we put $x_1 = z$ and $x_2 = 1$ throughout the work on binary quantics the general linear substitution may be written

$$z = \frac{az' + b}{cz' + d}, \text{ since } z = \frac{x_1}{x_2}.$$

Now put $z = x + iy$, and represent z as the real point x, y in the Argand diagram in the usual way; then the substitution

$$z = \frac{az' + b}{cz' + d}$$

is a point transformation.

Unless $c = 0$ the relation between z and z' may be reduced to the form

$$(z - \alpha)(z' - \alpha') = k,$$

wherein α, α', k are constants.

Suppose z, z', α, α' are the points P, P', A, A' respectively, then the geometrical meaning of the above is

$$AP \,.\, A'P' = \text{mod.}\,(k),$$

and the sum of the angles that AP and $A'P'$ make with any fixed line is constant. Hence the general linear substitution is equivalent to an inversion together with a change of origin and a reflexion of inclination of the line AP with respect to a fixed line.

If $c = 0$ the equation can be written

$$z' - \beta = m(z - \beta),$$

indicating that P' is derived from P by turning BP through a fixed angle and increasing it in a given ratio, B being the point which represents β.

Hence a binary form of order n represents n real points A in the plane, and a covariant of the form represents a group of points C whose relation to the points A is unaltered by a geometrical transformation of the types indicated.

In particular, the relation of the points C to the points A is unaltered by any inversion, because in the particular case in which A and A' coincide and k is real, the transformation is equivalent to an inversion with respect to A, and a reflexion with respect to the real axis through A; but the properties of the derived figure are evidently unaltered by a reflexion alone, and hence they are unaltered by an inversion alone.

174 A. Some of the simpler invariants and covariants can now be interpreted.

If $$az^2 + 2bz + c = 0;\quad a'z^2 + 2b'z + c' = 0$$

represent the points A, B and C, D respectively, then when

$$ac' + a'c - 2bb' = 0$$

A, B, C, D are four harmonic points on a circle.

In fact on changing the origin to O, the middle point of AB, the first quadratic becomes

$$z^2 - k^2 = 0,$$

and if the second be $$(z - z_1)(z - z_2) = 0,$$

then since the relation is invariantive we have

$$z_1 z_2 = k^2,$$

therefore $$OC \, . \, OD = OA^2 = OB^2$$

and OC, OD are equally inclined to OA.

If we produce CO to D' making $OD' = OD$, we have

$$OC \, . \, OD' = OA^2 = OA \, . \, OB;$$

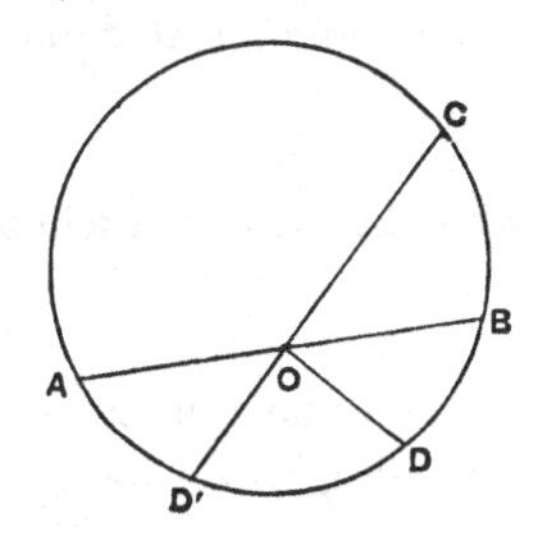

therefore $CAD'B$ are concyclic, and by symmetry D is on the circle.

Further as the pencil $D'\{ACBD\}$ is harmonic the four points AB and CD form two harmonic pairs on the circle. We shall call them harmonically concyclic.

Ex. Shew that there is one pair of points (P, P') harmonically concyclic to each of two given pairs (A, B) and (C, D).

Further if (Q, Q') be harmonically concyclic to (A, C), (B, D) and (R, R') to (A, D), (B, C), then any two of the three pairs P, P' ; Q, Q' ; R, R' are harmonically concyclic.

175. We shall now apply the complex variable to prove certain properties of the foci of conics.

If the tangential equation of a conic be

$$Al^2 + 2Hlm + Bm^2 + 2Gl + 2Fm + C = 0,$$

the axes of coordinates being rectangular, and $x_1 y_1$ be a focus, then the line

$$x + y\iota = x_1 + y_1\iota \text{ touches the curve.}$$

Hence the above equation is satisfied by $l = -\frac{1}{z_1}$, $m = -\frac{\iota}{z_1}$, and we have

$$A + 2H\iota - B - 2(G + F\iota) z_1 + Cz_1^2 = 0.$$

Consequently the two real foci z_1 and z_2 are given by the quadratic

$$Cz^2 - 2(G + F\iota) z + (A - B + 2H\iota) = 0.$$

Since $\frac{z_1 + z_2}{2} = \frac{G}{C} + \iota \frac{F}{C}$, we see that the centre is the point $\frac{G}{C}, \frac{F}{C}$.

Again, if O be the origin and S_1, S_2 the foci, we have

$$OS_1 . OS_2 = \frac{\sqrt{(A - B)^2 + 4H^2}}{C},$$

so that the origin can only be a focus when $A = B$ and $H = 0$.

If a system of conics be inscribed in the same quadrilateral, their tangential equations are of the form

$$\lambda\Sigma + \lambda'\Sigma' = 0,$$

and thus the real foci are given by the pencil of quadratics

$$\lambda f + \lambda' f' = 0.$$

All the quadratics are harmonic to the Jacobian (f, f'), and accordingly we have the theorem that the real foci of any conic inscribed in a quadrilateral are harmonically concyclic to a fixed pair of points J_1, J_2.

We leave the reader to prove that if T_1, T_2 be harmonically concyclic to S_1 and S_2, then the points of contact of the tangents drawn from T_1 and T_2 to any conic whose foci are S_1, S_2 lie on a circle through T_1 and T_2. Hence if tangents T_1P_1, T_1Q_1 be drawn to any one of these conics, the circles $T_1P_1Q_1$ all pass through another fixed point.

Thus the points J_1, J_2 are such that if tangents be drawn from them to any confocal to a conic inscribed in the quadrilateral, then the points of contact lie on a circle through J_1 and J_2.

176. *The binary cubic.*

Suppose the form is

$$az^3 + 3bz^2 + 3cz + d = 0,$$

and that the points A, B, C representing it are z_1, z_2, z_3.

The cubic covariant ϕ represents three points A', B', C' on the circle ABC such that A, A' are harmonically conjugate to B, C and so on. For these three points must be represented by a cubic covariant which is not f and therefore must be ϕ.

We have next to interpret the Hessian.

Let H_1, H_2 be the points representing h_1 and h_2 the roots of the Hessian, then we know that for real variables the range $\{ABCH_1\}$ is equianharmonic,

$$i.e. \quad \frac{(z_1 - z_2)(z_3 - h_1)}{1} = \frac{(z_2 - z_3)(z_1 - h_1)}{\omega} = \frac{(z_3 - z_1)(z_2 - h_1)}{\omega^2}.$$

Hence since mod $(\omega) = 1$, we have on equating moduli

$$AB \,.\, CH_1 = BC \,.\, AH_1 = CA \,.\, BH_1,$$

therefore the points H_1, H_2 are the points whose distances from the vertices are inversely proportional to the opposite sides.

To construct them we draw a circle having BC for inverse points, and passing through A, with analogous circles for CA and AB; then these three circles meet in the points H_1 and H_2.

It follows in addition that H_1 and H_2 are inverse points with respect to the circle ABC.

It is interesting to notice that the Hessian points of $A'B'C'$ are the same as those of ABC, a fact which gives rise to a curious geometrical theorem.

Another easily proved property of the Hessian points is that if H_1L, H_1M, H_1N be drawn perpendicular to the sides, then the triangle LMN is equilateral.

CHAPTER XI.

APOLARITY AND ELEMENTARY GEOMETRY OF RATIONAL CURVES.

177. Two binary forms of the same order are said to be *apolar* when the joint invariant which is linear in the coefficients of both is zero.

Suppose that the two forms in question are

$$f \equiv a_0 x_1^n + n a_1 x_1^{n-1} x_2 + \dots + a_n x_2^n = a_x^n,$$

$$\phi \equiv b_0 x_1^n + n b_1 x_1^{n-1} x_2 + \dots + b_n x_2^n = b_x^n,$$

then the only lineo-linear invariant is

$$(ab)^n = a_0 b_n - n a_1 b_{n-1} + \dots + (-1)^n a_n b_0$$

and this vanishes when the forms are apolar. An immediate consequence is that a form of odd order is always apolar to itself.

Thus the discussion of apolar forms may be regarded as the development of the theory of the simplest type of invariant; the fact that each set of coefficients only occur to the first degree in the invariant renders such a discussion simple and accounts for the relative importance of the allied geometrical properties. If two quadratic forms are apolar they are harmonic so that we may regard apolarity as being, in a certain way, the generalisation of harmonic properties.

The condition for apolarity may be written in other useful forms.

In fact, if the linear factors of ϕ are $\beta_x^{(1)}, \beta_x^{(2)}, \dots \beta_x^{(n)}$, we have

$$0 = \{a_x^n, \beta_x^{(1)} \beta_x^{(2)} \dots \beta_x^{(n)}\}^n = (a\beta^{(1)})(a\beta^{(2)}) \dots (a\beta^{(n)}).$$

Again, if ϕ vanishes for the values

$$x_1 = y_1^{(r)},\ x_2 = y_2^{(r)},\ r = 1, 2, \ldots n,$$

we have as far as ratios are concerned

$$\beta_1^{(r)} = -y_2^{(r)},\ \beta_2^{(r)} = +y_1^{(r)},$$

and hence the condition is

$$a_{y^{(1)}}\, a_{y^{(2)}} \ldots a_{y^{(n)}} = 0.$$

This form of the condition at once shews the connection with polar forms; given the form f, $n-1$ of the vanishing points of ϕ can be chosen arbitrarily and the remaining one is given by polarizing f with respect to each of the $n-1$ given values successively and equating the final result to zero.

We can at once find all perfect nth powers apolar to f, for if all the y's are the same we have

$$a_y^n = 0,$$

so that y must be a vanishing point of f. Hence f is apolar to the nth power of any one of its linear factors, and these determine the only nth powers apolar to f.

If f be apolar to each of the forms $\phi_1, \phi_2, \ldots \phi_n$ it is apolar to $\lambda_1\phi_1 + \lambda_2\phi_2 + \ldots + \lambda_k\phi_k$, where the λ's are any constants; for

$$\{f, \lambda_1\phi_1 + \lambda_2\phi_2 + \ldots + \lambda_k\phi_k\}^n = \lambda_1(f\phi_1)^n + \lambda_2(f\phi_2)^n + \ldots + \lambda_k(f\phi_k)^n = 0,$$

which establishes the result.

This result also follows at once from the fact that the equations of condition are all linear.

Again if f be apolar to each of the $(n+1)$ forms

$$\phi_1, \phi_2, \ldots \phi_{n+1},$$

then on elimination of the a's from the equations which they satisfy, we find that the determinant of the coefficients of the ϕ's is zero; but this is precisely the condition that there should be an identical relation of the form

$$\lambda_1\phi_1 + \lambda_2\phi_2 + \ldots + \lambda_{n+1}\phi_{n+1} = 0 \ldots\ldots\ldots\ldots\ldots\ldots\text{(A)},$$

and hence being given n linearly independent forms apolar to f any other apolar form is a linear combination of these. It is easy to prove directly that n linearly independent forms can be found which are apolar to a given form of order n, and this fact will appear independently from the special system of apolar forms to be constructed in the next article.

178. *To determine n linearly independent forms apolar to a given one.*

I. Suppose that the factors of the given form f are all different and that except for numerical multiples they are

$$\alpha_1^{(r)}x_1 + \alpha_2^{(r)}x_2, \quad r = 1, 2, \ldots n,$$

then the nth power of each factor is an apolar form, and they are linearly independent.

For if there be a relation of the type

$$\sum_1^n \lambda_r (\alpha_1^{(r)}x_1 + \alpha_2^{(r)}x_2)^n = 0,$$

where λ_r is a constant, then the $(n+1)$ determinants of the array

$$\left| \alpha_1^{(r)n}, \; \alpha_1^{(r)n-1}\alpha_2^{(r)}, \; \alpha_1^{(r)n-2}\alpha_2^{(r)}, \ldots \alpha_2^{(r)n} \right| \quad r = 1, 2, \ldots n,$$

vanish identically.

Hence the determinant

$$\begin{vmatrix} \alpha_1^{(1)n}, & \alpha_1^{(1)n-1}\alpha_2^{(1)}, & \alpha_1^{(1)n-2}\alpha_2^{(1)2}, & \ldots \alpha_2^{(1)n} \\ \cdots & \cdots & \cdots & \cdots \\ \cdots & \cdots & \cdots & \cdots \\ \alpha_1^{(n)n}, & \alpha_1^{(n)n-1}\alpha_2^{(n)}, & \alpha_1^{(n)n-2}\alpha_2^{(n)2}, & \ldots \alpha_2^{(n)n} \\ p_1^n, & p_1^{n-1}p_2, & p_1^{n-2}p_2^2, & \ldots p_2^n \end{vmatrix}$$

vanishes for all values of p_1 and p_2.

But this determinant being homogeneous in each set of symbols is equal to

$$\pm \Pi \left(\alpha_1^{(r)}\alpha_2^{(s)} - \alpha_1^{(s)}\alpha_2^{(r)}\right) \prod_{r=1}^{r=n} \left(\alpha_1^{(r)}p_2 - \alpha_2^{(r)}p_1\right),$$

where in the first product r and s have all the values $1, 2, \ldots n$, but are different.

Thus if we choose p_1, p_2 such that $\dfrac{p_1}{p_2} \neq \dfrac{\alpha_1^{(r)}}{\alpha_2^{(r)}}$, $r = 1, 2, \ldots n$, the determinant can only vanish when for some pair of values of r and s

$$\alpha_1^{(r)}\alpha_2^{(s)} = \alpha_2^{(r)}\alpha_1^{(s)},$$

in which case the two factors $\alpha_1^{(r)}x_1 + \alpha_2^{(r)}x_2$ and $\alpha_1^{(s)}x_1 + \alpha_2^{(s)}x_2$ only differ by a numerical multiplier, contrary to the hypothesis that the factors are all different.

Consequently the n forms

$$(\alpha_1^{(r)}x_1 + \alpha_2^{(r)}x_2)^n, \quad r = 1, 2, \dots n,$$

are linearly independent when the factors of f are all different.

We are thus led at once to the interesting result that the necessary and sufficient condition that a form ϕ can be represented as a linear combination of the nth powers of the factors of f is that the two forms should be apolar—the condition is necessary, because as each nth power is apolar to f a linear combination of them is also, and it is sufficient, because the nth powers are linearly independent in virtue of the foregoing, supposing always that f has no multiple factors.

Hence reciprocally, if ϕ have no multiple factors, f can be expressed as a linear combination of the nth powers of the factors of ϕ.

This may be regarded as the extension of the elementary theorem that all quadratics harmonic to $ax^2 + 2bx + c$ are of the form $\lambda(x-\alpha)^2 + \mu(x-\beta)^2$, where α, β are the roots of

$$ax^2 + 2bx + c = 0.$$

II. We have still to construct the apolar system when f has multiple factors.

Suppose that f has the factor $(\alpha_1 x_1 + \alpha_2 x_2)^r$ replacing r different linear factors, and that ϕ is apolar to f. Since the relation is invariantive, we may change the variables so that the multiple factor is simply x_2^r, and then

$$a_0, \; a_1, \; a_2, \dots a_{r-1} \text{ all vanish.}$$

Hence recalling the condition

$$a_0 b_n - n a_1 b_{n-1} + \dots + (-1)^n a_n b_0 = 0,$$

we see that it is satisfied for any values of

$$b_n, \; b_{n-1}, \dots b_{n-r+1},$$

provided that all the other b's vanish.

Thus f is now apolar to any form for which

$$b_0, b_1, b_2, \dots b_{n-r}$$

are all zero, that is to any form which contains the factor x_2^{n-r+1}. Among such we have the r forms

$$x_2^{n-r+1}x_1^{r-1},\ x_2^{n-r+1}x_1^{r-2}x_2, \ldots\ x_2^{n-r+1}x_2^{r-1},$$

which are obviously linearly independent.

Hence, in general, when a form has a linear factor of multiplicity r it is apolar to any form containing that factor $(n-r+1)$ times, and among these apolar forms it is possible to choose r which are linearly independent. If each multiple factor be treated in the same way we obtain in all n apolar forms, viz. r from each factor of multiplicity r; those derived from the same factor are linearly independent; it remains to shew that all the n forms are so independent.

Thus for the sextic $x_1^3(x_1+x_2)^2 x_2$ we have six apolar forms in three sets

$$x_1^6,\ x_1^5x_2,\ x_1^4x_2^2 \text{ containing the factor } x_1^4,$$

$$(x_1+x_2)^5x_1, (x_1+x_2)^5x_2 \text{ containing the factor } (x_1+x_2)^5,$$

$$\text{and } x_2^6 \text{ derived from the single factor } x_2.$$

The forms in each set are linearly independent, but it has not been shewn that the different sets are independent *inter se.*

The general proof that the sets so derived are independent presents no difficulty but is rather tedious owing to the complicated notation. (See Appendix II.)

Let $$f = A(x_1+\alpha_1 x_2)^{r_1}(x_1+\alpha_2 x_2)^{r_2}\ldots(x_1+\alpha_p x_2)^{r_p},$$

the α's being all different, $r_1+r_2+\ldots+r_p=n$, and A a constant. This assumes that in f the coefficient of x_1^n is not zero, if it be zero we can transform f by a linear substitution into one in which x_1^n actually occurs.

A form apolar to f is

$$\chi_1 = (x_1+\alpha_1 x_2)^n,$$

and $$\frac{\partial\chi_1}{\partial\alpha_1},\ \frac{\partial^2\chi_1}{\partial\alpha_1^2},\ \ldots\ \frac{\partial^{r_1-1}\chi_1}{\partial\alpha_1^{r_1-1}}$$

are all apolar forms since each involves the factor $(x_1+\alpha_1 x_2)^{n-r_1+1}$. We shall prove that this set and the corresponding ones derived from the other factors are linearly independent.

In fact if they are not independent the determinant

$$\begin{vmatrix} f_0(\alpha_1), & f_1(\alpha_1), & \dots & f_n(\alpha_1) \\ f_0'(\alpha_1), & f_1'(\alpha_1), & \dots & f_n'(\alpha_1) \\ \dots & \dots & \dots & \dots \\ f_0^{r_1-1}(\alpha_1), & f_1^{r_1-1}(\alpha_1), & \dots & f_n^{r_1-1}(\alpha_1) \\ \dots & \dots & \dots & \dots \\ f_0^{r_p-1}(\alpha_p), & \dots & \dots & f_n^{r_p-1}(\alpha_p) \\ f_0(q), & f_1(q), & \dots & f_n(q) \end{vmatrix},$$

where in general $$f_s(\alpha_r) = \alpha_r^s$$

and $$f_s^t(\alpha_r) = \frac{\partial^t \alpha_r^s}{\partial \alpha_r^t},$$

must vanish for all values of q.

Except for a non-vanishing numerical multiplier the above determinant is the limit of

$$\begin{vmatrix} f_0(\alpha_1), & f_1(\alpha_1), & \dots f_n(\alpha_1) \\ f_0(\alpha_1+t), & f_1(\alpha_1+t), & \dots f_n(\alpha_1+t) \\ \dots & \dots & \dots \\ f_0(\alpha_1+\overline{r_1-1}\,t), & f_1(\alpha_1+\overline{r_1-1}\,t), & \dots f_n(\alpha_1+\overline{r_1-1}\,t) \\ \dots & \dots & \dots \\ \dots & \dots & \dots \\ f_0(\alpha_p+\overline{r_p-1}\,t), & f_1(\alpha_p+\overline{r_p-1}\,t), & \dots f_n(\alpha_p+\overline{r_p-1}\,t) \\ f_0(q), & f_1(q), & \dots f_n(q) \end{vmatrix}$$

$$\div\, t^{(1+2+\dots+\overline{r_1-1})+(1+2+\dots+\overline{r_2-1})+\dots(1+2+\dots+\overline{r_p-1})},$$

when $t = 0$.

But the determinant last written is equal to

$$\pm\,\Pi\,(\alpha_\rho + \rho' t - \alpha_\sigma - \sigma' t),$$
$$\times\,\Pi\,(\alpha_\varpi + \varpi' t - q),$$

where in the first product $\rho = 1, 2, \dots p$,

$$\rho' = 0, 1, 2, \dots r_\rho - 1,$$
$$\sigma = 1, 2, \dots p,$$
$$\sigma' = 0, 1, 2, \dots r_\sigma - 1,$$

except that one of the inequalities $\rho \neq \sigma$, $\rho' \neq \sigma'$ must be satisfied, and in the second product

$$\varpi = 1, 2, \dots p,$$
$$\varpi' = 0, 1, \dots r_\varpi - 1.$$

A linear factor is a multiple of t when $\rho = \sigma$, hence t occurs as a factor of the determinant to a power equal to

$$\sum_{\rho} \frac{r_\rho (r_\rho - 1)}{2}, \quad \rho = 1, 2, \ldots p,$$

and this is precisely the power of t in the denominator.

To find the limit we put $t = 0$ when $\rho \neq \sigma$ and take the multiplier of t in the remaining factors.

Besides numerical factors which are certainly not zero, the limit is the product of a number of factors of the types $\alpha_\rho - \alpha_\sigma$ and $\alpha_\rho - q$, and in fact it is easily seen to be

$$N \prod_{\rho \neq \sigma} (\alpha_\rho - \alpha_\sigma)^{r_\rho r_\sigma} \prod_{\rho} (\alpha_\rho - q)^{r_\rho},$$

where N is an integer.

Now the quantities $\alpha_1, \alpha_2, \ldots \alpha_p$ are all unequal and we can choose q to be different from each of them, hence the determinant does not vanish for all values of q, and consequently the n apolar forms are linearly independent.

Ex. Evaluate the determinant

$$\begin{vmatrix} a^5, & a^4, & a^3, & a^2, & a, & 1 \\ 5a^4, & 4a^3, & 3a^2, & 2a, & 1, & 0 \\ 20a^3, & 12a^2, & 6a, & 2, & 0, & 0 \\ \beta^5, & \beta^4, & \beta^3, & \beta^2, & \beta, & 1 \\ 5\beta^4, & 4\beta^3, & 3\beta^2, & 2\beta, & 1, & 0 \\ \gamma^5, & \gamma^4, & \gamma^3, & \gamma^2, & \gamma, & 1 \end{vmatrix}.$$

179. *Forms apolar to two or more given forms.*

Suppose we have s linearly independent forms

$$f_r \equiv a_0^{(r)} x_1^n + n a_1^{(r)} x_1^{n-1} x_2 + \ldots + a_n^{(r)} x_2^n, \quad r = 1, 2, \ldots s,$$

then the determinants of the array formed by the coefficients cannot all vanish, because in that case values of λ which satisfy $s - 1$ of the $(n + 1)$ equations

$$\lambda_1 a_p^{(1)} + \lambda_2 a_p^{(2)} + \ldots + \lambda_s a_p^{(s)} = 0$$

would satisfy all these equations, and hence

$$\lambda_1 f_1 + \lambda_2 f_2 + \ldots + \lambda_s f_s = 0,$$

which is contrary to hypothesis.

If
$$b_0x_1^n + nb_1x_1^{n-1}x_2 + \dots + b_nx_2^n$$
be a form apolar to each of the f's, then

$$a_0^{(r)}b_n - na_1^{(r)}b_{n-1} + \frac{n(n-1)}{2}a_2^{(r)}b_{n-2} + \dots + (-1)^n a_n^{(r)}b_0 = 0,$$

$$r = 1, 2, \dots s.$$

These s equations connecting the b's can be solved for s of the b's in terms of the others, because, as we have seen, not all the determinants of s columns formed by the coefficients are zero. Having solved the equations we obtain s of the b's expressed as linear functions of the others, and as the remaining $(n-s+1)$ b's are arbitrary, the general form apolar to all the f's involves $(n-s+1)$ constants linearly, *i.e.* there are exactly $(n-s+1)$ linearly independent forms apolar to each of s given forms.

In particular it follows that there is a unique form apolar to each of n given linearly independent forms.

Further if
$$\phi_1, \phi_2, \dots \phi_{n-s+1}$$
be $(n-s+1)$ linearly independent forms apolar to the f's, it is clear that every ϕ is apolar to every f, and that the most general form apolar to each of the ϕ's is a linear combination of the f's; thus the relation between the two sets of forms is a reciprocal one.

Some interesting results follow from this reasoning. For example, given three independent cubics, there is one form apolar to each of them, and, if its factors be all different, each of the given cubics can be expressed as a linear combination of the same three cubes, viz. the cubes of the factors of the apolar cubic. A like result applies to forms of any order and constitutes the generalisation of the problem of expressing two quadratics each as linear combinations of the same two squares.

The form apolar to the three cubics a_x^3, b_x^3, c_x^3, is
$$(bc)(ca)(ab)\,a_x\,b_x\,c_x,$$
for this is not zero if the cubics are linearly independent and it is apolar to d_x^3 if
$$(bc)(ca)(ab)(ad)(bd)(cd) = 0,$$

or, as can be readily seen, if

$$\begin{vmatrix} a_0 & a_1 & a_2 & a_3 \\ b_0 & b_1 & b_2 & b_3 \\ c_0 & c_1 & c_2 & c_3 \\ d_0 & d_1 & d_2 & d_3 \end{vmatrix} = 0,$$

which is certainly true if d_x^3 is the same as any of the three original cubics.

The reader may verify that the equation of the apolar form may be written

$$\begin{vmatrix} a_0x_1 + a_1x_2, & a_1x_1 + a_2x_2, & a_2x_1 + a_3x_2 \\ b_0x_1 + b_1x_2, & b_1x_1 + b_2x_2, & b_2x_1 + b_3x_2 \\ c_0x_1 + c_1x_2, & c_1x_1 + c_2x_2, & c_2x_1 + c_3x_2 \end{vmatrix} = 0,$$

which can be easily done by expressing this determinant symbolically.

The extension of these results to n forms of order n will present no difficulty.

180. We may illustrate some of the foregoing results and anticipate some of the developments to come by reference to the geometry of the rational plane cubic curve.

Let ξ, η, ζ be homogeneous coordinates, then for all points on the curve they are rational integral functions of one parameter t and of the third order. To apply our results more directly we shall replace t by two variables which occur homogeneously, so that we have

$$\left.\begin{aligned} \rho\xi &= a_0x_1^3 + 3a_1x_1^2x_2 + 3a_2x_1x_2^2 + a_3x_2^3 \equiv f_1 \\ \rho\eta &= b_0x_1^3 + 3b_1x_1^2x_2 + 3b_2x_1x_2^2 + b_3x_2^3 \equiv f_2 \\ \rho\zeta &= c_0x_1^3 + 3c_1x_1^2x_2 + 3c_2x_1x_2^2 + c_3x_2^3 \equiv f_3 \end{aligned}\right\},$$

and to find the point equation of the curve we must eliminate x_1 and x_2 so as to obtain a result homogeneous in ξ, η, ζ. But the properties of the curve are naturally more easily obtained by using the parametric expressions.

We remark in the first place that the cubics f_1, f_2, f_3 must be linearly independent, otherwise all such points ξ, η, ζ lie on a straight line; next the points in which the line

$$l\xi + m\eta + n\zeta = 0$$

meets the curve are given by the cubic

$$lf_1 + mf_2 + nf_3 = 0,$$

which determines their parameters; hence any straight line meets the curve in three points and the curve is therefore of the third order.

Now there is a unique cubic ϕ apolar to f_1, f_2, f_3, and ϕ is apolar to any cubic giving the parameters of three collinear points. Conversely if a cubic be apolar to ϕ the three points whose parameters are determined by it are collinear because it is of the form $lf_1 + mf_2 + nf_3$.

Thus the three points whose parameters are (x_1, x_2), (y_1, y_2), (z_1, z_2) are in a straight line if

$$\phi_x \phi_y \phi_z = 0.$$

181. *Points of Inflexion.* At a point of inflexion three successive points on the curve are in a straight line, and hence the parameters of the points of inflexion are determined by the perfect cubes apolar to ϕ, that is by

$$\phi = 0.$$

Hence there are at most three points of inflexion, and, since a cubic is apolar to itself, when there are three they are collinear.

But there are other singularities for which three consecutive points on the curve are collinear, *e.g.* cusps, and accordingly we shall examine the equation $\phi = 0$ a little more closely.

If $\alpha_x, \beta_x, \gamma_x$ be the linear factors of ϕ we have

$$\xi = \lambda \alpha_x^3 + \mu \beta_x^3 + \nu \gamma_x^3,$$

with similar expressions for η and ζ; hence on replacing ξ, η, ζ by suitable linear combinations—which is tantamount to changing the triangle of reference—we shall have

$$\xi = \alpha_x^3, \ \eta = \beta_x^3, \ \zeta = \gamma_x^3,$$

from which it is readily seen that the straight lines $\xi = 0$, $\eta = 0$, $\zeta = 0$ are inflexional tangents, and therefore in this case there are three distinct points of inflexion.

If $\phi_x^3 = 0$ has a double factor β_x we have

$$\xi = \lambda \alpha_x^3 + \mu \beta_x^3 + \nu \beta_x^2 x_1 \text{ etc.},$$

and hence by a similar transformation we can reduce ξ, η, ζ to the forms

$$\xi = \alpha_x^3, \quad \eta = \beta_x^3, \quad \zeta = \beta_x^2 x_1.$$

In the neighbourhood of the point $\eta = 0$, $\zeta = 0$, whose parameter is given by $\beta_x = 0$, we see that $\eta^2 \propto \zeta^3$, and hence this point is a cusp, while the line $\xi = 0$ is an inflexional tangent. The case in which $\phi = 0$ has a treble factor may be rejected because under these conditions f_1, f_2, f_3 have a common factor and, as will be readily seen, the curve breaks up into a straight line and a conic.

We shall confine the further discussion to the case in which the factors of ϕ are distinct.

182. *Double point.* To each value of the ratio $x_1 : x_2$ corresponds a point on the curve. The same ratio cannot give rise to two different points, but the same point may be obtained from two different values of the ratio and then it will be a double point on the curve, because every straight line through it meets the curve in only one other point. We might find the double points directly by developing this idea, but the search is best conducted in a different manner.

If x, y, z be the parameters of three collinear points we have

$$\phi_x \phi_y \phi_z = 0,$$

and in general this determines z uniquely when x and y are given. When x and y give rise to the same point, and only then, the above condition does not determine z but is satisfied for all values of z.

Now the equation $\qquad \phi_x \phi_y \phi_z = 0$

indicates that the quadratic whose vanishing points are x, y is apolar to

$$\phi_x^2 \phi_z,$$

the first polar of z with respect to ϕ_x^3.

Hence if x, y be the parameters of a double point the quadratic giving them must be apolar to all first polars of ϕ_x^3.

Two such polars are

$$\phi_x^2 \phi_\alpha, \quad \phi'^2_x \phi'_\beta$$

and the quadratic apolar to each of these is their Jacobian,

i.e. $$(\phi\phi')\,\phi_x\phi'_x\,\phi_\alpha\phi'_\beta = -(\phi\phi')\,\phi_x\phi'_x\,\phi'_\alpha\phi_\beta,$$

ϕ and ϕ' being equivalent symbols.

This is

$$\tfrac{1}{2}(\phi\phi')\,\phi_x\phi'_x\,(\phi_\alpha\phi'_\beta - \phi'_\alpha\phi_\beta) = \tfrac{1}{2}(\phi\phi')\,\phi_x\phi'_x\,(\phi\phi')\,(\alpha\beta)$$

and the quadratic required is therefore

$$(\phi\phi')^2\,\phi_x\phi'_x = 0,$$

namely the Hessian of ϕ.

Thus a rational cubic curve has always one double point, the parameters of which are given by the Hessian of the cubic giving the parameters of the points of inflexion.

It will be noticed how readily properties of points on the curve are expressed by means of the form ϕ, and this is natural since ϕ being given we can write down three forms apolar to it for f_1, f_2, f_3 and thence find the ordinary equation of the curve.

As a further example, we remark that the points of contact of the tangents drawn from the point z on the curve to the curve are given by

$$\phi_x^2\phi_z = 0,$$

that is to say there are two such tangents, and the parameters of the points of contact are given by the first polar of z. Hence the quadratic giving them is apolar to the quadratic giving the parameters of the double points.

Ex. (i). Prove that the cross ratio of the pencil formed by joining the double point to four points on the curve is equal to the cross ratio of the parameters of those four points.

Ex. (ii). If the parameters of the points of inflexion be given by $a_x = 0$, $\beta_x = 0$, $\gamma_x = 0$, the point of contact of the remaining tangent to the curve from the first is given by

$$(a\beta)\,\gamma_x + (a\gamma)\,\beta_x = 0\,;$$

hence if $f = 0$ give the points of inflexion, the cubic covariant gives the parameters of the three points L, M, N, here indicated.

Ex. (iii). Prove that the six points in which any conic meets the cubic are given by a sextic apolar to a given sextic ψ. (ψ is apolar to the squares and products f_1, f_2, f_3.) Thence shew that $\psi = 0$ gives the parameters of the points where a conic can be drawn having six-point contact with the curve, and that inasmuch as these points are the points of inflexion and L, M, N, ψ is the product of f and its cubic covariant.

183. *Apolarity of forms of different orders.* In the foregoing discussion we have seen that if x and y be the vanishing points of the Hessian of a cubic f, then

$$a_x a_y a_z = 0$$

for all values of z, so that the Hessian although only a quadratic satisfies, in a manner, the condition of apolarity to the cubic.

If α_x, β_x be the factors of the Hessian, we have

$$(a\alpha)(a\beta)\,a_x = 0,$$

that is its second transvectant with f vanishes identically (cf. § 91).

Generalising this we shall call two forms

$$f = a_x^m,\ \phi = b_x^n,\quad m > n$$

apolar when the nth transvectant

$$(ab)^n\,a_x^{m-n}$$

vanishes identically.

Two important facts follow at once from this definition.

(i) *The form f is apolar to any form of order $n' \leqslant m$, having ϕ for factor.*

For if the new form be $\phi\psi$ we have

$$(f, \phi\psi)^{n'} = ((f, \phi)^n\ \psi)^{n'-n},$$

since ψ is of order $n' - n$ and the right-hand side vanishes by hypothesis.

A special result is that f is apolar to any m-ic containing the factor ϕ.

(ii) *ϕ is apolar to any polar form of f whose order is n.*

For let $$a_x^n\,a_y^{m-n}$$

be the apolar form, then

$$\{a_x^n a_y^{m-n},\ b_x^n\}^n = (ab)^n\,a_y^{m-n},$$

and this is zero since $$(ab)^n\,a_x^{m-n}$$

vanishes identically.

Cor. ϕ is apolar to any polar form of f whose order is $\geqslant n$ but $\leqslant m$.

The proof is as above.

184. The conditions of apolarity of

$$f = (a_0, a_1, \ldots a_m \between x_1, x_2)^m = a_x^m$$

and

$$\phi = (b_0, b_1, \ldots b_n \between x_1, x_2)^n = b_x^n$$

are equivalent to $m - n + 1$ linear homogeneous relations among the a's.

In fact, equating to zero the several coefficients in

$$(ab)^n a_x^{m-n}$$

we have the following equations:

$$(ab)^n a_1^{m-n} = 0, \ (ab)^n a_1^{m-n-1} a_2 = 0, \ \ldots (ab)^n a_2^{m-n} = 0.$$

On being expressed in terms of the actual coefficients the first of these relations involves

$$a_0, a_1, \ldots a_n,$$

the second

$$a_1, a_2, \ldots a_{n+1},$$

and so on, the last containing

$$a_{m-n}, a_{m-n+1}, \ldots a_n,$$

hence if, as without loss of generality we may do, we assume that none of the coefficients b are zero, these

$$m - n + 1$$

relations among the a's are obviously linearly independent.

It follows that by means of them we can express

$$(m - n + 1)$$

of the coefficients a linearly in terms of the remaining coefficients and thence that there are

$$(m + 1) - (m - n + 1) = n$$

linearly independent forms of order m apolar to any given form of order n less than m.

185. *Construction of a linearly independent set of forms of order m ($> n$) apolar to a given form ϕ of order n.*

I. Let the factors of the given form be

$$\beta_1^{(r)} x_1 + \beta_2^{(r)} x_2 = \beta_x^{(r)}, \quad r = 1, 2, \ldots n,$$

and all different.

Then the typical form β_x^m is apolar to b_x^n, for

$$(\beta_x^m, b_x^n)^n = (\beta b)^n \beta_x^{m-n}$$

and since
$$(\beta b)^n = 0,$$
when β_x is a factor of b_x^n, the result follows at once.

Next the system of forms

$$\beta_x^{(r)m}, \quad r = 1, 2, \dots n$$

is linearly independent, for if there were a relation of the type

$$\Sigma\lambda_r \beta_x^{(r)m} = 0,$$

on polarizing $(m-n)$ times with respect to y, we should have

$$\Sigma\lambda_r \beta_y^{(r)m-n} \beta_x^{(r)n} = 0,$$

which is contrary to the established fact that the system of forms

$$\beta_x^{(r)n}, \quad r = 1, 2, \dots n$$

is linearly independent.

II. Suppose next that the factors are not all different but that

$$\phi = \beta_x^{(1)\mu_1} \beta_x^{(2)\mu_2} \dots \beta_x^{(s)\mu_s}$$

Then since the factor $\quad \beta_x^{(1)}$

for example occurs μ_1 times in ϕ we know that ϕ is apolar to the n-ic

$$\beta_x^{(1)n-\mu_1+1} C,$$

where C is any form of order $\mu_1 - 1$.

In like manner ϕ is apolar to the m-ic

$$\beta_x^{(1)m-\mu_1+1} \Gamma,$$

Γ being any form of order $\mu_1 - 1$, for

$$(\beta_x^{(1)m-\mu_1+1} \Gamma, \phi)^n$$
$$= (\beta_x^{(1)m-\mu_1+1} \Gamma, b_x^n)^n$$

and this latter is an aggregate of forms each involving the factor

$$(b\beta^{(1)})^{n-\mu_1+1}.$$

But since the factor $\beta_x^{(1)}$ occurs μ_1 times in ϕ any form involving the factor

$$(b\beta^{(1)})^{n-\mu_1+1}$$

vanishes identically.

Hence choosing any s forms

$$\Gamma_1, \Gamma_2, \ldots \Gamma_s$$

of orders $\mu_1 - 1,\ \mu_2 - 1,\ \ldots\ \mu_s - 1$

respectively, forms $\beta_x^{(t)\,m-\mu_t+1}\Gamma_t$

are all apolar to ϕ.

Next there cannot be a linear relation between them because if there were, on polarizing it $m-n$ times with respect to y we should obtain a linear relation of the type

$$\Sigma\lambda_r\beta_x^{(t)\,n-\mu_t+1}\,\Gamma_t' = 0,$$

Γ_t' being of order $\mu_t - 1$. This is contrary to what was proved in constructing the apolar set of order n.

Now the form $\beta_x^{(t)\,m-\mu_t+1}\Gamma_t$

contains μ_t arbitrary constants and so we have a form apolar to ϕ involving

$$\mu_1 + \mu_2 + \ldots + \mu_s = n$$

arbitrary constants.

The coefficients of the various constants are each apolar to ϕ and they are n in number.

The discovery of forms of order n apolar to a form of order m $(> n)$ is a problem quite distinct from the foregoing.

Suppose f is the given form of order m, then a form of order n which is apolar to

$$a_x^n a_y^{m-n}$$

for all values of y will be apolar to f.

This condition has been shewn to be necessary and it is clearly sufficient because if

$$\phi = b_x^n,$$

then $(ab)^n\, a_y^{m-n}$

vanishes for all values of y.

Hence the form ϕ sought is apolar to the $m-n+1$ forms of order n

$$a_x^n a_1^{m-n},\ a_x^n a_1^{m-n-1} a_2,\ \ldots\ a_x^n a_2^{m-n},$$

and the problem is reduced to one in apolar forms of the same order.

If the forms just written down be linearly independent there are

$$n+1-(m-n+1)$$

linearly independent forms apolar to each of them, and thus there are at least $(2n-m)$ linearly independent forms of order n apolar to a given form of order m. There may be more owing to the subsidiary system of forms not being linearly independent and we shall discuss this question more fully in the sequel.

It is clear that if $n>\frac{m}{2}$ there is at least one form of order n apolar to f.

186. The latter theory finds its natural illustration in the problem of representing one or more given forms of order n as the sum of a number of perfect nth powers. We shall discuss the case of a single quintic at length as an example.

The general cases present no difficulties—they arise for forms of special character.

Suppose the quintic is

$$f=a_0x_1^5+5a_1x_1^4x_2+10a_2x_1^3x_2^2+10a_3x_1^2x_2^3+5a_4x_1x_2^4+a_5x_2^5$$
$$\equiv a_x^5=b_x^5=\ldots.$$

The second polars are linear combinations of

$$\frac{\partial^2 f}{\partial x_1^2},\quad \frac{\partial^2 f}{\partial x_1\partial x_2},\quad \frac{\partial^2 f}{\partial x_2^2}.$$

If these are linearly independent there is a unique cubic apolar to them, and being apolar to all second polars it is apolar to the quintic itself.

On referring to § 179 we may write this cubic in the form

$$\begin{vmatrix} \frac{\partial^4 f}{\partial x_1^4}, & \frac{\partial^4 f}{\partial x_1^3\partial x_2}, & \frac{\partial^4 f}{\partial x_1^2\partial x_2^2} \\ \frac{\partial^4 f}{\partial x_1^3\partial x_2}, & \frac{\partial^4 f}{\partial x_1^2\partial x_2^2}, & \frac{\partial^4 f}{\partial x_1\partial x_2^3} \\ \frac{\partial^4 f}{\partial x_1^2\partial x_2^2}, & \frac{\partial^4 f}{\partial x_1\partial x_2^3}, & \frac{\partial^4 f}{\partial x_2^4} \end{vmatrix}$$

or

$$(bc)^2(ca)^2(ab)^2a_xb_xc_x,$$

so that it is the covariant denoted by j.

Now suppose
$$j = \alpha_x \beta_x \gamma_x,$$
then
$$f = \lambda\alpha_x^5 + \mu\beta_x^5 + \nu\gamma_x^5,$$
$\alpha_x \beta_x \gamma_x$ being all different and λ, μ, ν numerical.

If $j = \alpha_x^2 \beta_x$ we have for an apolar quintic
$$f = \lambda\alpha_x^5 + \mu\alpha_x^4 x_1 + \nu\beta_x^5, \text{ or } \alpha_x^4 (px_1 + qx_2) + r\beta_x^5.$$

If $j = \alpha_x^3$ we have
$$f = \lambda\alpha_x^5 + \mu\alpha_x^4 x_1 + \nu\alpha_x^3 x_1^2, \text{ or } \alpha_x^3 (px_1^2 + 2qx_1x_2 + rx_2^2).$$

This exhausts the cases in which j is not identically zero.

If j be identically zero the forms $\frac{\partial^2 f}{\partial x_1^2}, \frac{\partial^2 f}{\partial x_1 \partial x_2}, \frac{\partial^2 f}{\partial x_2^2}$ are not linearly independent, because if they were they would determine a unique non-zero apolar form.

Let
$$p\frac{\partial^2 f}{\partial x_1^2} + q\frac{\partial^2 f}{\partial x_1 \partial x_2} + r\frac{\partial^2 f}{\partial x_2^2} = 0$$
be the relation.

Then we have
$$p\frac{\partial^3 f}{\partial x_1^3} + q\frac{\partial^3 f}{\partial x_1^2 \partial x_2} + r\frac{\partial^3 f}{\partial x_1 \partial x_2^2} = 0,$$
and
$$p\frac{\partial^3 f}{\partial x_1^2 \partial x_2} + q\frac{\partial^3 f}{\partial x_1 \partial x_2^2} + r\frac{\partial^3 f}{\partial x_2^3} = 0,$$
and all third polars can be expressed as linear combinations of
$$\frac{\partial^3 f}{\partial x_1^2 \partial x_2}, \frac{\partial^3 f}{\partial x_1 \partial x_2^2}.$$

If these are linearly independent they determine a unique quadratic apolar to both and being apolar to all third polars it is apolar to the quintic.

Suppose this quadratic to be

(i) $\alpha_x \beta_x$, then $f = \lambda\alpha_x^5 + \mu\beta_x^5$;

(ii) α_x^2, then $f = \alpha_x^4 (px_1 + qx_2)$;

(iii) identically zero, then
$$\frac{\partial^3 f}{\partial x_1^2 \partial x_2}, \frac{\partial^3 f}{\partial x_1 \partial x_2^2}$$
must be identical, hence in this case all third polars are identical.

Now all fourth polars are linear combinations of

$$\frac{\partial^4 f}{\partial x_1^3 \partial x_2}, \quad \frac{\partial^4 f}{\partial x_1^2 \partial x_2^2} \quad \ldots\ldots\ldots\ldots\ldots\ldots\ldots \text{(A)}.$$

But since $\frac{\partial^3 f}{\partial x_1^2 \partial x_2}$ and $\frac{\partial^3 f}{\partial x_1 \partial x_2^2}$ are identical all the fourth polars are identical with either of the forms (A).

Hence the fourth polar is apolar to the quintic, for being of odd order it is apolar to itself.

In this case the quintic is a perfect fifth power unless the linear apolar form is zero identically, in which case the quintic is also identically zero.

This completes the discussion and leaves us with six canonical forms for a quintic, viz.

(i) $f = \lambda \alpha_x^5 + \mu \beta_x^5 + \nu \gamma_x^5,$

(ii) $f = \lambda \alpha_x^5 + \mu \alpha_x^4 x_1 + \nu \beta_x^5,$

(iii) $f = \lambda \alpha_x^5 + \mu \alpha_x^4 x_1 + \nu \alpha_x^3 x_1^2,$

(iv) $f = \lambda \alpha_x^5 + \mu \beta_x^5,$

(v) $f = \lambda \alpha_x^5 + \mu \alpha_x^4 x_1,$

(vi) $f = \lambda \alpha_x^5.$

The discussion for any other single form can be conducted in an exactly similar manner.

187. The reader will have no difficulty in applying the method explained for the quintic to any binary form; in particular it will be easily seen that, whereas a form of odd order $(2n+1)$ always has at least one apolar form of order $(n+1)$, a form of even order $2n$ has not an apolar form of order n unless the determinant formed by the coefficients of its nth derivatives with respect to x_1 and x_2 be zero.

Thus a form of order $(2n+1)$ can in general be expressed in one way as the sum of $(n+1)$ $(2n+1)$th powers, but a form of order $2n$ cannot be expressed as the sum of n $2n$th powers unless a certain function of the coefficients—manifestly an invariant—be zero. For example, in the sextic

$$f \equiv a_0 x_1^6 + 6a_1 x_1^5 x_2 + \ldots + a_6 x_2^6 \equiv a_x^6 = b_x^6 = c_x^6 = d_x^6,$$

there is an apolar cubic only when

$$\begin{vmatrix} a_0 & a_1 & a_2 & a_3 \\ a_1 & a_2 & a_3 & a_4 \\ a_2 & a_3 & a_4 & a_5 \\ a_3 & a_4 & a_5 & a_6 \end{vmatrix} = 0.$$

The expression of this as a symbolical form is an instructive exercise. There must be a linear relation between

$$\frac{\partial^3 f}{\partial x_1^3}, \quad \frac{\partial^3 f}{\partial x_1^2 \partial x_2}, \quad \frac{\partial^3 f}{\partial x_1 \partial x_2^2}, \quad \frac{\partial^3 f}{\partial x_2^3},$$

i.e. between $\quad a_x^3 a_1^3, \quad b_x^3 b_1^2 b_2, \quad c_x^3 c_1 c_2^2, \quad d_x^3 d_2^3,$

and hence referring to § 179 we must have

$$I = (bc)(ca)(ab)(ad)(bd)(cd)\, a_1^3 b_1^2 b_2 c_1 c_2^2 d_2^3 = 0.$$

Interchanging the letters in every possible way we find that

$$I = \tfrac{1}{24}(bc)(ca)(ab)(ad)(bd)(cd) \begin{vmatrix} a_1^3, & a_1^2 a_2, & a_1 a_2^2, & a_2^3 \\ b_1^3, & b_1^2 b_2, & b_1 b_2^2, & b_2^3 \\ c_1^3, & c_1^2 c_2, & c_1 c_2^2, & c_2^3 \\ d_1^3, & d_1^2 d_2, & d_1 d_2^2, & d_2^3 \end{vmatrix}.$$

And hence the condition is

$$(bc)^2 (ca)^2 (ab)^2 (ad)^2 (bd)^2 (cd)^2 = 0.$$

The invariant I is called the catalecticant and it will be easily seen that a similar symbolical expression holds for the catalecticant of any form of even order.

188. It has been shewn that when

$$j = 0$$

identically the quintic can in general be expressed as the sum of two fifth powers, and in the course of the work we found the conditions under which it can be expressed as the sum of a smaller number of fifth powers. A similar process would of course apply to any form, but we shall now give a direct answer to the question as to what is the smallest number of nth powers in terms of which a given binary n-ic can be expressed*.

* See Gundelfinger, *Crelle*, Bd. c. 413—424.

189. If a binary n-ic can be expressed as the sum of r nth powers it must have an apolar r-ic whose factors are all different, so in the first place we proceed to find the necessary and sufficient conditions that the form should possess an apolar r-ic.

If $r > \frac{n}{2}$ there is always at least one apolar r-ic.

Suppose, then, that $r \not> \frac{n}{2}$ and that there is an apolar r-ic, namely ϕ. Then ϕ is apolar to all derivatives of f whose order is equal to or greater than r, and since there are $(r+1)$ derivatives of order $n-r$, viz.

$$\frac{\partial^r f}{\partial x_1^r}, \quad \frac{\partial^r f}{\partial x_1^{r-1}\partial x_2}, \quad \dots \quad \frac{\partial^r f}{\partial x_2^r},$$

these cannot be linearly independent.

Hence there is an identical relation of the form

$$\lambda_0 \frac{\partial^r f}{\partial x_1^r} + \lambda_1 \frac{\partial^r f}{\partial x_1^{r-1}\partial x_2} + \dots + \lambda_r \frac{\partial^r f}{\partial x_2^r} = 0;$$

on differentiating this r times with respect to x_1 and x_2 in the $(r+1)$ different ways possible and eliminating the λ's, we have

$$\begin{vmatrix} \dfrac{\partial^{2r} f}{\partial x_1^{2r}}, & \dfrac{\partial^{2r} f}{\partial x_1^{2r-1}\partial x_2}, & \dots, & \dfrac{\partial^{2r} f}{\partial x_1^r \partial x_2^r} \\ \dfrac{\partial^{2r} f}{\partial x_1^{2r-1}\partial x_2}, & \dfrac{\partial^{2r} f}{\partial x_1^{2r-2}\partial x_2^2}, & \dots, & \dfrac{\partial^{2r} f}{\partial x_1^{r-1}\partial x_2^{r+1}} \\ \dots & \dots & \dots & \dots \\ \dfrac{\partial^{2r} f}{\partial x_1^r \partial x_2^r}, & \dfrac{\partial^{2r} f}{\partial x_1^{r-1}\partial x_2^{r+1}}, & \dots, & \dfrac{\partial^{2r} f}{\partial x_2^{2r}} \end{vmatrix} = 0,$$

or say $$G_r = 0,$$

and it is easy to see that G_r is a covariant of f.

Conversely when $$G_r = 0$$

by the well-known theorem of Wronski* there is a linear relation between

$$\frac{\partial^r f}{\partial x_1^r}, \quad \frac{\partial^r f}{\partial x_1^{r-1}\partial x_2}, \quad \dots \quad \frac{\partial^r f}{\partial x_2^r}.$$

By differentiating this we obtain two independent relations between the $(r+1)$th derivatives, three between the $(r+2)$th

* See Appendix II.

derivatives and in general $(p+1)$ between the $(r+p)$th derivatives.

Now there are $(r+p+1)$ derivatives of the $(r+p)$th class and hence of these only r are linearly independent.

Hence in particular there are only r linearly independent $(n-r)$th derivatives and as these are of order r there is one form of order r apolar to them and therefore apolar to the form f.

Hence when $G_r = 0$ there is an apolar form of order r.

Thus forming the successive covariants

$$G_0, \; G_1, \; G_2, \; \ldots,$$

the necessary and sufficient condition for an apolar r-ic is $G_r = 0$.

If $G_{r-1} \neq 0$ there is no apolar form of order less than r, for if there were any such apolar forms there would be at least one of order $(r-1)$ and G_{r-1} would vanish.

Hence if G_r be the first of the covariants G which vanishes the lowest order of an apolar form is r.

Finally if $G_{r-1} \neq 0$, $G_r = 0$ there is only one apolar form of order r.

Suppose in fact there are two apolar forms of order r and that for simplicity their factors are all different in both cases.

Let $$(x_1 + \alpha_s x_2), \quad s = 1, 2, \ldots, r$$

be the factors of the first, and

$$(x_1 + \beta_s x_2), \quad s = 1, 2, \ldots, r$$

be the factors of the second, then we have

$$f \equiv \sum_1^r \lambda_s (x_1 + \alpha_s x_2)^n \equiv \sum_1^r \mu_s (x_1 + \beta_s x_2)^n;$$

therefore there is an identical relation of the type

$$\sum_1^r \lambda_s (x_1 + \alpha_s x_2)^n - \sum_1^r \mu_s (x_1 + \beta_s x_2)^n = 0.$$

Now $2r$ is less than n, hence by § 178 such a relation is only possible when the coefficients of the various nth powers severally vanish; thus since the α's are all different and the β's are all different it follows that either every λ and every μ is zero or else for a certain number t of values of s

$$\alpha_s = \beta_s;\ \lambda_s = \mu_s,$$

while for other values of s

$$\alpha_s \neq \beta_s,\ \lambda_s = 0,\ \mu_s = 0.$$

Consequently f can be expressed as the sum of t nth powers where $t < r$, hence there is an apolar form of order t. But in this case we must have $G_{r-1} = 0$ contrary to hypothesis, hence there is only one apolar form of order r.

The reader will easily establish the fact that if there are only two apolar r-ics they must have $(r-1)$ common factors and that these factors multiplied together give the apolar $(r-1)$-ic. Further the extension of the above to the case in which two or more α's are equal will present no difficulty.

Cor. Since G_1 is the Hessian of f the necessary and sufficient condition that f should be a perfect nth power is that its Hessian should vanish identically.

Ex. (i). If f is apolar to ϕ then f is apolar to every form having ϕ for factor.

Ex. (ii). If a binary form of order n have an apolar r-ic $(r < n)$, then it has at least $(s+1)$ independent apolar forms of order $r+s$.

Ex. (iii). Shew that the argument of § 188 can be extended to any number of binary forms, and construct a table of canonical forms of a simultaneous system consisting of a cubic and a quartic.

Ex. (iv). Shew that in general two forms of orders n_1 and n_2 can be expressed as linear combinations of powers of p linear forms if

$$3p - 2 = n_1 + n_2;$$

find the p-ic giving these linear forms and extend the results to any number of forms.

Ex. (v). From the symbolical form of a catalecticant deduce the symbolical forms of the covariants G.

190. We shall conclude this chapter with a few geometrical illustrations of the foregoing theory.

Binary Quadratics in connection with the Geometry of a Conic.

If we have the equations

$$\xi = a_0x_1^2 + 2a_1x_1x_2 + a_2x_2^2 \equiv a_x^2 = f,$$
$$\eta = b_x^2 = \phi,$$
$$\zeta = c_x^2 = \psi,$$

then the point ξ, η, ζ lies on a fixed conic.

The equation of the conic is easy to find. For the line

$$\lambda\xi + \mu\eta + \nu\zeta = 0$$

touches the curve when

$$\text{Disct.}\ (\lambda f + \mu\phi + \nu\psi) = 0,$$

i.e. if $\lambda^2 i_{11} + \mu^2 i_{22} + \nu^2 i_{33} + 2\mu\nu i_{23} + 2\nu\lambda i_{31} + 2\lambda\mu i_{12} = 0,$

where $i_{12} = (f, \phi)^2$ etc.

This being the tangential equation the point equation is

$$\begin{vmatrix} i_{11} & i_{12} & i_{13} & \xi \\ i_{12} & i_{22} & i_{23} & \eta \\ i_{13} & i_{23} & i_{33} & \zeta \\ \xi & \eta & \zeta & 0 \end{vmatrix} = 0,$$

or $I_{11}\xi^2 + I_{22}\eta^2 + I_{33}\zeta^2 + 2I_{23}\eta\zeta + 2I_{31}\zeta\xi + 2I_{12}\xi\eta = 0,$

where I_{rs} is the minor of i_{rs} in the determinant

$$\begin{vmatrix} i_{11} & i_{12} & i_{13} \\ i_{12} & i_{22} & i_{23} \\ i_{13} & i_{23} & i_{33} \end{vmatrix}.$$

In particular the equation of the conic gives an identical relation between three quadratic forms and their invariants.

191. An immediate inference from the parametric expressions for ξ, η, ζ is that the cross-ratio of the pencil joining a variable point P on the conic to four fixed points x, y, z, ω on the curve is equal to the cross-ratio of the parameters of those four points, *i.e.*

$$\frac{(xy)(z\omega)}{(x\omega)(zy)}.$$

For let the equations of two lines through P be $X=0$, $Y=0$ and let $X+tY=0$ be the equation of the line joining P to the point x on the curve. Then there is an algebraic relation connecting t with $\frac{x_1}{x_2}$ and since to one value of t corresponds one value of $\frac{x_1}{x_2}$ and *vice versa* we must have

$$t = \frac{Ax_1 + Bx_2}{Cx_1 + Dx_2},$$

where A, B, C, D are constants.

Hence the cross-ratio of four values of t is equal to the cross-ratio of the corresponding four values of the parameter $\frac{x_1}{x_2}$.

192. In connection with this conic there is a simple correspondence between binary quadratics and straight lines in the plane, for a binary quadratic χ equated to zero gives two points P, Q on the conic, so that if we make χ correspond to the line PQ when either is given the other is uniquely determined.

The quadratic χ can be written in one way in the form

$$\lambda f + \mu\phi + \nu\psi,$$

and then $\lambda\xi + \mu\eta + \nu\zeta = 0$ is the equation of the corresponding straight line.

Accordingly if the line pass through a fixed point in the plane, say ξ_0, η_0, ζ_0, we have

$$\begin{aligned} \chi &= \lambda f + \mu\phi + \nu\psi \\ &= \lambda f + \mu\phi - \frac{\lambda\xi_0 + \mu\eta_0}{\zeta_0}\psi \\ &= \lambda\left(f - \frac{\xi_0}{\zeta_0}\psi\right) + \mu\left(\phi - \frac{\eta_0}{\zeta_0}\psi\right), \end{aligned}$$

and χ is apolar to (*i.e.* harmonic to) the fixed quadratic apolar to

$$f - \frac{\xi_0}{\zeta_0}\psi \text{ and } \phi - \frac{\eta_0}{\zeta_0}\psi.$$

Hence if the line PQ passes through a fixed point T, the corresponding quadratic χ is harmonic to a fixed quadratic τ.

Now the perfect squares apolar to τ correspond to lines touching the conic, since in this case the points P and Q coincide, hence these perfect squares determine the points of contact of the tangents drawn from T to the conic.

If these points are R, S, then the quadratic giving R and S is apolar to the square of the linear forms giving R and S respectively, and since J is apolar to these latter squares, it follows that J corresponds to RS, as can be seen in many ways.

Thus if PQ pass through T, and RS be the polar of T, the quadratics corresponding to PQ, RS are harmonic, or in other words, when two quadratics are harmonic, the corresponding lines are conjugate with respect to the conic.

This can be easily verified by using the tangential equation of the conic.

Consider a binary quartic representing four points A, B, C, D on the conic, and let BC, AD meet in E, CA, BD in F, and AB, CD in G.

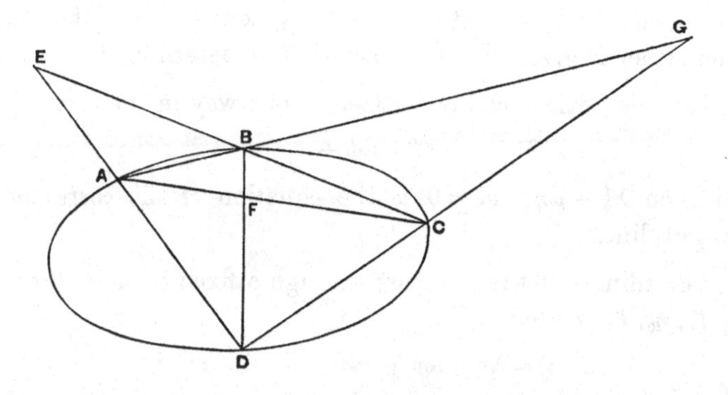

Then the quadratic corresponding to the polar of E is harmonic to the two quadratics corresponding to AD and BC, *i.e.* this polar meets the conic in two points which are the double points of the involution having B, C and A, D for conjugate elements.

Thus the polars of E, F, G meet the conic in the double points of the three involutions determined by the four points A, B, C, D; but EFG being a self-conjugate triangle of the conic, the polar of E is FG, and the lines EF, FG for example are conjugate lines with respect to the triangle, hence the pairs of double points of any two of the three involutions are harmonically conjugate.

[This corresponds to the fact that the sextic covariant of a quartic can be written as the product of three quadratics which are mutually harmonic.]

193. As another example, let us prove that a triangle and its polar triangle with respect to the conic are in perspective.

Suppose the sides of the triangle $A_1A_2A_3$ correspond to the quadratics f_1, f_2, f_3 respectively, and that the sides of the polar triangle $B_1B_2B_3$ correspond to ϕ_1, ϕ_2, ϕ_3 respectively. Then since B_1B_2 is conjugate to A_1A_3 and A_2A_3, it follows that ϕ_3 is harmonic to f_2 and f_1, so that ϕ_3 is the Jacobian of f_1 and f_2.

Hence the sides of $B_1B_2B_3$ correspond to

$$J_{23},\ J_{31}, \text{ and } J_{12} \text{ where } J_{12}=(f_1f_2).$$

Now let B_2B_3 meet A_1A_3 in P_1, then the polar of P_1 is conjugate to both these lines, and therefore corresponds to

$$(\phi_1,\ f_1) \text{ or to } (f_1,\ J_{23}).$$

The polars of the analogous points P_2, P_3 correspond to (f_2, J_{31}), (f_3, J_{12}) respectively, and P_1, P_2, P_3 will be collinear if their polars are concurrent, *i.e.* if the quadratics

$$(f_1,\ J_{23}),\ (f_2,\ J_{31}),\ (f_3,\ J_{12})$$

are harmonic to the same quadratic.

To prove that this is so, let us calculate the quadratic harmonic to the first two.

Representing

$$f_1,\ f_2,\ f_3 \text{ by } a_x^2,\ b_x^2,\ c_x^2,$$

we have

$$J_{23}=(bc)\,b_xc_x, \text{ etc.}$$

$$\begin{aligned}\therefore\ (f_1,\ J_{23})' &= \{a_x^2,\ (bc)\,b_xc_x\} = \tfrac{1}{2}(ac)(bc)\,a_xb_x + \tfrac{1}{2}(ab)(bc)\,a_xc_x\\ &= \tfrac{1}{4}\{(ac)^2 b_x^2 + (bc)^2 a_x^2 - (ab)^2 c_x^2\} - \tfrac{1}{4}\{(ab)^2 c_x^2 + (bc)^2 a_x^2 \\ &\qquad - (ac)^2 b_x^2\}\\ &= \tfrac{1}{2}(ac)^2 b_x^2 - \tfrac{1}{2}(ab)^2 c_x^2.\end{aligned}$$

Similarly

$$(f_2,\ J_{31}) = \tfrac{1}{2}(ab)^2 c_x^2 - \tfrac{1}{2}(bc)^2 a_x^2,$$

and the quadratic harmonic to these two is

$$\{(ac)^2 b_x^2 - (ab)^2 c_x^2,\ \ (ab)^2 c_x^2 - (bc)^2 a_x^2\},$$

or

$$(ab)^2(ac)^2 J_{23} - (ac)^2(bc)^2 J_{21} + (ab)^2(bc)^2 J_{31},$$

or

$$(ab)^2(ac)^2 J_{23} + (ba)^2(bc)^2 J_{31} + (ca)^2(cb)^2 J_{12},$$

and by symmetry this is harmonic to $(f_3,\ J_{12})$ also.

Thus the polars of $P_1P_2P_3$ meet in a point whose polar corresponds to

$$\frac{J_{23}}{i_{23}} + \frac{J_{31}}{i_{31}} + \frac{J_{12}}{i_{12}} = 0,$$

where $i_{23} = (f_2, f_3)^2 = (bc)^2$.

And hence $P_1P_2P_3$ lie on the straight line corresponding to the quadratic

$$\frac{J_{23}}{i_{23}} + \frac{J_{31}}{i_{31}} + \frac{J_{12}}{i_{12}} = 0.$$

The reader will find it interesting to shew that the above quadratic may be written

$$\frac{f_1}{I_{23}} + \frac{f_2}{I_{31}} + \frac{f_3}{I_{12}} = 0,$$

where the I's are derived from the i's as before.

Ex. (i). ABC is a triangle inscribed in a conic, and the tangents at A, B, C meet BC, CA, AB in L, M, N. Determine the parameters of the points in which the polar of L meets the conic in terms of those of A, B, C. Hence shew that L, M, N are collinear, and that the line LMN meets the conic in the Hessian points of ABC.

Ex. (ii). Deduce Pascal's Theorem from the parametric representation of a conic.

Ex. (iii). Prove that six conics can be drawn through four fixed points to touch a given conic, and that their points of contact are given by the Jacobian of the quartics determined by any two conics through the four points.

194. Twisted Cubic. The relation of the rational cubic space curve to the binary cubic is exactly the same as that of the conic to the binary quadratic.

The line of argument is similar to that used in the previous case.

We have

$$\begin{aligned}
\xi &= a_0x_1^3 + 3a_1x_1^2x_2 + 3a_2x_1x_2^2 + a_3x_2^3 = a_x^3 = f,\\
\eta &= b_x^3 = \phi,\\
\zeta &= c_x^3 = \psi,\\
\varpi &= d_x^3 = \chi,
\end{aligned}$$

and the curve represented is a cubic because the plane

$$\lambda\xi + \mu\eta + \nu\zeta + \rho\varpi = 0$$

meets it in three points.

We have now an exact correspondence between planes and binary cubics. If a plane pass through a fixed point T the corresponding cubic is apolar to a fixed cubic which determines P, Q, R the points of contact of the osculating planes of the cubic through T. Cf. § 192.

Hence there are three such points, and since a cubic is apolar to itself, the plane joining them passes through T. We shall call T the pole of the plane PQR; and we observe that when two cubics are apolar, each of the corresponding planes passes through the pole of the other.

If TLM be a chord of the cubic through T, then any plane through this line is apolar to the cubic giving PQR, so that L, M are determined by a quadratic apolar to the cubic, *i.e.* by the Hessian of the cubic.

195. We shall illustrate the above remarks by the discussion of an algebraical problem intimately connected with the cubic curve.

The planes passing through a fixed line l correspond to cubics of the form

$$\lambda_1 f_1 + \lambda_2 f_2,$$

where λ_1 and λ_2 are variable numbers. We call such a linear system of cubics a *pencil*, and then we see that the Jacobian of two members

$$\lambda_1 f_1 + \lambda_2 f_2, \quad \mu_1 f_1 + \mu_2 f_2$$

of the pencil is

$$J = (\lambda_1\mu_2 - \lambda_2\mu_1)(f_1 f_2),$$

so that except for a numerical factor it is the same for every pair of cubics in the pencil. Hence we may call this quartic the Jacobian of the pencil and the question arises—*Given a binary quartic J, can a pencil of cubics be found of which it is the Jacobian?*

To answer this question consider the geometrical meaning of the Jacobian with reference to the line l.

If the cubic

$$\lambda_1 f_1 + \lambda_2 f_2$$

have a double factor this factor occurs in the Jacobian, because it occurs in both

$$\lambda_1 \frac{\partial f_1}{\partial x_1} + \lambda_2 \frac{\partial f_2}{\partial x_1} \text{ and } \lambda_1 \frac{\partial f_1}{\partial x_2} + \lambda_2 \frac{\partial f_2}{\partial x_2}.$$

And further, as the discriminant of a cubic is of degree four there are only four members of the pencil which have a double factor; therefore the Jacobian of the forms contains these four double factors and no others.

But if a cubic $$\lambda_1 f_1 + \lambda_2 f_2$$
have a double factor the corresponding plane touches the curve, and hence through a given line four planes can be drawn touching the curve, and their points of contact are given by $J = 0$; in other words there are four tangent lines to the curve which intersect a given line l, and they are tangents at the points given by

$$J = 0.$$

Now our problem is equivalent to the following: *Being given the four points of contact, to construct the line l.* But the four tangents being given there are two lines meeting them, and, as each of these corresponds to a pencil of cubics having J for Jacobian, it follows that the question is always soluble and that there are two solutions which may coincide in particular cases.

Ex. (i). Prove that three osculating planes and one chord of a cubic can be drawn through any point. The chord determines the Hessian of the cubic giving the points of contact of the planes.

Ex. (ii). The plane joining the points of contact of the three osculating planes from a point O passes through O and is called the polar plane of O. If the polar plane of O passes through O' the polar plane of O' passes through O and the corresponding cubics are apolar.

Ex. (iii). By using line coordinates prove that the tangents to the cubic belong to a linear complex and that any two planes corresponding to apolar cubics meet in a line belonging to this complex.

Ex. (iv). If a line l belong to the above complex the points of contact of the tangents to the cubic which meet l are given by a quartic for which the invariant i vanishes. In this case the two pencils of cubics having the quartic for Jacobian coincide.

Ex. (v). If two pencils of cubics have the same Jacobian, then any member of the first pencil is apolar to any member of the second pencil.

196. The twisted quartic. The rational space curve of the fourth degree furnishes the most convenient geometrical representation of a binary quartic and its concomitants. Suppose that

$$\xi = p_x^4, \quad \eta = q_x^4, \quad \zeta = r_x^4, \quad \varpi = t_x^4$$

is the parametric representation of such a curve. Then since ξ, η, ζ, ϖ are not connected by a linear relation the four binary quartics are linearly independent and are therefore all apolar to a unique quartic f which we shall denote by

$$a_x^4 \equiv b_x^4 \equiv c_x^4 \equiv d_x^4.$$

It is evident that four points on the curve are coplanar when and only when the quartic determining their parameters is apolar to f—in fact such quartics are linear combinations of ξ, η, ζ, ϖ. Hence the four points $\alpha, \beta, \gamma, \delta$ are coplanar when

$$a_\alpha a_\beta a_\gamma a_\delta = 0.$$

Thus $a_x^4 = 0$ gives the four points of superosculation, *i.e.* the four points at which the osculating plane contains four consecutive points of the curve.

Through a point δ on the curve can be drawn three osculating planes other than that at δ and their points of contact are given by

$$a_x^3 a_\delta = 0,$$

i.e. by the first polar of δ.

By varying δ we obtain a pencil of cubics each of which determines three points on the curve such that the osculating planes at them meet in a point on the curve.

Now if a member of this pencil have a double factor we have two consecutive osculating planes whose line of intersection meets the curve, and hence the double factor gives a point the tangent at which meets the curve again. But the double factors of members of the pencil are precisely the factors of the Jacobian of the pencil, which in this case is the Hessian of f. Hence $H = 0$ gives four points on the curve the tangents at which meet the curve again.

197. We have still to interpret the sextic covariant which gives six points on the curve.

For this purpose let us seek for points P, Q on the curve such that the osculating plane at P passes through Q and the osculating plane at Q passes through P.

If λ, μ are the parameters of P and Q we have

$$a_\lambda^3 a_\mu = 0,$$

$$b_\lambda b_\mu^3 = 0.$$

To eliminate λ we note that from the second equation

$$\lambda_2 = -b_1 b_\mu^3, \ \lambda_1 = b_2 b_\mu^3,$$

and as λ occurs to degree three in the first equation we must use three different equations of this type.

Hence we have

$$(ab)(ac)(ad)\, a_\mu b_\mu^3 c_\mu^3 d_\mu^3 = 0$$

as the equation for μ.

Now

$$\begin{aligned}(ab)(ac)(ad)\, a_x b_x^3 c_x^3 d_x^3 &= \tfrac{1}{2}(ab)\, a_x b_x^3 c_x^2 d_x^2 \{(ac)^2 d_x^2 + (ad)^2 c_x^2 - (cd)^2 a_x^2\} \\ &= \tfrac{1}{2}(ab)(ac)^2 a_x b_x^3 c_x^2 \,.\, d_x^4 + \tfrac{1}{2}(ab)(ad)^2 a_x b_x^3 d_x^2 \,.\, c_x^4 \\ &\qquad - \tfrac{1}{2}(ab)\, a_x^3 b_x^3 \,.\, (cd)^2 c_x^2 d_x^2 \\ &= tf,\end{aligned}$$

where t is the sextic covariant.

For $f = 0$ the points P and Q coincide, hence there are three pairs of points such that each lies in the osculating plane of the other and they are given by the sextic covariant.

198. We shall now sketch some further investigations connected with the curve.

The triple secants of the curve are given by the pencil of cubics apolar to f, so that if one cubic give three points on a triple secant and another three points whose osculating planes meet on the curve the two cubics are apolar. The two pencils have the same Jacobian, as follows from geometry and analysis.

The four points in which the tangents at the Hessian points meet the curve again are given by

$$(ab)^2 (ac)(bd)(cd)^2 a_x b_x c_x d_x = 0,$$

for the discriminant of their first polars must vanish.

This reduces to $\qquad iH - jf = 0.$

If $i = 0$ the four points of superosculation are coplanar. If $j = 0$ there is an actual double point on the curve, because in that case f has an apolar quadratic—the quartic is now the complete intersection of two quadrics which touch, whereas in general it is the partial intersection of a quadric and a cubic.

There are two other tangents meeting any given tangent to the curve and when $j=0$ there exist pairs of chords such that the tangents at the extremities of either intersect those at the extremities of the other.

Between the parameters of the points of contact of two intersecting tangents there is a symmetrical (2, 2) relation, hence by comparison with Poncelet's porism we infer that if one twisted polygon with n sides can be circumscribed to the curve there is an infinite number of such circumscribing polygons.

By forming the expressions for the six coordinates of the tangent at any point, which are the Jacobians of the fundamental quartics taken in pairs, we see that the condition that six tangents should belong to the same linear complex is that the sextic determining their point of contact should be apolar to a fixed sextic. This fixed sextic can be shewn to be the sextic covariant either geometrically or analytically. We can in fact establish the following theorem: If two quartics are apolar to a third then their Jacobian is apolar to the sextic covariant of the latter. Ex. (ii), p. 52.

The sextic covariant therefore bears the same relation to the line geometry of the curve as the fundamental quartic does to the point geometry.

By discussion of the quartics apolar to this sextic the reader will easily prove that the tangents at four points given by a quartic covariant (of the type $\lambda iH+\mu jf$) are generators of a quadric.

References to various memoirs of Reye, Rosanes and others on the subjects discussed in this chapter will be found in Meyer's *Berichte.* See also the same author's book *Apolarität und rationale Curve.* There is an interesting paper on the twisted cubic by Sturm in *Crelle's Journal*, Vol. LXXXVI., and a very full list of references for the twisted quartic in a paper by Richmond, *Camb. Phil. Trans.* 1900.

CHAPTER XII.

TERNARY FORMS.

199. In this chapter we shall extend the symbolical notation, which has been used throughout for binary forms, to the case of forms with a higher number of variables. It may be well to remark at the outset that the methods developed up to the present are nowhere else so effective as in the case of binary forms; accordingly, while we shall explain at some length the notation and methods for ternary forms, we shall content ourselves with a very brief indication of the extension to forms with four or more variables.

200. A ternary form in one set of variables x_1, x_2, x_3 will be written in the form

$$f = \Sigma \frac{n!}{p!\,q!\,r!} a_{pqr}\, x_1^p x_2^q x_3^r\,;\ p+q+r=n,$$

and the summation is of course extended to all possible values of the integers p, q, r satisfying this condition.

In agreement with the symbols previously introduced, we shall represent f by the umbral expression

$$(\alpha_1 x_1 + \alpha_2 x_2 + \alpha_3 x_3)^n = \alpha_x{}^n,$$

so that

$$\alpha_1{}^p \alpha_2{}^q \alpha_3{}^r = a_{pqr}.$$

Just as before an expression in the α's has no actual (as opposed to symbolical) significance unless the total degree in the α's is exactly n. To represent forms whose degree in the a's is greater than unity, we have to introduce other equivalent systems of symbols β and γ, so that

$$f = \alpha_x{}^n = \beta_x{}^n = \gamma_x{}^n \ldots$$

201. Let us at once point out how concisely polar forms are represented in this notation; the rth polar of $y_1y_2y_3$ with respect to f is

$$\frac{(n-r)!}{n!}\left(y_1\frac{\partial}{\partial x_1}+y_2\frac{\partial}{\partial x_2}+y_3\frac{\partial}{\partial x_3}\right)^r f,$$

the numerical factor being introduced for arithmetical convenience.

This expression is

$$\frac{(n-r)!}{n!}\left(y_1\frac{\partial}{\partial x_1}+y_2\frac{\partial}{\partial x_2}+y_3\frac{\partial}{\partial x_3}\right)^r (\alpha_1x_1+\alpha_2x_2+\alpha_3x_3)^n$$

$$=\frac{(n-r)!}{n!}\,\frac{n!}{(n-r)!}\,\alpha_x^{n-r}\alpha_y^{r},$$

i.e. the rth polar is represented by

$$\alpha_x^{n-r}\alpha_y^{r}.$$

Thus for example the equation

$$\alpha_x^{n}=0$$

represents a curve of order n, and if y be a point on it the equation of the tangent at that point is

$$\alpha_x\alpha_y^{n-1}=0.$$

Again, the points of contact of the tangents drawn from a point y to the curve lie on the first polar

$$\alpha_x^{n-1}\alpha_y=0.$$

Next let us consider the effect of a linear substitution on a form such as f.

We shall write such a transformation in the form

$$\left.\begin{aligned} x_1&=\xi_1X_1+\eta_1X_2+\zeta_1X_3\\ x_2&=\xi_2X_1+\eta_2X_2+\zeta_2X_3\\ x_3&=\xi_3X_1+\eta_3X_2+\zeta_3X_3\end{aligned}\right\}.$$

The effect of this change of variables on α_x is to change it into

$$\alpha_1(\xi_1X_1+\eta_1X_2+\zeta_1X_3)+\alpha_2(\xi_2X_1+\eta_2X_2+\zeta_2X_3)+\alpha_3(\xi_3X_1+\eta_3X_2+\zeta_3X_3),$$

or

$$\alpha_\xi X_1+\alpha_\eta X_2+\alpha_\zeta X_3.$$

Hence f which is α_x^n becomes

$$(\alpha_\xi X_1 + \alpha_\eta X_2 + \alpha_\zeta X_3)^n,$$

or $$\Sigma \frac{n!}{p!q!r!} \alpha_\xi^p \alpha_\eta^q \alpha_\zeta^r X_1^p X_2^q X_3^r, \text{ where } p+q+r=n.$$

Thus the coefficient of X_1^n is α_ξ^n or the result of putting the ξ's for the x's in f, and subsequent coefficients may be deduced from this by means of the polarizing operators

$$\left(\eta_1 \frac{\partial}{\partial \xi_1} + \eta_2 \frac{\partial}{\partial \xi_2} + \eta_3 \frac{\partial}{\partial \xi_3}\right) \text{ and } \left(\zeta_1 \frac{\partial}{\partial \xi_1} + \zeta_2 \frac{\partial}{\partial \xi_2} + \zeta_3 \frac{\partial}{\partial \xi_3}\right)$$

operating on the first.

202. The above will shew how very convenient the symbolical notation is for dealing with constantly occurring functions like polars, but, as in the case of binary forms, its great value lies in the application to the theory of invariants and covariants.

Let us recall the significance of these terms especially for ternary forms and illustrate them by some examples.

Suppose that after the application of a linear transformation

$$x_r = \xi_r X_1 + \eta_r X_2 + \zeta_r X_3, \qquad r = 1, 2, 3,$$

a ternary quantic

$$f \equiv \Sigma a_{pqr} \frac{n!}{p!q!r!} x_1^p x_2^q x_3^r$$

becomes $$F = \Sigma A_{pqr} \frac{n!}{p!q!r!} X_1^p X_2^q X_3^r,$$

then a rational integral function I of the coefficients is said to be an invariant when

$$I(A_{pqr}) = I(a_{pqr}) M,$$

where M depends only on the transformation.

A covariant of f possesses a similar property, but involves the variables as well as the coefficients, and similar definitions apply to invariants and covariants of two or more forms.

After what has been said on the subject in dealing with binary forms, we need not stop to prove that it suffices to consider invariants and covariants which are homogeneous in the coefficients of each form involved, and covariants which are homogeneous in the variables—further it may be easily proved that the factor M referred to in the definition must be a power of the determinant of the transformation, and we shall assume henceforth that it is so. The reader may regard this as a simplification of the definition, or supply the necessary proof on the lines of that given for binary forms. See § 24.

As examples, we may notice that the discriminant

$$abc + 2fgh - af^2 - bg^2 - ch^2$$

is an invariant of the ternary quadric

$$ax_1^2 + bx_2^2 + cx_3^2 + 2fx_2x_3 + 2gx_3x_1 + 2hx_1x_2.$$

In this case the power of the determinant which occurs as multiplying factor is two.

Again the Hessian

$$\begin{vmatrix} \frac{\partial^2 f}{\partial x_1^2} & \frac{\partial^2 f}{\partial x_1 \partial x_2} & \frac{\partial^2 f}{\partial x_1 \partial x_3} \\ \frac{\partial^2 f}{\partial x_2 \partial x_1} & \frac{\partial^2 f}{\partial x_2^2} & \frac{\partial^2 f}{\partial x_2 \partial x_3} \\ \frac{\partial^2 f}{\partial x_3 \partial x_1} & \frac{\partial^2 f}{\partial x_3 \partial x_2} & \frac{\partial^2 f}{\partial x_3^2} \end{vmatrix}$$

is a covariant of any ternary form f.

As a further example, the Jacobian of three forms is a covariant of the three forms.

We shall not stop to verify that these expressions actually are invariants or covariants as the case may be. They are well known in the theory of curves and we mention them here to remind the reader that such invariant functions as defined above do actually exist. By the use of the symbolical notation we shall be able to verify our assertions much more easily, and also to construct any number of invariants and covariants.

203. Just as in the theory of binary forms we denoted the determinant

$$\begin{vmatrix} \alpha_1 & \alpha_2 \\ \beta_1 & \beta_2 \end{vmatrix} \text{ by } (\alpha\beta),$$

so now we shall denote

$$\begin{vmatrix} \alpha_1 & \alpha_2 & \alpha_3 \\ \beta_1 & \beta_2 & \beta_3 \\ \gamma_1 & \gamma_2 & \gamma_3 \end{vmatrix} \text{ by } (\alpha\beta\gamma).$$

We are now in a position to enunciate and prove the first theorem connecting invariants and symbolical notation, namely: *Every expression represented symbolically by factors of the type* $(\alpha\beta\gamma)$ *is an invariant, and every expression represented by factors of the types* $(\alpha\beta\gamma)$ *and* α_x *is a covariant.*

For by a linear substitution α_x becomes

$$\alpha_\xi X_1 + \alpha_\eta X_2 + \alpha_\zeta X_3,$$

and hence $(\alpha\beta\gamma)$ becomes

$$\begin{vmatrix} \alpha_\xi & \alpha_\eta & \alpha_\zeta \\ \beta_\xi & \beta_\eta & \beta_\zeta \\ \gamma_\xi & \gamma_\eta & \gamma_\zeta \end{vmatrix} \quad i.e. \ (\alpha\beta\gamma)(\xi\eta\zeta).$$

So that the expression formed from the new coefficients is equal to that formed from the old coefficients multiplied by a power of $(\xi\eta\zeta)$; and the power of $(\xi\eta\zeta)$ that occurs is equal to the number of symbolical factors of the type $(\alpha\beta\gamma)$ that occur in the expression.

204. The preceding proof, depending only on the way in which the symbolical letters occur in the expression under consideration, applies equally well to an invariant or covariant of any number of ternary forms. The only further condition that such a symbolical expression should actually represent a covariant is that it should have a meaning when expressed in terms of the original coefficients, *i.e.* each symbolical letter must occur to the requisite degree in every term in which it appears. It is easy to verify that the three examples already given are actually invariants or covariants as the case may be.

Thus if

$$a_{200}x_1^2 + a_{020}x_2^2 + a_{002}x_3^2 + 2a_{110}x_1x_2 + 2a_{101}x_3x_1 + 2a_{011}x_2x_3$$
$$= \alpha_x^2 = \beta_x^2 = \gamma_x^2,$$

we have by direct multiplication

$$(\alpha\beta\gamma)^2 = (\Sigma \pm \alpha_1\beta_2\gamma_3)^2 = 6 \begin{vmatrix} a_{200}, & a_{110}, & a_{101} \\ a_{110}, & a_{020}, & a_{011} \\ a_{101}, & a_{011}, & a_{002} \end{vmatrix},$$

since $\alpha_1^2 = \beta_1^2 = \gamma_1^2 = a_{200}$, etc.

Thus the discriminant of the ternary quadric is an invariant.

Again, if we have three forms

$$f_1 = a_x^{n_1}, \quad f_2 = \beta_x^{n_2}, \quad f_3 = \gamma_x^{n_3},$$

then

$$(\alpha\beta\gamma) a_x^{n_1-1} \beta_x^{n_2-1} \gamma_x^{n_3-1} = \begin{vmatrix} \alpha_1 \alpha_x^{n_1-1}, & \alpha_2 \alpha_x^{n_1-1}, & \alpha_3 \alpha_x^{n_1-1} \\ \beta_1 \beta_x^{n_2-1}, & \beta_2 \beta_x^{n_2-1}, & \beta_3 \beta_x^{n_2-1} \\ \gamma_1 \gamma_x^{n_3-1}, & \gamma_2 \gamma_x^{n_3-1}, & \gamma_3 \gamma_x^{n_3-1} \end{vmatrix}$$

$$= \frac{1}{n_1 n_2 n_3} \begin{vmatrix} \frac{\partial f_1}{\partial x_1} & \frac{\partial f_1}{\partial x_2} & \frac{\partial f_1}{\partial x_3} \\ \frac{\partial f_2}{\partial x_1} & \frac{\partial f_2}{\partial x_2} & \frac{\partial f_2}{\partial x_3} \\ \frac{\partial f_3}{\partial x_1} & \frac{\partial f_3}{\partial x_2} & \frac{\partial f_3}{\partial x_3} \end{vmatrix} = \frac{1}{n_1 n_2 n_3} \frac{\partial (f_1 f_2 f_3)}{\partial (x_1 x_2 x_3)},$$

which shews at once that the Jacobian is a covariant.

The fact that $(\alpha\beta\gamma)^2 \alpha_x^{n-2} \beta_x^{n-2} \gamma_x^{n-2}$ represents the Hessian of the form $f = \alpha_x^n = \beta_x^n = \gamma_x^n$ may be easily verified in the same way. The deduction of the equation of the Hessian from the property that the polar conic of any point on it degenerates into two straight lines affords an instructive example of the symbolical calculus.

The polar conic of the point y_1, y_2, y_3 with respect to the curve

$$f = \alpha_x^n = \beta_x^n = \gamma_x^n = 0$$

is represented by the equation

$$\alpha_y^{n-2} \alpha_x^2 \equiv \beta_y^{n-2} \beta_x^2 \equiv \gamma_y^{n-2} \gamma_x^2 = 0,$$

and the discriminant of this is

$$(\alpha\beta\gamma)^2 \alpha_y^{n-2} \beta_y^{n-2} \gamma_y^{n-2}.$$

Hence changing y into x we get the Hessian as the locus of points possessing the property in question.

205. Geometrical Meaning of a Linear Transformation. To obtain a geometrical representation of ternary forms we naturally suppose the variables x_1, x_2, x_3 to be the homogeneous coordinates of a point in a plane referred to a certain triangle. The equation to zero of a ternary form will then represent a curve in the plane.

A linear transformation

$$x_r = \xi_r X_1 + \eta_r X_2 + \zeta_r X_3, \qquad r = 1, 2, 3,$$

may be regarded from two points of view, for if P be the point

x_1, x_2, x_3 we may either suppose P to be unaltered and the triangle of reference changed or we may suppose P to be changed and the triangle of reference unaltered, in other words X_1, X_2, X_3 may be the coordinates of the original point referred to a new fundamental triangle or they may be the coordinates of a new point (into which P is changed) referred to the original triangle of reference.

The general linear transformation as written above can be, in fact, represented as a change in the triangle of reference together with a multiplication of each coordinate by a suitable quantity—the equations of the sides of the old triangle of reference referred to the new triangle are $x_r = 0$, that is

$$\xi_r X_1 + \eta_r X_2 + \zeta_r X_3 = 0, \qquad r = 1, 2, 3,$$

and on solving for X_1, X_2, X_3 in terms of x_1, x_2, x_3 we can easily find the equations of the sides of the new triangle of reference in the old coordinates.

From this point of view, if I be an invariant of a ternary form f, when I vanishes for one triangle of reference it vanishes for any other triangle of reference; that is to say, $I = 0$ expresses a property of the curve itself and not a relation between the curve and some particular triangle.

In like manner the curve represented by a covariant is connected with the original curve by relations which are quite independent of the triangle of reference.

From the second point of view the point P is changed into a point P' by a homographic transformation, *i.e.* a transformation possessing the following properties:

(i) any point P is changed into a point P';

(ii) any straight line p is changed into a straight line p';

(iii) if P lies on p then P' lies on p';

(iv) the cross-ratio of four collinear points is equal to the cross-ratio of the four corresponding points; and the cross-ratio of four concurrent straight lines is equal to the cross-ratio of the four corresponding lines.

Of these properties (iv) is the only one we need prove.

Let $p_1 = 0$ and $p_2 = 0$ be the equations of two straight lines and $p_1' = 0$, $p_2' = 0$ the equations of the corresponding lines, then

$p_1' + \lambda p_2' = 0$ corresponds to $p_1 + \lambda p_2 = 0$ for all values of λ, and inasmuch as the cross-ratio of the four lines

$$p_1 + \lambda_r p_2 = 0, \quad r = 1, 2, 3, 4$$

is $-\dfrac{(\lambda_2 - \lambda_3)(\lambda_1 - \lambda_4)}{(\lambda_3 - \lambda_1)(\lambda_2 - \lambda_4)}$ the result follows at once.

Again, suppose that a figure in one plane is projected into another plane and that P' is the projection of P, then if x_1, x_2, x_3 be the coordinates of P referred to a fixed triangle in the first plane and X_1, X_2, X_3 be the coordinates of P' referred to a triangle in the second plane, it is very easy to see that x_1, x_2, x_3 are linear functions of X_1, X_2, X_3 and *vice versâ*. On the connection between projection and linear transformation see § 157.

Thus projection affords an example of linear transformation and in this case the four properties enunciated above are self-evident.

Hence if I be an invariant of a form f, then $I = 0$ expresses a property of the curve $f = 0$ that is unaltered by any homographic transformation and in particular by projection. A like remark applies to the connection between a covariant curve and the original curve—it is undisturbed by projection.

206. The connection between these two different points of view is interesting.

The linear transformation

$$x_r = \xi_r X_1 + \eta_r X_2 + \zeta_r X_3$$

leaves three points of the plane unaltered in general. In fact to find the points whose position is unchanged we have merely to put $X_r = \rho x_r$ in the equations above and then elimination of x_1, x_2, x_3 gives a cubic for $\dfrac{1}{\rho}$, viz.

$$\begin{vmatrix} \xi_1 - \dfrac{1}{\rho}, & \eta_1, & \zeta_1 \\ \xi_2, & \eta_2 - \dfrac{1}{\rho}, & \zeta_2 \\ \xi_3, & \eta_3, & \zeta_3 - \dfrac{1}{\rho} \end{vmatrix} = 0.$$

In general this equation has three unequal roots and to each root corresponds a single determination of the ratios $x_1 : x_2 : x_3$. If we take the triangle formed by these three points for triangle of reference the linear transformation takes the simple form

$$x_1 = k_1 X_1, \quad x_2 = k_2 X_2, \quad x_3 = k_3 X_3.$$

And hence in general by a suitable choice of the triangle of reference a linear transformation can be reduced to this form in which the coordinates are only multiplied by constants.

Ex. (i). Prove that by suitably choosing the coordinate system any four points may be reduced to the form

$$x_2^{(1)} = 0, \quad x_3^{(1)} = 0, \quad x_1^{(2)} = 0, \quad x_3^{(2)} = 0, \quad x_1^{(3)} = 0, \quad x_2^{(3)} = 0,$$
$$x_1^{(4)} = x_2^{(4)} = x_3^{(4)}.$$

And hence prove that a linear transformation can be found which changes any four points into four given points.

Ex. (ii). Given four points and their four corresponding points, shew that the point corresponding to any fifth point can be constructed by means of the ruler only. (Use the cross-ratio property.)

Ex. (iii). Hence or otherwise shew that any linear transformation is equivalent to a ruler construction.

Ex. (iv). If the linear transformation leave an isolated point and every point on a fixed line unaltered, shew that the equation of § 206 has a double root which is a root of every first minor. Give an equivalent geometrical construction in this case.

207. Starting with the definitions of invariants and covariants already given we might proceed to prove that every invariant and covariant can be represented symbolically in the form indicated in a previous theorem, and then go on to develop a theory of invariants and covariants as far as possible on the lines of binary forms. However it is essential to remember that the real and primary importance of such a theory lies in its application to geometry, and as in geometry line coordinates and tangential equations occur quite at the beginning of the analytical exposition, we are led here to introduce line coordinates as well as point coordinates from the first. Indeed it is not difficult to shew on purely analytical grounds that line coordinates are essential for the proper treatment of the algebraical questions that arise; such an explanation together with a comparison with the theory of binary forms will be given later.

The equation of a straight line being

$$u_1x_1 + u_2x_2 + u_3x_3 = 0,$$

u_1, u_2, u_3 are as usual called the coordinates of the line. Following Clebsch it is preferable to regard the equation just written as neither the point equation of the line nor the line equation of the point, but as the condition that the line (u_1, u_2, u_3) and the point (x_1, x_2, x_3) should bear a certain relation to one another.

The first fact to notice about (u_1, u_2, u_3) is that they are contragredient to x_1, x_2, x_3, for if u_1, u_2, u_3 become U_1, U_2, U_3 and x_1, x_2, x_3 become X_1, X_2, X_3, then

$$u_1x_1 + u_2x_2 + u_3x_3 \text{ must become } U_1X_1 + U_2X_2 + U_3X_3,$$

which is precisely the condition for contragredience (§ 39).

Further, the u's being contragredient to the x's are cogredient with the α's, and in fact by the linear transformation of § 201 we see at once that u_1 becomes u_ξ, u_2 becomes u_η and u_3 becomes u_ζ.

It is now evident that by a linear transformation the factor $(\alpha\beta u)$ merely becomes itself multiplied by the determinant of transformation, so that every form whose symbolical expression consists exclusively of factors of the types

$$(\alpha\beta\gamma),\ (\alpha\beta u),\ \alpha_x$$

possesses the property of invariance.

There are four classes of such invariant functions.

(i) Containing neither x's nor u's. These as already mentioned are called invariants.

(ii) Containing x's but not u's. These are covariants.

(iii) Containing u's but not x's. These are new introductions and are called contravariants.

(iv) Containing both u's and x's. These are called mixed concomitants.

Further there is the identical invariant form u_x which does not contain the coefficients of the original forms at all.

The degree to which the coefficients occur in a given invariant form is called simply its *degree*, the degree in the x's is called its *order* and the degree in the u's is called its *class*.

Examples of invariants and covariants have already been given.

As an example of a contravariant we may mention

$$\begin{vmatrix} a_{200} & a_{110} & a_{101} & u_1 \\ a_{110} & a_{020} & a_{011} & u_2 \\ a_{101} & a_{011} & a_{002} & u_3 \\ u_1 & u_2 & u_3 & 0 \end{vmatrix}$$

as a contravariant of the conic. In fact equated to zero it represents the line equation, and the reader may verify by direct multiplication that it is equivalent to

$$(\alpha\beta u)^2,$$

and therefore actually is an invariant form.

As an example of calculation similar to that given for the Hessian consider the locus of a point whose polar conic with respect to the cubic

$$f \equiv \alpha_x^3 \equiv \beta_x^3 = 0$$

touches a given straight line u.

The polar conic of the point y is

$$\alpha_x^2 \alpha_y \equiv \beta_x^2 \beta_y = 0,$$

and the condition that this should touch the given line is

$$(\alpha\beta u)^2 \alpha_y \beta_y = 0.$$

Thus the equation of the required locus is

$$(\alpha\beta u)^2 \alpha_x \beta_x = 0.$$

This is a mixed concomitant of the cubic. Equated to zero it represents the point equation of a curve when the u's are taken as constants, viz. the point equation of the locus mentioned. When the x's are taken to be constant it represents the line equation of a curve, viz. the line equation of the polar conic of the point x.

Similar remarks apply to mixed concomitants in general.

208. Principle of Duality. The fact that the condition that a straight line whose equation is $\alpha_x = 0$ should pass through a given point is symmetrical in the coordinates of the line and the point leads to an important remark. For if the proof of any theorem relating to a figure containing m straight lines and n

points is thrown into an analytical form, then by interchanging point and line coordinates throughout the investigation we obtain a correlative theorem for n straight lines and m points—if in the first figure one of the straight lines pass through one of the points, then in the second figure the corresponding point lies on the corresponding straight line. It must be clearly understood that such theorems as are here contemplated depend only on lines passing through points and points lying on lines, so that the process is exactly analogous to the reciprocation of descriptive properties. Naturally we can pass to point loci and line envelopes by the introduction of an infinite number of points and lines into the figures.

A fuller explanation of this principle of duality will be found in works on geometry, but the above will suffice as an indication of some general ideas which will underlie much of our subsequent work, especially on the theory of two conics.

Ex. (i). Shew that in every case the theorem dual to a given one may be obtained by reciprocating the given one (supposed descriptive) with respect to a conic.

Ex. (ii). If D, E, F be three points in the sides BC, CA, AB of a triangle, and AD, BE, CF be concurrent, then the lines EF, FD, DE meet BC, CA, AB respectively in three collinear points. What is the dual theorem?

209. Methods for transforming symbolical expressions. As in the case of binary forms there are two different methods by which a symbolical expression may be transformed:

(i) the interchange of equivalent symbols,

(ii) the use of identities among elementary symbolical expressions.

As an illustration of (i) we may notice that for the ternary quadratic $\alpha_x^2 = \beta_x^2 =$ etc. the mixed concomitant $(\alpha\beta u)\,\alpha_x\beta_x$ vanishes identically.

In fact, we have

$$(\alpha\beta u)\,\alpha_x\beta_x = (\beta\alpha u)\,\alpha_x\beta_x$$

since α, β are equivalent and

$$(\alpha\beta u)\,\alpha_x\beta_x = -(\beta\alpha u)\,\alpha_x\beta_x$$

by the properties of determinants, hence $(\alpha\beta u)\,\alpha_x\beta_x = 0$.

The fundamental identity for ternary forms is

$$(\beta\gamma\delta)\,\alpha_x-(\gamma\delta\alpha)\,\beta_x+(\delta\alpha\beta)\,\gamma_x-(\alpha\beta\gamma)\,\delta_x=0,$$

deduced from

$$\begin{vmatrix} \alpha_1 & \beta_1 & \gamma_1 & \delta_1 \\ \alpha_2 & \beta_2 & \gamma_2 & \delta_2 \\ \alpha_3 & \beta_3 & \gamma_3 & \delta_3 \\ \alpha_x & \beta_x & \gamma_x & \delta_x \end{vmatrix}=0.$$

On replacing δ by u

$$(\beta\gamma u)\,\alpha_x-(\gamma u\alpha)\,\beta_x+(u\alpha\beta)\,\gamma_x=(\alpha\beta\gamma)\,u_x$$

or

$$(\beta\gamma u)\,\alpha_x+(\gamma\alpha u)\,\beta_x+(\alpha\beta u)\,\gamma_x=(\alpha\beta\gamma)\,u_x,$$

an easily remembered result.

Again, if in the first identity on replacing x by $(\epsilon\zeta)$, *i.e.*

$$x_1,\ x_2,\ x_3 \text{ by } \epsilon_2\zeta_3-\epsilon_3\zeta_2,\ \epsilon_3\zeta_1-\epsilon_1\zeta_3,\ \epsilon_1\zeta_2-\epsilon_2\zeta_1$$

respectively, we deduce

$$(\alpha\beta\gamma)(\delta\epsilon\zeta)-(\delta\alpha\beta)(\gamma\epsilon\zeta)+(\gamma\delta\alpha)(\beta\epsilon\zeta)-(\beta\gamma\delta)(\alpha\epsilon\zeta)=0.$$

Another important identity is

$$\alpha_x\beta_y-\alpha_y\beta_x=(\alpha\beta u),$$

where u_1, u_2, u_3 are the coordinates of the line joining the points x and y so that

$$u_1=x_2y_3-x_3y_2,\quad u_2=x_3y_1-x_1y_3,\quad u_3=x_1y_2-x_2y_1.$$

In exactly the same way we have

$$v_\alpha w_\beta-v_\beta w_\alpha=(\alpha\beta x),$$

where x is the point of intersection of the lines v and w.

Ex. For the ternary quadratic $a_x^2=\beta_x^2=$etc. shew that

$$(\alpha\beta\gamma)(\alpha\beta\delta)\,\gamma_x\delta_x=\tfrac{1}{3}(\alpha\beta\gamma)^2\,.\,\delta_x^2,$$

$$(\alpha\beta u)(\alpha\gamma u)\,\beta_x\gamma_x=\tfrac{1}{2}(\alpha\beta u)^2\,.\,\gamma_x^2-\tfrac{1}{6}(\alpha\beta\gamma)^2\,.\,u_x^2.$$

210. We now come to the fundamental theorem in the present calculus, namely, that every invariant form can be represented symbolically by three types of factors, viz.

$$(\alpha\beta\gamma),\ (\alpha\beta u) \text{ and } \alpha_x$$

together with the identical invariant form u_x.

Some preliminary observations will enable us to simplify the proof.

In fact an invariant form containing the u's may be regarded as a pure covariant of the original form or forms together with the linear form

$$u_1x_1 + u_2x_2 + u_3x_3,$$

and hence if the theorem be proved for any number of ground forms for concomitants in which the u's are absent the general case follows at once by adjoining a linear form to the original system.

Let us then proceed to prove the theorem for invariants and covariants.

In the proof we need two lemmas regarding the operator Ω denoted by

$$\begin{vmatrix} \frac{\partial}{\partial \xi_1} & \frac{\partial}{\partial \xi_2} & \frac{\partial}{\partial \xi_3} \\ \frac{\partial}{\partial \eta_1} & \frac{\partial}{\partial \eta_2} & \frac{\partial}{\partial \eta_3} \\ \frac{\partial}{\partial \zeta_1} & \frac{\partial}{\partial \zeta_2} & \frac{\partial}{\partial \zeta_3} \end{vmatrix}$$

which is the natural extension to three variables of the operator so important in the binary theory.

211. I. *If D denote the determinant*

$$\begin{vmatrix} \xi_1 & \xi_2 & \xi_3 \\ \eta_1 & \eta_2 & \eta_3 \\ \zeta_1 & \zeta_2 & \zeta_3 \end{vmatrix}$$

then $\Omega^r D^n$ is a numerical multiple of D^{n-r}, where n, r are positive integers and $r \ngtr n$.

This result can be proved for $r = 1$ by straightforward differentiation and then the general result follows. We may shorten the proof by using properties of the minors of the determinant.

For a moment denote the minor of ξ_1 by X_1, that of η_1 by Y_1, of ζ_1 by Z_1, etc.

Then

$$\frac{\partial D^n}{\partial \eta_2} = nD^{n-1} Y_2$$

$$\frac{\partial^2 D^n}{\partial \eta_2 \partial \zeta_3} = n(n-1) D^{n-2} Y_2 Z_3 + nD^{n-1} \xi_1$$

and

$$\frac{\partial^2 D^n}{\partial \eta_3 \partial \zeta_2} = n(n-1) D^{n-2} Y_3 Z_2 - nD^{n-1} \xi_1.$$

Hence

$$\left(\frac{\partial^2}{\partial \eta_2 \partial \zeta_3} - \frac{\partial^2}{\partial \eta_3 \partial \zeta_2}\right) D^n = n(n-1) D^{n-2} (Y_2 Z_3 - Y_3 Z_2) + 2nD^{n-1} \xi_1$$

and

$$Y_2 Z_3 - Y_3 Z_2 = \xi_1 D$$

by the properties of minors, hence

$$\left(\frac{\partial^2}{\partial \eta_2 \partial \zeta_3} - \frac{\partial^2}{\partial \eta_3 \partial \zeta_2}\right) D^n = n(n+1) D^{n-1} \xi_1.$$

Consequently

$$\frac{\partial}{\partial \xi_1}\left(\frac{\partial^2}{\partial \eta_2 \partial \zeta_3} - \frac{\partial^2}{\partial \eta_3 \partial \zeta_2}\right) D^n = n(n+1)(n-1) D^{n-2} \xi_1 X_1 + n(n+1) D^{n-1},$$

and adding to this two like terms we have

$$\begin{aligned} \Omega D^n &= n(n+1)(n-1) D^{n-2} (\xi_1 X_1 + \xi_2 X_2 + \xi_3 X_3) + 3n(n+1) D^{n-1} \\ &= n(n+1)(n+2) D^{n-1}, \end{aligned}$$

since

$$D = \xi_1 X_1 + \xi_2 X_2 + \xi_3 X_3.$$

Operating with Ω again we have

$$\Omega^2 D^2 = (n-1) n^2 (n+1)^2 (n+2) D^{n-2},$$

and proceeding in this way we see that $\Omega^r D^n$ is a numerical multiple of D^{n-r} and, in particular, $\Omega^n D^n$ is a numerical constant.

The reader will find it instructive to extend this property of the operator Ω to four or more variables.

212. II. *If S be a product of m factors of the type α_ξ, n factors of the type β_η, and p factors of the type γ_ζ, then $\Omega^r S$ is the sum of a number of terms each containing r factors of the type $(\alpha\beta\gamma)$, $m-r$ factors of the type α_ξ, $n-r$ of the type β_η and $p-r$ of the type γ_ζ.*

The proof of this result is exactly similar to that of the corresponding theorem in binary forms. In fact let $S=PQR$, where

$$P=\alpha_\xi^{(1)}\alpha_\xi^{(2)}\dots\alpha_\xi^{(m)},$$

$$Q=\beta_\eta^{(1)}\beta_\eta^{(2)}\dots\beta_\eta^{(n)},$$

$$R=\gamma_\zeta^{(1)}\gamma_\zeta^{(2)}\dots\gamma_\zeta^{(p)},$$

then
$$\frac{\partial^3 S}{\partial\xi_1\partial\eta_2\partial\zeta_3}=\sum_{\mu,\nu,\pi}\alpha_1^{(\mu)}\beta_2^{(\nu)}\gamma_3^{(\pi)}\frac{P}{\alpha_\xi^{(\mu)}}\frac{Q}{\beta_\eta^{(\nu)}}\frac{R}{\gamma_\zeta^{(\pi)}},$$

the sum being taken for

$$\mu=1,2,\dots m;$$

$$\nu=1,2,\dots n;$$

$$\pi=1,2,\dots p.$$

Writing down all six such terms and adding we have

$$\Omega\,.\,S=\sum_{\mu,\nu,\pi}(\alpha^{(\mu)}\beta^{(\nu)}\gamma^{(\pi)})\frac{P}{\alpha_\xi^{(\mu)}}\frac{Q}{\beta_\eta^{(\nu)}}\frac{R}{\gamma_\zeta^{(\pi)}},$$

which proves the theorem for $r=1$, since $\dfrac{P}{\alpha_\xi^{(\mu)}}$ contains $m-1$ factors, and so on.

Now Ω has no effect on a factor $(\alpha\beta\gamma)$, hence applying the same result to each term in $\Omega\,.\,S$ we obtain the result for $r=2$ and so on for any value of r. In particular if $r=m=n=p$ the result is expressed exclusively in terms of factors of the type $(\alpha\beta\gamma)$.

213. Our proof that any invariant or covariant can be represented by factors of the two types

$$(\alpha\beta\gamma),\quad \alpha_x$$

follows exactly the lines of the second proof of the corresponding theorem for binary forms. As the proof is exactly the same for any number of forms, we give it explicitly for one only.

First suppose that $I(a)$ is an *invariant* of the form

$$\Sigma a_{pqr}\frac{n!}{p!\,q!\,r!}x_1^p x_2^q x_3^r,$$

then if $\Sigma A_{pqr}\dfrac{n!}{p!\,q!\,r!}X_1^p X_2^q X_3^r$ be the transformed expression, we have

$$I(A)=I(a)\,.\,(\xi\eta\zeta)^w\ \text{identically.}$$

Now express the left-hand side of this equation in symbols. We have

$$a_{pqr} = \alpha_1^p \alpha_2^q \alpha_3^r, \text{ and hence } A_{pqr} = \alpha_\xi^p \alpha_\eta^q \alpha_\zeta^r.$$

Thus on the left we have the sum of a number of terms. Each contains w factors of the type α_ξ, w of the type α_η, and w of the type α_ζ.

Hence operating on both sides of the equation with Ω^w, the left-hand side becomes an aggregate of terms each made up entirely of factors of the type $(\alpha\beta\gamma)$ and the right-hand side is a numerical multiple of $I(a)$, so that we have expressed $I(a)$ in the required manner.

The proof for a covariant is similar in form to the above, but as in the case of binary forms, a little care is necessary. Of course, here as always, we confine ourselves to covariants that are homogeneous in the variables x_1, x_2, x_3.

Suppose that $C(a, x)$ is a covariant of order m, then by definition

$$C(A, X) = (\xi\eta\zeta)^w \times C(a, x).$$

Now on solving the equations of transformation

$$x_1 = \xi_1 X_1 + \eta_1 X_2 + \zeta_1 X_3, \text{ etc.}$$

we have

$$X_1 = \frac{x_1(\eta_2\zeta_3 - \eta_3\zeta_2) + x_2(\eta_3\zeta_1 - \eta_1\zeta_3) + x_3(\eta_1\zeta_2 - \eta_2\zeta_1)}{(\xi\eta\zeta)}, \text{ etc.}$$

On replacing x_1 by $(v_2w_3 - v_3w_2)$, x_2 by $(v_3w_1 - v_1w_3)$ and x_3 by $(v_1w_2 - v_2w_1)$—a set of equations which may be written $x = (vw)$—we have

$$X_1 = \frac{v_\eta w_\zeta - v_\zeta w_\eta}{(\xi\eta\zeta)},$$

$$X_2 = \frac{v_\zeta w_\xi - v_\xi w_\zeta}{(\xi\eta\zeta)},$$

$$X_3 = \frac{v_\xi w_\eta - v_\eta w_\xi}{(\xi\eta\zeta)}.$$

Again, $A_{pqr} = \alpha_\xi^p \alpha_\eta^q \alpha_\zeta^r$, and hence on substituting for the A's and x's their values as given above in $C(A, X)$, and multiplying across by $(\xi\eta\zeta)^m$, the left-hand side becomes an aggregate of products of factors having the suffixes ξ, η, ζ, and the right-hand side becomes $(\xi\eta\zeta)^{w+m} \times C(a, x)$. Moreover each term on the left-hand side must contain $(w+m)$ factors with each suffix.

Now if we operate on the left-hand side once with Ω, we obtain an aggregate of terms each containing one determinantal factor and $(w+m-1)$ factors with each suffix.

The determinantal factors are of three types:

$$\text{(i)}\quad (\alpha\beta\gamma),$$

$$\text{(ii)}\quad (\alpha v w) = \alpha_x,$$

$$\text{(iii)}\quad (\alpha\beta v),$$

where the first two are of the form we require in the result, but the third is not.

Now a term involving a factor $(\alpha\beta v)$ must have arisen from operating with Ω on a term containing a factor such as $\alpha_\xi\beta_\eta v_\zeta$ and this term must therefore contain a further factor w_η (or w_ξ). Accordingly let the term be in the original expression

$$G\alpha_\xi\beta_\eta v_\zeta w_\eta,$$

then from the mode in which v and w occur in the expression on the left-hand side there must also be present a term

$$-G\alpha_\xi\beta_\eta v_\eta w_\zeta,$$

and on operating with Ω this gives

$$-G(\alpha\beta w)v_\eta,$$

whereas the former term is

$$+G(\alpha\beta v)w_\eta.$$

But by the fundamental identity

$$(\alpha\beta v)w_\eta + (\beta w v)\alpha_\eta + (w\alpha v)\beta_\eta = (w\alpha\beta)v_\eta,$$

or $$(\alpha\beta v)w_\eta - (\alpha\beta w)v_\eta = (\beta v w)\alpha_\eta - (\alpha v w)\beta_\eta = \beta_x\alpha_\eta - \alpha_x\beta_\eta,$$

and the sum of the two terms mentioned is

$$G(\beta_x\alpha_\eta - \alpha_x\beta_\eta).$$

Thus although factors like $(\alpha\beta v)$ do appear explicitly after operating with Ω, the terms containing them may be paired in such a way that the aggregate may be expressed entirely in terms of factors of the types $(\alpha\beta\gamma)$ and α_x together with factors having the suffixes ξ, η, or ζ.

Hence if we operate with Ω, $w+m$ times on the left and transform at each step as indicated above, we finally have an aggregate of terms each containing $(w+m)$ factors of the types $(\alpha\beta\gamma)$ and α_x only; also after performing the same operations on the right-hand side, we are left with a numerical multiple of $C(a, x)$. We have therefore expressed the covariant as an aggregate of terms each of which is the product of a number of factors of the types $(\alpha\beta\gamma)$ and α_x, and since the order must be m, it follows that in each term there are w factors of the type $(\alpha\beta\gamma)$ and m of the type α_x.

214. Leading Coefficients of Concomitants. In § 33 it was shewn that a covariant of a binary form could be deduced from its leading coefficient; we shall in what follows consider the extension of this idea to ternary forms, but as the results are easily obtained and not necessary for present purposes our remarks will be somewhat brief; the reader who desires further information is referred to a memoir by Forsyth, *Am. Journal of Mathematics*, 1889.

If a mixed concomitant be of class m' and order n', the coefficient of $x_1^{n'}u_1^{m'}$ is called the leading coefficient.

Given the concomitant, the leading coefficient is unique of course, but the reverse is not true, because if we multiply a concomitant by any power of u_x we obtain another concomitant with the same leading coefficient.

However, save as to a power of u_x, a concomitant can be found when its leading coefficient is given, as we proceed to shew.

The leading coefficient is an aggregate of products of factors of the types

$$(a\beta\gamma), \quad (a_2\beta_3 - a_3\beta_2), \quad a_1;$$

and on replacing

$$a_1 \text{ by } a_x,$$
$$a_2 \text{ by } a_y,$$
$$a_3 \text{ by } a_z,$$

and so on, the above factors become

$$(a\beta\gamma)\,u_x, \quad (a\beta u), \quad a_x$$

respectively, where $u=(yz)$. Cf. § 209.

The symbolical substitution is equivalent to replacing the coefficient a_{pqr} by

$$\frac{p\,!}{n\,!}\left(y\frac{\partial}{\partial x}\right)^q\left(z\frac{\partial}{\partial x}\right)^r a_x^n,$$

so that if in a leading coefficient we replace a_{pqr} by the value just given we obtain the corresponding concomitant multiplied by a power of u_x—the exponent of u_x is in fact equal to the number of factors of the type $(a\beta\gamma)$ that occur in the symbolical expression of the concomitant.

Hence, as in binary forms, it follows that "If there be an identical relation between a number of leading coefficients, the same relation exists among the corresponding concomitants."

We leave the reader to establish this on the lines of § **33**; the concomitants corresponding to the various leading coefficients must be chosen so as to make the whole expression of uniform order and class.

On multiplication by a sufficiently high power of u_x a concomitant can be completely expressed by means of factors of the types $(\alpha\beta u)$ and α_x.

The leading coefficient can therefore be expressed in terms of factors of the types

$$(\alpha_2\beta_3 - \alpha_3\beta_2) \text{ and } \alpha_1.$$

It follows immediately that the leading coefficient is an invariant of the binary forms

$$(a_2x_2 + a_3x_3)^n, \quad (a_2x_2 + a_3x_3)^{n-1} a_1, \quad \ldots\ldots \quad (a_2x_2 + a_3x_3)\, a_1{}^{n-1},$$

which are n in number.

These binary forms are the coefficients of the various powers of x_1 in the original ternary form, hence they can be obtained at once from f.

This is the fundamental result of Forsyth on algebraically complete sets of ternariants and from it readily flow all the results in the memoir cited. Similar methods may be applied to obtain the results of another paper of Forsyth's*.

215. We shall now explain an important principle due to Clebsch which establishes a connection between the invariants of binary forms and the contravariants of ternary forms. To facilitate the discussion we shall commence with a special case.

Suppose we require to find the condition that the line $u_x = 0$ should touch the conic $a_x{}^2 = b_x{}^2 = 0$. Let the points on the line be $y_1y_2y_3$ and $z_1z_2z_3$, so that

$$u_1 : u_2 : u_3 = y_2z_3 - y_3z_2 : y_3z_1 - y_1z_3 : y_1z_2 - y_2z_1.$$

The coordinates x_1, x_2, x_3 of any point on this line are of the form

$$\xi_1y_1 + \xi_2z_1, \quad \xi_1y_2 + \xi_2z_2, \quad \xi_1y_3 + \xi_2z_3,$$

and ξ_1, ξ_2 may clearly be regarded as the coordinates of a variable point on the line referred to the points y and z as base points.

We have

$$a_x = a_1x_1 + a_2x_2 + a_3x_3 = \xi_1 a_y + \xi_2 a_z,$$

* See *London Math. Soc. Proc.*, 1898.

and the points in which the line meets the conic are given by

$$(a_y\xi_1 + a_z\xi_2)^2 = (b_y\xi_1 + b_z\xi_2)^2 = \ldots = 0.$$

As the line touches the conic this expression regarded as a quadratic $\alpha_\xi^2 = \beta_\xi^2 = \ldots$ in ξ must have its invariant $(\alpha\beta)^2$ zero.

Hence $$(a_y b_z - a_z b_y)^2 = 0,$$

and by an identity already given

$$(a_y b_z - a_z b_y) = (abu);$$

hence we have $$(abu)^2 = 0$$

as the required condition.

Ex. If the line $u_x = 0$ cuts the conics $a_x^2 = 0$, $b_x^2 = 0$ in harmonic points, prove that $(abu)^2 = 0$.

Thus when the points on the line satisfy the invariant relation $(\alpha\beta)^2 = 0$, the line satisfies the relation $(abu)^2 = 0$.

This principle is true in general, that is to say, in order to pass from the invariant relation satisfied by the points in which the line meets any number of curves to the condition satisfied by the coordinates of the line, we have merely to change every factor of the type $(\alpha\beta)$, in the expression of the binary invariant, into a factor (abu).

The proof is very simple—suppose the equations of the curves are

$$a_x^m = 0, \quad b_x^n = 0, \text{ etc.},$$

and that y, z are two points on the line, then the points in which the line meets the curves are given by the binary forms

$$(a_y\xi_1 + a_z\xi_2)^m = 0, \quad (b_y\xi_1 + b_z\xi_2)^n = 0, \text{ etc.}$$

But an invariant of these forms is expressible in terms of factors of the type

$$(a_y b_z - a_z b_y)$$

entirely, and inasmuch as

$$a_y b_z - a_z b_y = (abu)$$

the result follows at once.

Thus, knowing the discriminant of a binary form of order n, we can find the tangential equation of a curve of the nth degree, for the discriminant being equated to zero gives the condition that two points of ntersection should coincide, and in this case the line touches the curve; *e.g.* the discriminant of a binary cubic being

$$(ab)^2 (ac) (bd) (cd)^2,$$

the tangential equation of the curve

$$a_x^3 = b_x^3 = c_x^3 = d_x^3 = 0$$

is

$$(abu)^2 (acu) (bdu) (cdu)^2 = 0.$$

216. The principle can be extended to covariants and mixed concomitants, as we shall explain by an example.

A line $u_x = 0$ cuts two conics $a_x^2 = 0$, $b_x^2 = 0$ in pairs of points PQ, RS; to find the pair of points harmonic to both P, Q and R, S.

We know that the pair of points harmonic to those given by

$$\alpha_\xi^2 = 0, \quad \beta_\xi^2 = 0,$$

are given by equating the Jacobian $(\alpha\beta)\,\alpha_\xi \beta_\xi$ to zero.

Now if u be the join of the points y and z, then ξ_1, ξ_2 are given by

$$(a_y \xi_1 + a_z \xi_2)^2 = 0, \quad (b_y \xi_1 + b_z \xi_2)^2 = 0,$$

hence the common harmonic pair are given by

$$(a_y b_z - a_z b_y)(a_y \xi_1 + a_z \xi_2)(b_y \xi_1 + b_z \xi_2) = 0,$$

or they satisfy

$$(abu)\, a_x b_x = 0,$$

and are therefore the points in which this conic meets the given line.

The extension to covariants in general will now be sufficiently obvious.

217. There are also dual methods which enable us to determine, for example, the locus of a point such that the pencils of tangents drawn from it to a number of fixed curves possess a given projective property.

After the study of a simple example it will be easy for the reader to enunciate and prove the necessary theorems.

Let us find the locus of a point such that the pairs of tangents drawn from it to two conics

$$u_\alpha^2 = 0, \quad u_\beta^2 = 0,$$

are harmonically conjugate.

If any two lines through the point are $v_x=0$, $w_x=0$, then the coordinates of a variable line through it are

$$\xi_1 v_1+\xi_2 w_1,\quad \xi_1 v_2+\xi_2 w_2,\quad \xi_1 v_3+\xi_2 w_3,$$

where ξ_1, ξ_2 may be regarded as the coordinates of a variable line in the pencil.

The tangents to the two conics are given by the binary quadratics

$$(u_1\alpha_1+u_2\alpha_2+u_3\alpha_3)^2,\ \text{or}\ (v_\alpha\xi_1+w_\alpha\xi_2)^2=0,$$

and

$$(v_\beta\xi_1+w_\beta\xi_2)^2=0.$$

And since these are harmonic

$$(v_\alpha w_\beta-v_\beta w_\alpha)^2=0,$$

but

$$v_\alpha w_\beta-v_\beta w_\alpha=(\alpha\beta x),$$

where x is the point of intersection of the lines v and w.

Hence the required locus is

$$(\alpha\beta x)^2=0.$$

If α and β are equivalent symbols, this gives the point equation of the conic whose tangential equation is

$$u_\alpha^2=u_\beta^2=0.$$

Ex. (i). Prove that if three binary quadratics a_x^2, b_x^2, c_x^2 are in involution, then $(ab)(bc)(ca)=0$; deduce the locus of a point such that the tangents from it to three conics form a pencil in involution, and state the correlative theorem.

Ex. (ii). Interpret the envelope loci $(abu)^4=0$, $(abu)^2(bcu)^2(cau)^2=0$ in connection with the quartic curve $a_x^4=b_x^4=c_x^4=0$; shew that the inflexional tangents touch each of the above curves, and are therefore in general completely determined as the common tangents.

Ex. (iii). A straight line p meets a cubic curve in P, Q, R, shew that

(i) there are two points H, H' on the line such that their polar conics touch the line;

(ii) the polar conic of H touches the line in H' and that of H' touches the line in H;

(iii) H and H' are the Hessian points of P, Q, R;

(iv) if p pass through a fixed point y the locus of H and H' is the quartic obtained by putting $u=(xy)$ in $(abu)^2 a_x b_x=0$;

(v) the equation of the quartic is $fP-C^2=0$, where $f=0$ is the cubic, C the first and P the second polar of y.

218. Quaternary Forms. The general quaternary form of order n is

$$\Sigma \frac{n!}{p!\,q!\,r!\,s!} x_1^p x_2^q x_3^r x_4^s,$$

where
$$p + q + r + s = n.$$

In accordance with the methods used in this work, we write this in the umbral form

$$(\alpha_1 x_1 + \alpha_2 x_2 + \alpha_3 x_3 + \alpha_4 x_4)^n,$$

and this in turn we denote by α_x^n.

To represent expressions of degree higher than unity in the coefficients, we have to introduce equivalent symbols β, γ, etc., so that

$$f = \alpha_x^n = \beta_x^n = \gamma_x^n = \text{etc.}$$

Denoting the determinant

$$\begin{vmatrix} \alpha_1 & \alpha_2 & \alpha_3 & \alpha_4 \\ \beta_1 & \beta_2 & \beta_3 & \beta_4 \\ \gamma_1 & \gamma_2 & \gamma_3 & \gamma_4 \\ \delta_1 & \delta_2 & \delta_3 & \delta_4 \end{vmatrix}$$

by $(\alpha\beta\gamma\delta)$ and extending the definitions of invariants and covariants, we can shew that every expression completely represented by factors of the type $(\alpha\beta\gamma\delta)$ is an invariant, and every expression represented by factors of the types $(\alpha\beta\gamma\delta)$ and α_x is a covariant; of course in a symbolical expression each symbol must occur to degree n.

Then using plane coordinates u, such that

$$u_1 x_1 + u_2 x_2 + u_3 x_3 + u_4 x_4 = 0$$

is the equation of the plane u, we have mixed concomitants containing the three types of factors

$$\alpha_x,\ (\alpha\beta\gamma\delta),\ (\alpha\beta\gamma u).$$

Finally, introducing a second plane v, we have a more comprehensive type of mixed concomitant expressed by factors of the four types

$$\alpha_x,\ (\alpha\beta\gamma v),\ (\alpha\beta\gamma u),\ (\alpha\beta u v).$$

There is no need to introduce a third set of plane coordinates w, because

$$(\alpha u v w)$$

is really of the form α_x, where x is the point of intersection of the planes u, v, w.

The reader will not have more than purely algebraical difficulties in extending the methods of §§ 211, 212 to four variables.

219. In illustration of the foregoing we may mention that if

$$f = \alpha_x^2 = \beta_x^2 = \gamma_x^2 = \delta_x^2$$

be a quaternary quadratic, then

(i) $(\alpha\beta\gamma\delta)^2$ is its discriminant;

(ii) $(\alpha\beta\gamma u)^2 = 0$ is the condition that the plane u should touch the quadric $f = 0$;

(iii) $(\alpha\beta uv)^2 = 0$ is the condition that the line of intersection of the planes u and v should touch the quadric;

(iv) $(\alpha uvw)^2 = 0$ is the condition that the point of intersection of the three planes u, v, and w should be on the quadric.

220. It is clear that when any invariant form expresses a relation of a projective nature between a straight line in space and a surface, factors of the type

$$(\alpha\beta uv)$$

must occur in its symbolical form, because the straight line is given in the first place as the intersection of two planes, which are u and v in this case.

It is easy to see that

$$(\alpha\beta uv) = \Sigma\,(\alpha_1\beta_2 - \alpha_2\beta_1)\,(u_3v_4 - u_4v_3);$$

the six quantities

$$u_2v_3 - u_3v_2,\ u_3v_1 - u_1v_3,\ u_1v_2 - u_2v_1,\ u_1v_4 - u_4v_1,\ u_2v_4 - u_4v_2,\ u_3v_4 - u_4v_3$$

are called the six coordinates of the line of intersection of the planes u and v. It can be verified that they are altered in the same ratio, when instead of the planes u, v we take any other two planes through the line.

Further, if x, y be any two points on the line, the quantities

$$x_1y_4 - x_4y_1,\ x_2y_4 - x_4y_2,\ x_3y_4 - x_4y_3,\ x_2y_3 - x_3y_2,\ x_3y_1 - x_1y_3,\ x_1y_2 - x_2y_1$$

are altered all in the same ratio, when instead of x, y we use two other points on the same line.

Finally, from the equations

$$u_x = 0, \quad u_y = 0, \quad v_x = 0, \quad v_y = 0,$$

it follows that

$$x_1 y_4 - x_4 y_1 = \rho\,(u_2 v_3 - u_3 v_2) = p_{14},$$
$$x_2 y_4 - x_4 y_2 = \rho\,(u_3 v_1 - u_1 v_3) = p_{24},$$
$$x_3 y_4 - x_4 y_3 = \rho\,(u_1 v_2 - u_2 v_1) = p_{34},$$
$$x_2 y_3 - x_3 y_2 = \rho\,(u_1 v_4 - u_4 v_1) = p_{23},$$
$$x_3 y_1 - x_1 y_3 = \rho\,(u_2 v_4 - u_4 v_2) = p_{31},$$
$$x_1 y_2 - x_2 y_1 = \rho\,(u_3 v_4 - u_4 v_3) = p_{12},$$

in other words, that the six coordinates of a line are either the six determinants of the array

$$\begin{vmatrix} x_1 & x_2 & x_3 & x_4 \\ y_1 & y_2 & y_3 & y_4 \end{vmatrix}$$

or those of the array

$$\begin{vmatrix} u_1 & u_2 & u_3 & u_4 \\ v_1 & v_2 & v_3 & v_4 \end{vmatrix}.$$

The expression $(\alpha\beta uv)$ is written $(\alpha\beta p)$ for convenience, but it must not be confused with the similar expression for a three-rowed determinant.

The discussion of concomitants involving line coordinates is complicated by the existence of the relation

$$p_{23}p_{14} + p_{31}p_{24} + p_{12}p_{34} = 0$$

between the six coordinates of a line.

221. As a simple illustration of the use of the above methods, we shall take the extension of the principle of Clebsch explained in § 215.

To explain this extension we take a particular problem, viz. to find the condition that the plane $u_x = 0$ should touch the quadric surface represented by

$$\alpha_x^2 = \beta_x^2 = \gamma_x^2 = 0.$$

Let x, y, z be three points in the plane, then any other point (X_1, X_2, X_3, X_4) in the plane is given by

$$X_r = \xi_1 x_r + \xi_2 y_r + \xi_3 z_r, \quad r = 1, 2, 3, 4,$$

and ξ_1, ξ_2, ξ_3 may be regarded as the coordinates of (X_1, X_2, X_3, X_4) in the plane referred to x, y, z as fundamental triangle.

Now
$$\sum_1^4 \alpha_r X_r = \alpha_x \xi_1 + \alpha_y \xi_2 + \alpha_z \xi_3,$$

hence the equation of the conic in which the plane meets the quadric may be written

$$(\alpha_x \xi_1 + \alpha_y \xi_2 + \alpha_z \xi_3)^2 = 0.$$

If the plane touch the quadric, this conic must be a pair of straight lines; the condition for this is

$$\begin{vmatrix} \alpha_x & \alpha_y & \alpha_z \\ \beta_x & \beta_y & \beta_z \\ \gamma_x & \gamma_y & \gamma_z \end{vmatrix}^2 = 0,$$

and inasmuch as the determinant here written is equal to

$$(\alpha\beta\gamma u),$$

the condition required is

$$(\alpha\beta\gamma u)^2 = 0.$$

The reader will have no difficulty in applying the same method to find the envelope of a plane cutting a surface in a curve which has a definite projective property, of which the invariant equivalent is given.

222. We can in like manner solve problems leading to line coordinates, *e.g.* to find the condition that a line should cut the quadrics

$$\alpha_x^2 = 0, \quad \beta_x^2 = 0,$$

in pairs of points harmonically conjugate.

Let x, y be two points on the line, then any other point X is given by

$$X_r = \xi_1 x_r + \xi_2 y_r, \quad r = 1, 2, 3, 4,$$

and
$$\alpha_X = \alpha_x \xi_1 + \alpha_y \xi_2,$$

so that the quadratics giving the ratio in which the line is divided by the quadrics are

$$(\alpha_x \xi_1 + \alpha_y \xi_2)^2 = 0, \quad (\beta_x \xi_1 + \beta_y \xi_2)^2 = 0.$$

The condition that the pairs should be harmonically conjugate is

$$(\alpha_x\beta_y - \alpha_y\beta_x)^2 = 0,$$

or

$$(\alpha\beta p)^2 = 0,$$

where p is a typical coordinate of the line.

The construction of other examples and the discovery of the dual principles will be easy for the reader who has grasped the corresponding results relating to ternary forms.

CHAPTER XIII.

TERNARY FORMS (*continued*).

223. In this chapter we shall give some further theorems relating to ternary forms; mainly such as arise in obtaining the irreducible systems in the simpler cases. In connection with Hilbert's proof of Gordan's theorem it has already been remarked that this proof can be extended without much difficulty to the case of ternary forms; we are therefore certain that for any number of such forms the irreducible system is finite, *i.e.* there exists a finite number of invariant forms in terms of which all others can be expressed as rational integral algebraic functions. But although the existence of the finite system is thus established, no clue is given as to the method of discovering such systems and, as a matter of fact, very little is known in this branch of the subject. The more important complete systems known up to the present are those for one, two* and three† ternary quadratics, that for a single ternary cubic‡ form, and quite recently Gordan has given the complete system for two quaternary quadratics§.

224. In his paper on the ternary cubic Gordan gave a systematic method of searching for all irreducible forms by proceeding from those of degree $r-1$ to those of degree r in the manner of Chapter V.; in any but the simplest cases the application of this method requires uncommon skill and patience. The reader will find an introductory sketch of the method at the end of this chapter; we shall content ourselves with obtaining the complete systems for one and two quadratics by an elegant and ingenious

* Clebsch, *Geometrie*, p. 174.

† Ciamberlini, *Battaglini*, Vol. XXIV.

‡ Gordan, *Math. Ann.*, Band I.

§ *Math. Ann.*, Band LVI.

process, also due to Gordan, which has been applied with success to three, and would probably be equally successful in dealing with any number of, quadratics. The complete systems are in all cases, except for linear forms or one quadratic form, so large that any method for their discovery will involve a great deal of labour.

Ex. (i). For a single linear form a_x the complete system is a_x.

Ex. (ii). For two linear forms a_x and b_x the complete system is

$$a_x, \quad b_x, \quad (abu).$$

Ex. (iii). For any number of linear forms $a_x, b_x, c_x, \ldots$ the complete system consists of

(α) Forms of the type a_x,

(β) Contravariants of the type (abu),

(γ) Invariants of the type (abc).

These results all follow immediately from the symbolical expressions for invariant forms. Cf. § 84.

225. One Quadratic Form. Let the form be

$$f = a_x^2 = b_x^2 = c_x^2 = \text{etc.},$$

then the typical invariant factors are

$$(abc), \quad (abu), \quad a_x,$$

and there are three invariant forms

$$(abc)^2, \quad (abu)^2, \quad a_x^2;$$

we proceed to shew that these constitute the complete irreducible system.

I. *Any invariant form P containing the factor (abc) can be transformed so as to contain the factor $(abc)^2$.*

First suppose that no two of the letters a, b, c occur again in the same factor, then

$$P = (abc)\, a_p b_q c_r\, M,$$

where M does not contain a, b or c, and

$$p = (de), \ (du) \text{ or } x \text{ etc.}$$

Hence by interchanging a, b, c in all possible ways

$$\begin{aligned} 6P &= (abc)\, M\, (a_p b_q c_r - a_p b_r c_q - a_q b_p c_r + a_q b_r c_p - a_r b_q c_p + a_r b_p c_q) \\ &= (abc)\, M\, (abc)\, (pqr), \\ &= (abc)^2\, M\, (pqr). \end{aligned}$$

Thus P involves the factor $(abc)^2$ which is an invariant, hence the other factor must be an invariant form* and P is therefore reducible.

Secondly, suppose that

$$P = (abc)(abp)\, c_q N,$$

where

$$p = d \text{ or } u,$$

$$q = (de), \quad (du) \text{ or } x.$$

Then

$$\begin{aligned} 3P &= (abc)\, N \{(abp)\, c_q + (bcp)\, a_q + (cap)\, b_q\} \\ &= (abc)\, N\, (abc)\, p_q \\ &= (abc)^2\, N p_q, \end{aligned}$$

and as before it follows that P is reducible.

Thus any form containing the factor (abc) is reducible, and we may neglect such forms in future.

II. *A form Q containing the factor (abu) can be expressed in terms of forms containing the factor $(abu)^2$ and reducible forms.*

Rejecting forms containing factors of the type (cde), we have the three following possibilities:

(i) $Q = (abu)(acu)(bdu)\, M,$

(ii) $Q = (abu)(acu)\, b_x M,$

(iii) $Q = (abu)\, a_x b_x M.$

As regards (i), on interchanging a and b,

$$\begin{aligned} 2Q &= (abu)\, M \{(acu)(bdu) + (cbu)(adu)\} \\ &= (abu)\, M \{-(bau)(cdu)\} \\ &= (abu)^2 (cdu)\, M, \end{aligned}$$

and Q is reducible.

(ii) Here

$$\begin{aligned} 2Q &= (abu)\, M \{(acu)\, b_x - (bcu)\, a_x\} \\ &= (abu)\, M \{(abu)\, c_x - (abc)\, u_x\} \\ &= (abu)^2 M c_x - (abc)\, u_x (abu)\, M, \end{aligned}$$

and the latter part is reducible because it contains the factor (abc). Thus again Q is reducible.

* We shall return to this point later, see § 226.

(iii) In this case Q vanishes, as we see by interchanging a and b.

Now every symbolical expression representing an invariant form contains a factor of one of the types (abc) or (abu) unless it be a_x^2; it follows that all such are reducible except

$$(abc)^2, \quad (abu)^2, \quad a_x^2,$$

and these form the complete irreducible system.

Of course, to be strictly accurate we should add to this and every complete system the identical concomitant u_x.

Ex. (i). Express $(abu)(acu)\,b_x c_x$ in terms of the irreducible system.

Ex. (ii). Prove the symbolical identity

$$\{ab\,(pq)\} = a_p b_q - a_q b_p.$$

226. The foregoing discussion leads to some general remarks that will be useful in the sequel.

The artifice of replacing (abc) by a_p, or, what is the same thing, writing

$$(bc) = p,$$

and more explicitly

$$(b_2c_3 - b_3c_2) = p_1; \quad (b_3c_1 - b_1c_3) = p_2; \quad (b_1c_2 - b_2c_1) = p_3,$$

will be frequently used. (It is clear that if b and c be two straight lines then p is their point of intersection.)

In particular the contravariant $(abu)^2$ will be written u_α^2, thus

$$\alpha = (ab),$$

and the invariant of the quadratic is c_α^2.

The α's may be regarded as expressible in terms of the original symbols, but we shall more often regard them as independent symbols; the contravariant above will thus be

$$u_\alpha^2 = u_\beta^2 = u_\gamma^2 = \ldots,$$

the original form is

$$f = a_x^2 = b_x^2 = c_x^2 = \ldots,$$

and the invariant is

$$a_\alpha^2 = a_\beta^2 = b_\alpha^2 = \text{etc.}$$

Of course in replacing an α by symbols belonging to f, we must use no symbol already occurring in the expression because then that symbol would occur more than twice in the transformed expression.

The α's being formed from the a's and b's, as line coordinates are formed from point coordinates, it follows that the α's are contragredient to a and u but cogredient with x.

We shall now return to a point that arose in the discussion of a single quadratic, where we deduced from first principles that if a_p, a_q, a_r be all invariant factors then (pqr) is an invariant factor —it will be shewn here how to express it in terms of factors of the types (abc), (abu) and a_x.

The possible forms for p, q, r are (de), (du) and x; if each of them were x, (pqr) would vanish, so we may suppose that p for example is not x.

Let $p = (dv)$, where v is either e or u, then

$$(pqr) = \{(dv)\, qr\} = d_q v_r - d_r v_q, \qquad (\S\ 209)$$

but since b_q is an invariant factor, d_q and u_q are both invariant factors; for if in such a factor we replace a letter of the type a by another of the same type the factor is still invariant, while if we replace it by u the resulting factor either vanishes or is an invariant factor.

Hence (pqr) can be expressed entirely in terms of factors of the types (abc), (abu) and a_x; it is therefore an invariant factor.

227. To make the matter clearer we shall illustrate it by some examples.

Suppose a quadratic is

$$a_x^2 = b_x^2 = c_x^2 = \dots,$$

and its contravariant is

$$u_\alpha^2 = u_\beta^2 = u_\gamma^2 = \dots,$$

then a_α, a_β, b_α, b_β etc. are invariant factors.

Hence $(\alpha\beta x)$ must be an invariant factor and $(\alpha\beta x)^2$ an invariant form of the quadratic. We proceed to express it in terms of the members of the irreducible system as follows:

$$(\alpha\beta x)^2 = (\overline{ab}\ \beta x)^2 = (a_\beta b_x - a_x b_\beta)^2$$
$$= a_\beta^2 \,.\, b_x^2 + a_x^2 \,.\, b_\beta^2 - 2a_x b_x a_\beta b_\beta,$$

and we have now to deal with the term $a_x b_x a_\beta b_\beta$.

As a guide to the further reduction we remark that since a_β is a factor the form must involve the factor a_β^2 after suitable transformation.

Thus

$$a_x b_x a_\beta b_\beta = a_x b_x (acd)(bcd) = (bcd)\, a_x (acd)\, b_x,$$

and $$3a_x b_x a_\beta b_\beta = (bcd)\, a_x \{(acd)\, b_x - (abd)\, c_x - (acb)\, d_x\},$$

or since $$(acd)\, b_x - (abd)\, c_x - (acb)\, d_x = a_x (bcd),$$

$$3a_x b_x a_\beta b_\beta = (bcd)^2 a_x^2.$$

Consequently

$$(\alpha\beta x)^2 = 2a_\alpha^2 \,.\, b_x^2 - \tfrac{2}{3} a_\alpha^2 \,.\, b_x^2 = \tfrac{4}{3} a_\alpha^2 \,.\, b_x^2,$$

and except for a numerical factor the invariant form is equal to the product of the original form and its invariant.

The following example will shew the advantage of introducing symbols like α, β, γ above in geometrical investigations.

Let $a_x = 0$, $b_x = 0$, $c_x = 0$ be the equations of three straight lines, then if $\alpha = (bc)$, $\beta = (ca)$, $\gamma = (ab)$,

$$(\beta\gamma) = (abc)\, a.$$

This is not surprising, because α, β, γ are the points of intersection of b, c; c, a; a, b respectively and $(\beta\gamma)$ must therefore give the line joining β and γ, that is a*.

The equation of the straight line joining the point y to the point of intersection of $a_x = 0$ and $b_x = 0$ is

$$(\overline{ab}, xy) = 0 \text{ or } a_x b_y - a_y b_x = 0,$$

which is the well-known form.

The line joining the point (a, b) to the point (c, d) is

$$(\overline{ab}, \overline{cd}, x) = 0, \text{ or } (acd)\, b_x - (bcd)\, a_x = 0.$$

This may also be written $(abc)\, d_x - (abd)\, c_x = 0$, and the two forms are equivalent in virtue of the fundamental identity from which both can, in fact, be at once inferred.

Now let a', b', c' be the sides of a second triangle and α', β', γ' its angular points.

* The similarity of these ideas to the methods of Grassmann's *Ausdehnungslehre* will not escape the notice of the reader who is acquainted with the Calculus.

The lines joining corresponding vertices of the two triangles are $(\alpha\alpha')$, $(\beta\beta')$, $(\gamma\gamma')$, and these are concurrent if

$$(\overline{\alpha\alpha'},\ \overline{\beta\beta'},\ \overline{\gamma\gamma'})=0,$$

i.e. if
$$(\alpha\beta\beta')(\alpha'\gamma\gamma')-(\alpha'\beta\beta')(\alpha\gamma\gamma')=0.$$

But since $(\alpha\beta)=(abc)\,c$ etc. this condition is

$$(abc)(a'b'c')\,\{c_{\beta'}b'_{\gamma}-c'_{\beta}b_{\gamma'}\}=0,$$

or
$$(abc)(a'b'c')\,\{(a'c'c)(abb')-(acc')(a'b'b)\}=0,$$

or
$$(abc)(a'b'c')\,\{(abb')(a'cc')-(acc')(a'bb')\}=0,$$

or
$$(abc)(a'b'c')\,(\overline{aa'},\ \overline{bb'},\ \overline{cc'})=0.$$

Now $(\overline{aa'},\ \overline{bb'},\ \overline{cc'})=0$ is the condition that the points of intersection of corresponding sides should be collinear.

Hence when the joins of corresponding vertices are concurrent the intersections of corresponding sides are collinear—a well-known theorem. A direct analytical proof by the ordinary methods without using a particular triangle of reference is by no means easy.

Ex. (i). Prove that the cross ratio of the range in which $u_x=0$ is met by $a_x=0$, $b_x=0$, $c_x=0$, $d_x=0$ is $-\dfrac{(bcu)(adu)}{(abu)(cdu)}$; hence the general equation of a conic touching the four lines is

$$(abu)(cdu)+\lambda(bcu)(adu)=0.$$

If the six lines a, b, c, d, e, f touch a conic, then

$$(abe)(cde)(bcf)(adf)=(bce)(ade)(abf)(cdf).$$

Ex. (ii). State and prove the results dual to those of (i).

Ex. (iii). If the points $\alpha, \beta, \gamma, \delta, \epsilon, \zeta$ be on a conic and

$$(\alpha\beta)=p,\quad(\beta\gamma)=q,\quad(\gamma\delta)=r,\quad(\delta\epsilon)=p',\quad(\epsilon\zeta)=q',\quad(\zeta\alpha)=r',$$

then
$$(\overline{pp'},\ \overline{qq'},\ \overline{rr'})=0.$$

Deduce Pascal's theorem and prove Brianchon's theorem in the same way.

228. Two Quadratics. Suppose the quadratics are

$$f=a_x^2=b_x^2=c_x^2=\text{etc.},$$
$$f'=a'^2_x=b'^2_x=c'^2_x=\text{etc.}$$

and that their contravariants are

$$\phi=u_\alpha^2=u_\beta^2=u_\gamma^2=\text{etc.}$$
$$\phi'=u_{\alpha'}^2=u_{\beta'}^2=u_{\gamma'}^2=\text{etc.}$$

respectively, then their invariants are a_α^2 and $a'^2_{\alpha'}$ respectively.

The types of invariant factors are

$$(abc) = a_\alpha, \quad (aba') = a'_\alpha, \quad (ab'c') = a_{\alpha'},$$

$$(a'b'c') = a'_{\alpha'}, \quad (abu) = u_\alpha, \quad (a'b'u) = u_{\alpha'},$$

$$(aa'u), \quad a_x, \quad a'_x,$$

and in the course of the work we shall use the additional types

$$(\alpha\beta x), \quad (\alpha'\beta' x), \quad (\alpha\alpha' x)$$

which, as has been pointed out, certainly are invariant factors.

There are four types of symbols occurring in any expression, viz.

$$a, \; a', \; \alpha, \; \alpha',$$

and the leading idea of the investigation is to reduce the number of symbols in the expression for any invariant form to a minimum.

Thus $(abu)^2$ involves two symbols a, b, but it can be written u_α^2 and then only involves one.

Again $$2a_{\alpha'} b_{\alpha'} a_x b_x$$

is equivalent to $$(a_{\alpha'} b_x - a_x b_{\alpha'})^2$$

except for reducible terms; then since

$$a_{\alpha'} b_x - a_x b_{\alpha'} = (\alpha\alpha' x)$$

we see that $$a_{\alpha'} b_{\alpha'} a_x b_x$$

can be expressed in terms of reducible forms and forms containing a smaller number of symbols.

This example will make clear what we mean in the sequel by reducing the number of symbols.

I. *If a symbolical product contain a factor of the type* a_α *or* $a'_{\alpha'}$ *it is reducible.*

For let $$P = (abc)\, a_p b_q c_r M,$$

then, as in § 225,

$$6P = (abc)^2 . (pqr)\, M,$$

and (pqr) is an invariant factor; hence P is reducible.

This shews incidentally how factors of the type $(\alpha\beta x)$ arise naturally in the course of the work, for we might have

$$p = \alpha, \; q = \beta, \; r = x.$$

The case in which a and b occur again in the same factor, *e.g.* (abu), is treated as in § 225.

II. *If a symbolical product contain a factor of the type* (abv), *where* v *is of the type* a' *or is* u, *then it can be so transformed as to contain an additional factor* (abw).

Suppose in fact that

$$P = (abv)\, a_p b_q M,$$

then

$$2P = (abv)\,(a_p b_q - a_q b_p)\, M$$
$$= (abv)\,(\overline{ab},\, pq)\, M$$
$$= v_\alpha\,(\alpha pq)\, M,$$

and (αpq) can be expressed in terms of α and the symbols occurring in p and q.

Hence in the transformed expression there are two factors involving the symbol α.

A like theorem applies to α', and here we see a great advantage in introducing these symbols, because just as when one a occurs in a product there must be another, so when α occurs the expression can be transformed so as to be of degree two in α.

III. *If an expression contain a factor of the type* $(\alpha\beta y)$ *or* $(\alpha'\beta' y)$ *it is reducible.*

For let

$$P = (\alpha\beta y)\, M, \text{ where } y \text{ must be of the type } \alpha,\ \alpha' \text{ or } x.$$

Then since

$$(\alpha\beta y) = (\overline{ab},\, \beta y) = a_\beta b_y - b_\beta a_y,$$
$$P = a_\beta b_y M - b_\beta a_y M$$

and by I. the form P contains the invariant a_α^2 as a factor.

Summing up these results we infer that any concomitant other than a_x^2, a'^2_x, a_α^2, $a'^2_{\alpha'}$ must be composed of factors of the types

$$a_x,\ a'_x,\ a_{\alpha'},\ a'_\alpha,\ u_\alpha,\ u_{\alpha'},\ (\alpha\alpha' x),\ (aa'u),$$

and if one α occurs there must be another present after a suitable transformation.

IV. *Any expression containing two equivalent symbols can be expressed in terms of concomitants that are either reducible or contain a smaller number of symbols.*

There are two cases according as the equivalent symbols are of the type a or α, so that we have to consider the expressions

$$a_p a_q b_r b_s M \text{ and } \varpi_\alpha \rho_\alpha \sigma_\beta \tau_\beta M,$$

with the exactly similar ones

$$a'_p a'_q b'_r b'_s M \text{ and } \varpi_{\alpha'} \rho_{\alpha'} \sigma_{\beta'} \tau_{\beta'} M.$$

Since $$a_p b_r - a_r b_p = (apr),$$
and $$a_q b_s - a_s b_q = (aqs),$$
it follows that expressions derived from

$$a_p a_q b_r b_s M,$$

by permuting the symbols p, q, r, s, differ from the original expression by forms in which the two symbols a and b are replaced by the simple symbol α.

Hence if any one of the three expressions

$$a_p a_q b_r b_s M,\ a_p a_r b_q b_s M,\ a_p a_s b_q b_r M$$

can be expressed in terms of reducible forms and forms containing fewer symbols the same is true of the other two.

The same result can be easily established for the three expressions

$$\varpi_\alpha \rho_\alpha \sigma_\beta \tau_\beta M,\ \varpi_\alpha \sigma_\alpha \rho_\beta \tau_\beta M,\ \varpi_\alpha \tau_\alpha \rho_\beta \sigma_\beta M;$$

in fact the difference between any two is reducible by III.

Consider now the expression

$$a_p a_q b_r b_s M.$$

There are only five possible types for p, q, r, s, viz. α, α', (bu), $(a'u)$ and x; of these α may be neglected because if it occur the form is reducible, and (bu) may be neglected because if it occur we can replace a and b at once by the single symbol α.

Hence there are only three remaining possibilities, viz. α', $(a'u)$, x, and of p, q, r, s some two must certainly be of the same type.

First suppose that two are identical, say p and r, then

$$a_p a_r b_q b_s M = a_p^2 . b_q b_s M;$$

thus $a_p a_r b_q b_s M$ is reducible, and hence $a_p a_q b_r b_s M$ can be expressed in terms of reducible forms and forms containing fewer symbols.

Similar reasoning applies to the expression

$$\varpi_\alpha \rho_\alpha \sigma_\beta \tau_\beta M$$

when any two of the symbols ϖ, ρ, σ, τ are identical; here, as in the other case, there are three possibilities, viz. a, a' and u, so some two must be of the same type.

Proceed now to the general case in which no two of the letters p, q, r, s are identical.

In this case some two must be equivalent without being identical.

Suppose the equivalent symbols be p and r, then there are three cases to consider.

(i) Let $p = r = x$. Here p and r are identical and the result has been established already.

(ii) Let $p = \alpha'$, $r = \beta'$, then

$$a_p a_r b_q b_s M = a_{\alpha'} a_{\beta'} b_q b_s M.$$

And since the symbols α', β' each occur once the expression can be transformed by II. so that each occurs twice, hence

$$a_p a_r b_q b_s M = a_{\alpha'} a_{\beta'} \gamma_{\alpha'} \delta_{\beta'} N.$$

But $a_{\alpha'} a_{\beta'} \gamma_{\alpha'} \delta_{\beta'} N$ falls under that class of expression

$$\varpi_{\alpha'} \rho_{\alpha'} \sigma_{\beta'} \tau_{\beta'} M$$

in which two of the symbols ϖ, ρ, σ, τ are identical, for $\varpi = \sigma = a$; hence, as we have seen, it can be expressed in terms of simpler forms.

Thus $a_p a_r b_q b_s M$ can be expressed in terms of forms that are either reducible or contain fewer symbols, and the same is true of $a_p a_q b_r b_s M$.

(iii) Let $p = (a'u)$, $r = (b'u)$, then

$$a_p a_r b_q b_s M = (aa'u)(ab'u)\, b_q b_s M.$$

And the latter expression can be written

$$(a'au)(b'au)\, a'_q b'_s M',$$

or

$$a'_{p'} b'_{r'} a'_q b'_s M',$$

where $p' = (au)$, $r' = (au)$, *i.e.* p' and r' are identical.

Thus, as we have already shewn

$$a'_{p'}a'_{q'}b'_{r'}b'_{s'}M'$$

can be expressed in terms of simpler forms; hence $a_p a_r b_q b_s M$ can be expressed in terms of forms that are either reducible or contain fewer symbols, and the same is true of $a_p a_q b_r b_s M$.

This completely establishes the Theorem IV. when the equivalent symbols are of the type a or a'.

We have still to consider the general expression

$$\varpi_\alpha \rho_\alpha \sigma_\beta \tau_\beta M,$$

where ϖ, ρ, σ, τ are each of the types a', $(\alpha' x)$ or u. The case in which some two are identical has already been discussed and reasoning exactly similar to that of (i), (ii), (iii) above enables us to prove Theorem IV. for the general expression.

The theorem we have just proved simplifies vastly the evolution of the irreducible system inasmuch as it shews that in an irreducible form there cannot be more than one symbol of any of the types

$$a, \quad a', \quad \alpha, \quad \alpha'.$$

A further limitation is imposed by the following theorem.

V. *A form containing both the factors* $(aa'u)$ *and* $(\alpha\alpha' x)$ *may be rejected in constructing the irreducible system.*

In fact by direct multiplication

$$(aa'u) \times (\alpha\alpha' x) = \begin{vmatrix} a_\alpha & a'_\alpha & u_\alpha \\ a_{\alpha'} & a'_{\alpha'} & u_{\alpha'} \\ a_x & a'_x & u_x \end{vmatrix},$$

hence if a form involve both these factors it can be transformed so as to only contain the simpler factors

$$a_\alpha,\ a'_{\alpha'},\ a'_\alpha,\ a_{\alpha'},\ u_\alpha,\ u_{\alpha'},\ a_x,\ a'_x \text{ and } u_x.$$

With the aid of the above five theorems the problem of the complete system is reduced to a very simple one; in fact we have only to write down such products of factors of the types

$$a_\alpha,\ a'_{\alpha'},\ a_{\alpha'},\ a'_\alpha,\ u_\alpha,\ u_{\alpha'},\ a_x,\ a'_x,\ (aa'u),\ (\alpha\alpha' x),$$

as satisfy the conditions implied in the theorems.

Since no two equivalent symbols can occur in the same product we need introduce no more symbols; further a_α, $a'_{\alpha'}$ can only occur in the invariants a_α^2 and $a'_{\alpha'}{}^2$ respectively; therefore it only remains to write down all products of the factors

$$a_{\alpha'},\ a'_\alpha,\ u_\alpha,\ u_{\alpha'},\ a_x,\ a'_x,\ (aa'u),\ (\alpha\alpha'x),$$

in which every letter except u and x which appears at all appears twice and no more; besides the two last factors must not appear in the same product.

Hence we have the following forms:

(A) $a_\alpha^2,\ a'_{\alpha'}{}^2,\ a_{\alpha'}^2,\ a'_\alpha{}^2,\ u_\alpha^2,\ u_{\alpha'}^2,\ a_x^2,\ a'_x{}^2,\ (aa'u)^2,\ (\alpha\alpha'x)^2,$

obtained by squaring each factor.

(B) $a_{\alpha'}a_x u_{\alpha'},\ a_{\alpha'}a_x(\alpha\alpha'x)u_\alpha,\ a_{\alpha'}a_x(\alpha\alpha'x)a'_\alpha a'_x,$

$a_{\alpha'}(aa'u)a'_x,\ a_{\alpha'}(aa'u)a'_\alpha u_\alpha u_{\alpha'},$

containing the factor $a_{\alpha'}$.

(C) $a'_\alpha a'_x u_\alpha,\ a'_\alpha a'_x(\alpha\alpha'x)u_{\alpha'},\ a'_\alpha(aa'u)a_x,$

containing a'_α but not $a_{\alpha'}$.

(D) $u_\alpha(\alpha\alpha'x)u_{\alpha'},$

containing u_α but neither a'_α nor $a_{\alpha'}$.

(E) $a_x(aa'u)a'_x,$

containing a_x but neither a'_α nor $a_{\alpha'}$.

There are twenty forms in all, viz.:

Four Invariants,

$$a_\alpha^2 = A_{111},\ a'_{\alpha'}{}^2 = A_{222},\ a_{\alpha'}^2 = A_{122},\ a'_\alpha{}^2 = A_{112}.$$

Four Covariants,

$$a_x^2 = S_1,\ a'_x{}^2 = S_2,\ (\alpha\alpha'x)^2 = F,\ (\alpha\alpha'x)a_{\alpha'}a'_\alpha a_x a'_x.$$

Four Contravariants,

$$u_\alpha^2 = \Sigma_1,\ u_{\alpha'}^2 = \Sigma_2,\ (aa'u)^2 = \Phi,\ (aa'u)a_{\alpha'}a'_\alpha u_\alpha u_{\alpha'}.$$

Eight Mixed Concomitants,

$a_x a_{\alpha'} u_{\alpha'},\ a'_x a'_\alpha u_\alpha,\ (aa'u)a_x a'_x,\ (aa'u)a_{\alpha'}a'_x u_{\alpha'},$

$(aa'u)a'_\alpha a_x u_\alpha,\ (\alpha\alpha'x)u_\alpha a_{\alpha'}a_x,\ (\alpha\alpha'x)u_{\alpha'}a'_\alpha a'_x,$

$(\alpha\alpha'x)u_\alpha u_{\alpha'}.$

To these twenty should be added the identical concomitant u_x.

It should be noticed in conclusion that, although it follows from the five theorems of this article that every other invariant form is expressible in terms of these as a rational integral algebraic function, it has not been shewn that these twenty forms are themselves irreducible. Cf. note, p. 131.

Ex. (i). Prove that

$$(a\beta\gamma)^2 = \tfrac{4}{3} A_{111}^2, \quad (a\beta\gamma')^2 = \tfrac{4}{3} A_{111} A_{112}.$$

Ex. (ii). Prove that

$$a_\alpha a'_\alpha a_{\alpha'} a'_{\alpha'} = \tfrac{1}{3} A_{111} A_{222},$$

and thence that $$(\alpha\alpha', \overline{aa'})^2 = \tfrac{1}{3} A_{111} A_{222} + A_{112} A_{122}.$$

Ex. (iii). Prove that

$$a_x a'_x a_\alpha a'_\alpha = \tfrac{1}{3} A_{111} S_2.$$

Ex. (iv). Prove that

$$a_\alpha u_{\alpha'} a_x (\alpha\alpha' x) = 0.$$

229. A complete irreducible system of invariant forms may be regarded as giving rise to two inquiries.

(i) What is the geometrical meaning of each member of the irreducible system?

(ii) What is the expression in terms of the members of that system of the invariant forms which arises in the analytical treatment of a given problem?

To the first of these inquiries an answer can generally be given, provided a sufficiently complex geometrical apparatus be allowed, but it commonly happens that the significance of some members of the system is so remote as to render them of little geometrical importance.

The second inquiry is naturally unanswerable until the problem be named, and thus all we can do is to illustrate it by the discussion of some simple problems.

Before going further it may be well to add that the second inquiry is the really important one; in a manner it includes the first as a particular case, and in fact there being no direct method of proceeding from the invariant to the geometrical meaning, the answer of the first inquiry is obtained fortuitously in pursuing the second. If it be not obtained we should console ourselves with the reflexion that the uninterpreted forms are of little geometrical interest in the present state of knowledge; besides, if we regard the algebra as being merely helpful to geometry in the analytical formulation of results, it does not follow that everything in the algebra need be taken seriously from the geometrical point of view*.

In spite of what we have said, we shall begin the geometrical theory of two conics by interpreting the members of the irreducible system; it will be seen that they are all of importance in elementary geometry.

230. Geometrical Theory of two Quadratics. The irreducible system. The forms themselves when equated to zero represent two conics, viz.

$$a_x^2 = 0, \quad a'^2_x = 0,$$

which we write $S_1 = 0$, $S_2 = 0$, and call S_1 and S_2.

Invariants. The meaning of the invariant of a single conic is expressed by the fact that $a_\alpha^2 = 0$ is the condition for two straight lines.

Again, $a'^2_\alpha = 0$ is satisfied when the point equation of S_1 involves only product terms and the line equation S_2 only squared terms, or when the point equation of S_1 involves only squared terms, and the line equation of S_2 only product terms.

It is then the poristic condition† that there should be an infinite number of triangles inscribed in the first conic and self-conjugate in the second, or, what is the same thing, that there should be an infinite number of triangles self-conjugate to the first conic and inscribed in the second. We shall consider this type of invariant more fully in the next chapter.

Covariants, etc. The simple ones are well known, viz.

$a_x^2 = 0$ is the condition that (x) should be on the first conic,

$u_\alpha^2 = 0$ is the condition that (u) should touch the first conic.

* Reducibility itself is a purely algebraical idea and the reader will soon convince himself that it is generally hard to obtain any geometrical satisfaction from the fact that a covariant is reducible. See a curious remark of Clifford's, *Collected Papers*, p. 81.

† The sufficiency of the condition can easily be seen by taking a triangle inscribed in the first conic and having two pairs of vertices conjugate with respect to the second for fundamental triangle.

Of those not involving determinantal factors, there remain

$$a'_\alpha a'_x u_\alpha, \quad a_{\alpha'} a_x u_{\alpha'}.$$

It will be sufficient to consider one of these. Now $a'_x a'_\alpha u_\alpha = 0$ is the polar of the point $(\alpha_1 u_\alpha, \alpha_2 u_\alpha, \alpha_3 u_\alpha)$ with respect to $a'^2_x = 0$.

This point is the pole of u with respect to the first conic

$$u_\alpha^2 = 0 \text{ or } a_x^2 = 0$$

and hence $a'_x a'_\alpha u_\alpha = 0$, when u is constant, represents the polar with respect to the second conic of the pole of u with respect to the first. When x is constant, it is the tangential equation of the pole with respect to the first conic of the polar of x with respect to the second.

Again $(aa'u)^2 = 0$ is the equation of the envelope of the lines cutting the two conics in harmonic point pairs, and in like manner $(\alpha\alpha' x)^2$ is the locus of a point such that the tangents drawn to the two conics from it are harmonic line pairs.

Consider now the forms involving one determinant factor.

(i) $(aa'u) a_x a'_x = 0$ is the condition that the lines

$$a_x a_y = 0, \; a'_x a'_y = 0, \; u_y = 0,$$

should be concurrent, y being the current coordinate, *i.e.* when u is constant it is the locus of points whose polars intersect on the line u, and when x is constant it is the equation of the point of intersection of the polars of x with respect to the conics.

(ii) $(\alpha\alpha' x) u_\alpha u_{\alpha'}$ is dual to the last; when x is constant it represents the envelope of a line such that the line joining its poles passes through x, and when u is constant it represents the equation of the line joining the poles of u with respect to the two conics.

(iii) $(aa'u) a'_\alpha a_x u_\alpha$ is the Jacobian with respect to y of the quantities

$$a_x a_y, \; a'_\alpha u_\alpha a'_y, \; u_y.$$

Equated to zero, these represent three straight lines, namely, the polar of x with respect to the first conic, the polar with respect to the second conic of the pole of u with respect to the first, and the line u. The vanishing of the concomitant is the condition that the three lines should be concurrent; hence, for example, when u is constant, the equation represents a straight line constructed as follows: let P_1 be the pole of u with respect to the first conic, and v the polar of P_1 with respect to the second conic, then the line represented is the polar of the point (u, v) with respect to the first conic.

(iv) $(\alpha\alpha' x)\, a_{\alpha'} a_x u_\alpha$ is the Jacobian with respect to the v's of

$$u_\alpha v_\alpha, \quad v_{\alpha'} a_{\alpha'} a_x, \quad v_x,$$

and hence equated to zero is the condition that these three points should be collinear.

Now $u_\alpha v_\alpha = 0$ represents the pole of u with respect to the first conic, and $a_x a_{\alpha'} v_{\alpha'} = 0$ represents the pole with respect to the second conic of the polar of x with respect to the first. Thus we can interpret the concomitant geometrically. See also Ex. ix. p. 294.

(v) $(aa'u)\, a_{\alpha'} a'_\alpha u_\alpha u_{\alpha'}$ is a contravariant of the third class, and moreover the only such contravariant that can be built up from the members of the system. But the angular points of the common self-conjugate triangle must be given by equating to zero a contravariant of the third class. Hence the contravariant in question represents the vertices of the common self-conjugate triangle of the conics $a_x^2 = 0$ and $a'^2_x = 0$.

We may prove this as follows*: The three conics $u_\alpha^2 = 0$, $u_{\alpha'}^2 = 0$, and $(aa'u)^2 = 0$ have a common self-conjugate triangle if there be a proper triangle self-conjugate to the first two, as we see by taking it for triangle of reference. Further, when three conics are referred to their common self-conjugate triangle, the Jacobian of their tangential equations is the tangential equation of the vertices.

Hence the Jacobian of u_α^2, $u_{\alpha'}^2$, $(aa'u)^2$ represents the vertices of the common self-conjugate triangle.

Now it is

$$(aa'u)\, u_\alpha u_{\alpha'} \overline{(aa'\alpha\alpha')} = (aa'u)\, u_\alpha u_{\alpha'} (a_\alpha a'_{\alpha'} - a_{\alpha'} a'_\alpha).$$

But
$$(aa'u)\, u_\alpha u_{\alpha'} a_\alpha a'_{\alpha'} = 0,$$
for otherwise it would involve the factors a_α^2 and $a'^2_{\alpha'}$, whereas its total degree is only six, so that the remaining factor would be a contravariant of zero degree which is impossible.

* See also Ex. vi. p. 293.

Hence the Jacobian in question equated to zero is equivalent to

$$(aa'u)\, u_a u_{a'} a_{a'} a'_a = 0.$$

(vi) By exactly similar methods we can shew that

$$(\alpha\alpha' x)\, a_{\alpha'} a'_\alpha a'_x a_x = 0$$

represents the sides of the common self-conjugate triangle.

231. Let us now consider some problems bearing on two conics with a view to illustrating the second inquiry of § 229.

(i) To find the equation of the reciprocal of the conic S_2 with respect to S_1.

If y be a point on the reciprocal conic its polar with respect to $a_x^2 = 0$ must touch $u_{\alpha'}^2 = 0$.

The polar is

$$a_x a_y \equiv b_x b_y = 0,$$

and if $u_x = 0$ touches $u_{\alpha'}^2 = 0$

$$0 = u_{\alpha'}^2 = \tfrac{1}{4}\left(\frac{\partial^2}{\partial v_1 \partial x_1} + \frac{\partial^2}{\partial v_2 \partial x_2} + \frac{\partial^2}{\partial v_3 \partial x_3}\right)^2 \{u_x^2 v_{\alpha'}^2\}.$$

Hence in this case

$$0 = \left(\frac{\partial^2}{\partial v_1 \partial x_1} + \frac{\partial^2}{\partial v_2 \partial x_2} + \frac{\partial^2}{\partial v_3 \partial x_3}\right)^2 \{a_x a_y\, b_x b_y\, v_{\alpha'}^2\}$$

$$= 4 a_{\alpha'} b_{\alpha'}\, a_y b_y,$$

so that the reciprocal is

$$a_x b_x\, a_{\alpha'} b_{\alpha'} = 0 \quad \text{or} \quad -(a_x b_{\alpha'} - a_{\alpha'} b_x)^2 + a_x^2 b_{\alpha'}^2 + b_x^2 a_{\alpha'}^2 = 0$$

or $$0 = 2 a_x^2 b_{\alpha'}^2 - (\alpha\alpha' x)^2 = 2 A_{122} \,.\, S_1 - F,$$

which expresses the equation in terms of the irreducible system.

(ii) To find the point equation of the covariant conic

$$(aa'u)^2 = 0.$$

The point equation of $u_\alpha^2 = 0$ is $(\alpha\beta x)^2 = 0$, and hence in our case the point equation is

$$\{\overline{aa'}\,\overline{bb'}\,x\}^2 = 0,$$

or $$\{(abb')\, a'_x - (a'bb')\, a_x\}^2 = 0,$$

i.e. $$(abb')^2 a'^2_x + (a'bb')^2 a_x^2 - 2\,(abb')(a'bb')\, a_x a'_x = 0.$$

Now

$$(abb')(a'bb')\,a_x a'_x = -(abb')(a'ab')\,a'_x b_x$$
$$= \tfrac{1}{2}(abb')\,a'_x\{a_{\alpha'}b_x - b_{\alpha'}a_x\}, \text{ where } \alpha' = (a'b')$$
$$= \tfrac{1}{2}(\alpha\alpha' x)(abb')\,a'_x$$
$$= -\tfrac{1}{2}(\alpha\alpha' x)(aba')\,b'_x = -\tfrac{1}{4}(\alpha\alpha' x)\{a'_\alpha b'_x - a'_x b'_\alpha\} = \tfrac{1}{4}(\alpha\alpha' x)^2,$$

thus the point equation is

$$(abb')^2\,a'^2_x + (a'bb')^2\,a_x^2 - \tfrac{1}{2}(\alpha\alpha' x)^2 = 0,$$

or
$$A_{112}\,.\,S_2 + A_{122}\,S_1 - \tfrac{1}{2}F = 0.$$

(iii) To find the locus of the point of intersection of harmonic pairs of tangents to F and S.

The locus for $u_\alpha^2 = 0$ and $u_{\alpha'}^2 = 0$ is $(\alpha\alpha' x)^2 = 0$ and in our case the locus is accordingly

$$\{\overline{aa'}\,\alpha x\}^2 = 0$$

or $\quad (a_x a'_\alpha - a'_x a_\alpha)^2 = 0,$ *i.e.* $S_1 A_{112} + S_2 A_{111} - 2a_x a'_x a_\alpha a'_\alpha = 0.$

To reduce the last term we perceive that it must contain A_{111} and we write it

$$(abc)\,a'_x(a'bc)\,a_x = \tfrac{1}{3}(abc)\,a'_x\{(a'bc)\,a_x + (caa')\,b_x - (aa'b)\,c_x\}$$
$$= \tfrac{1}{3}(abc)\,a'_x(bca)\,a'_x = \tfrac{1}{3}A_{111}\,.\,S_2.$$

The equation required is

$$S_1 A_{112} + S_2 A_{111} - \tfrac{2}{3}S_2 A_{111} = 0,$$

or
$$A_{111} S_2 + 3\,S_1 A_{112} = 0.$$

The following gives an easy means of verifying the results of (i), (ii), (iii) above and of the examples which follow.

Taking the conics in the canonical forms

$$S_1 = a_1 x_1^2 + a_2 x_2^2 + a_3 x_3^2$$
$$S_2 = x_1^2 + x_2^2 + x_3^2$$

we have

$$\Sigma_1 = 2(a_2 a_3 u_1^2 + a_3 a_1 u_2^2 + a_1 a_2 u_3^2)$$
$$\Sigma_2 = 2(u_1^2 + u_2^2 + u_3^2),$$

$$A_{111} = 6a_1 a_2 a_3, \quad A_{222} = 6, \quad A_{112} = 2(a_2 a_3 + a_3 a_1 + a_1 a_2),$$
$$A_{122} = 2(a_1 + a_2 + a_3),$$

$$\Phi = (a_2 + a_3)\,u_1^2 + (a_3 + a_1)\,u_2^2 + (a_1 + a_2)\,u_3^2$$
$$F = 4\{a_1(a_2 + a_3)\,x_1^2 + a_2(a_3 + a_1)\,x_2^2 + a_3(a_1 + a_2)\,x_3^2\}.$$

Ex. (i). The locus of points whose polars with respect to S_2 cut S_1, S_2 in pairs of points harmonically conjugate is

$$(aa'b')(aa'c')b'_x c'_x = 0,$$

or

$$A_{222}S_1 - 3A_{122}S_2 = 0.$$

Ex. (ii). Prove that

$$(abu)^2 = \tfrac{1}{8}\Omega^2 (a_x^2 b_y^2 u_z^2)$$

where

$$\Omega = \begin{vmatrix} \frac{\partial}{\partial x_1} & \frac{\partial}{\partial x_2} & \frac{\partial}{\partial x_3} \\ \frac{\partial}{\partial y_1} & \frac{\partial}{\partial y_2} & \frac{\partial}{\partial y_3} \\ \frac{\partial}{\partial z_1} & \frac{\partial}{\partial z_2} & \frac{\partial}{\partial z_3} \end{vmatrix},$$

and hence that

$$(\overline{a\beta}, \overline{a'\beta'}, u)^2 = \tfrac{16}{9} A_{111} A_{222} (aa'u)^2.$$

Interpret this result geometrically.

Ex. (iii). The line equation of $F=0$ is

$$(\overline{aa'}\ \overline{\beta\beta'}\ u)^2 = 0,$$

or

$$(a\beta\beta')^2 u_{a'}^2 + (a'\beta\beta')^2 u_a^2 - \tfrac{1}{2}(\overline{a\beta}\ \overline{a'\beta'}\ u)^2.$$

Thence

$$(\overline{aa'}\ \overline{\beta\beta'}\ u)^2 = \tfrac{4}{3}\{3A_{111}A_{122}\Sigma_2 + 3A_{222}A_{112}\Sigma_1 - 2A_{111}A_{222}\Phi\}.$$

Cf. Ex. (ii).

Ex. (iv). Deduce from Ex. (iii) that the discriminant of F is

$$(\overline{aa'}\ \overline{\beta\beta'}\ \overline{\gamma\gamma'})^2 = \tfrac{8}{27}(9A_{112}A_{122} - A_{111}A_{222})A_{111}A_{222}.$$

Ex. (v). Deduce from Ex. (ii),

$$(\overline{aa'}\ \overline{bb'}\ \overline{cc'})^2 = \tfrac{1}{6}(9A_{112}A_{122} - A_{111}A_{222}).$$

Ex. (vi). Prove that the discriminant of $(aa'u)a_x a'_x$ is

$$(aa'u)(bb'u)(cc'u)(ab'c')(a'bc).$$

Hence if the conic $(aa'u)a_x a'_x = 0$ be two straight lines

$$(aa'u)a_{a'}a'_a u_a u_{a'} = 0.$$

Thence verify that this equation represents the angular points of the common self-conjugate triangle and work out the dual results.

Ex. (vii). Prove that if the point equation of a conic be

$$\lambda_1 S_1 + \lambda_2 S_2 = 0,$$

then its tangential equation is

$$\lambda_1^2\Sigma_1 + 2\lambda_1\lambda_2\Phi + \lambda_2^2\Sigma_2 = 0,$$

and its discriminant is

$$\lambda_1^3 A_{111} + 3\lambda_1^2\lambda_2 A_{112} + 3\lambda_1\lambda_2^2 A_{122} + \lambda_2^3 A_{222}.$$

Ex. (viii). Prove that the point equation of any covariant conic is of the form

$$\lambda_1 S_1 + \lambda_2 S_2 + \lambda F = 0,$$

where λ_1, λ_2, λ are invariants, and that its line equation is

$$\lambda_1{}^2\Sigma_1+2\lambda_1\lambda_2\Phi+\lambda_2{}^2\Sigma_2+\tfrac{2}{3}\lambda\lambda_1(A_{111}\Sigma_2+3A_{122}\Sigma_1)+\tfrac{2}{3}\lambda\lambda_2(A_{222}\Sigma_1+3A_{112}\Sigma_2)$$
$$+\lambda^2\tfrac{4}{9}(3A_{111}A_{122}\Sigma_2+3A_{222}A_{112}\Sigma_1-2A_{111}A_{222}\Phi)=0.$$

Ex. (ix). The locus of points whose polars with respect to S_1 and F meet on the line u is

$$(a\ \overline{aa'}\ u)\,a_x\,(aa'x)=0,$$

or

$$(aa'x)\,a_{\alpha'}u_{\alpha}a_x=0.$$

This gives a simpler interpretation of the irreducible form than that given in § 230.

Ex. (x). Use the method of Ex. (ix) to interpret

$$(aa'u)\,a'_{\alpha}a_x u_{\alpha}=0.$$

Examples (ix) and (x) enable us to interpret all the irreducible mixed concomitants very simply in connection with the conics S_1, S_2, F, Φ.

Ex. (xi). The locus of points whose polars with respect to F and Φ meet on u is

$$A_{112}(aa'x)\,a_x u_{\alpha}a_{\alpha'}+A_{122}(aa'x)\,a'_x u_{\alpha'}a_{\alpha'}=0.$$

(Use the point equation of Φ given in Ex. (ii).)

Ex. (xii). The equation of the line joining the poles of u with respect to F and Φ is

$$A_{222}A_{112}(aa'u)\,u_{\alpha}a_x a'_{\alpha}+A_{111}A_{122}(aa'u)\,u_{\alpha'}a'_x a_{\alpha'}=0.$$

(Use the result of Ex. (iii).)

Ex. (xiii). Calculate the four invariants of the conics S_1 and F.

We have

$$C_{222}=\tfrac{8}{27}(9A_{112}A_{122}-A_{111}A_{222})A_{111}A_{222}\text{ by Ex. (iv).}$$
$$C_{122}=(a\ \overline{aa'}\ \beta\beta')^2=\tfrac{4}{3}(3A_{111}A^2_{122}+A_{222}A_{112}A_{111}).$$
$$C_{112}=\tfrac{4}{3}A_{111}A_{122}.$$
$$C_{111}=A_{111}.$$

232. Gordan's general method. Consider a concomitant of any number of forms containing the r letters $a, b, c, \ldots h, k$. If we replace each factor of the type (aku) by a_x, each factor of the type (abk) by (abu) and delete all factors k_x the resulting expression is still an invariant form but only of degree $r-1$ because it only contains $r-1$ symbols.

Hence reversing the operation, *i.e.* replacing a proper number of factors of the type a_x by (aku), some of the type (abu) by (abk) and introducing a sufficient power of k_x we can deduce the form of degree r from one of degree $r-1$. Applying this process in all possible ways to all invariant forms of degree $r-1$ we certainly obtain all invariant forms of degree r. We pause to explain more precisely what we mean by applying the reverse process in all possible ways to an invariant ϕ. Suppose in fact that the newly introduced letter k belongs to a form of order n, then we replace any p factors of the type a_x by (aku) any q factors of the type (abu) by (abk) and multiply by k_x^r where of course we must have

$$p+q+r=n,$$

and we must take all values of p, q, r satisfying this equation subject, of course, to the condition that there exist p factors of the given type to alter and a like restriction for q.

For example, let the original form be

$$(abu)\,(acu)\,b_x c_x,$$

and suppose the new letter d like a, b, c belongs to a form of order 2.

We can only change 2 factors at most in this case and we have

$$p+q=1 \text{ or } 2,$$

there are five cases,

$$p=1,\ q=0;\ p=0,\ q=1;\ p=2,\ q=0;\ p=1,\ q=1;\ p=0,\ q=2;$$

and in following out the case $p=1$, $q=1$, $r=0$, for example, we deduce four forms from the given one, viz.

$$(abd)\,(acu)\,(bdu)\,c_x,$$
$$(abd)\,(acu)\,b_x\,(cdu),$$
$$(abu)\,(acd)\,(bdu)\,c_x,$$
$$(abu)\,(acd)\,b_x\,(cdu).$$

The above indicates the general method of procedure, but some introductory lemmas are necessary to render the method of any practical value—for example for all we know at present a form of degree $r-1$ which is identically zero might lead to irreducible forms of degree r, and we need hardly say that this would complicate matters enormously.

233. The reader will have observed some likeness between the above and the methods used in Chapter v. on binary forms to deduce the invariants of degree r from those of degree $r-1$. This analogy will be further exemplified in the rest of the argument.

Consider the effect of the operator Ω^p on the expression

$$\phi k_y^n u_z^p$$

ϕ being a covariant of degree $r-1$ and Ω being the operator

$$\begin{vmatrix} \dfrac{\partial}{\partial x_1} & \dfrac{\partial}{\partial x_2} & \dfrac{\partial}{\partial x_3} \\ \dfrac{\partial}{\partial y_1} & \dfrac{\partial}{\partial y_2} & \dfrac{\partial}{\partial y_3} \\ \dfrac{\partial}{\partial z_1} & \dfrac{\partial}{\partial z_2} & \dfrac{\partial}{\partial z_3} \end{vmatrix}.$$

As we saw in the last chapter, the result is the sum of a number of terms each containing the determinantal factors of ϕ together with p of the form (aku), there being in the end p fewer factors of the type a_x, $(n-p)$ factors k_y and no factors of the type u_z in each term.

Next operate on the complete result of which a typical term is ψk_y^{n-p}

$$Q \equiv \frac{\partial^2}{\partial u_1 \partial y_1} + \frac{\partial^2}{\partial u_2 \partial y_2} + \frac{\partial^2}{\partial u_3 \partial y_3}$$

q times in succession.

Since the effect of this operator on $(abu)\, k_y^{n-p}$ is to give a multiple of $(abk)\, k_y^{n-p-1}$, it is very easy to see that the effect of the operator q times is to give a number of terms each containing the same determinantal factors as ψ, with the exception that q factors of the type (abu) are replaced by (abk); the power of k_y remaining in each term is k_y^{n-p-q}, and k replaces u in q places in all possible ways.

Hence if we operate with $Q^q \Omega^p$ on the product $\phi k_y^n u_z^p$ and then put $y=x$ we obtain the sum of a number of terms each of which has the same determinantal factors as ϕ except that in any q of them u is replaced by k and p new ones of the type (aku) are introduced while p factors of the type a_x disappear and finally the factor k_x^{n-p-q} is introduced in each term.

Consequently in the resulting expression there will be contained every term derived from ϕ in the reverse process explained above with these definite values of p and q.

We may conveniently call

$$Q^q \Omega^p \{\phi,\ k_y^n u_z^p\}_{y=x}$$

the transvectant of ϕ and k_x^n whose indices are p, q, the order of the indices being essential, and we have the result that every concomitant of degree r is the sum of a number of terms each occurring as a term in a transvectant of a form of degree $r-1$ with k_x^n. Naturally when there are different forms we have to introduce in turn a symbol belonging to each.

234. We next require certain relations that exist among the terms of the same transvectant, and to establish them we shall alter our notation for a moment.

Suppose in fact that

$$\phi = a_x^{(1)} a_x^{(2)} \ldots a_x^{(p')} u_{\alpha_1} u_{\alpha_2} \ldots u_{\alpha_{q'}} M,$$

where M contains neither u nor x.

In each term of the transvectant

$$\{\phi,\ k_x^n\}^{p,q}$$

we have q of the u's replaced by k, p of the terms $a_x^{(r)}$ by (aku).

We shall call two terms N_1 and N_2 *adjacent* when $p+q-1$ of the factors affected in ϕ to obtain them are common, and two cases will arise according as the remaining factor affected is an a_x or a u_α.

In the first case we have, supposing that $a_x^{(r)}$ and $a_x^{(s)}$ are the additional altered factors in the two terms respectively,

$$\begin{aligned} N_1 - N_2 &= N\{(a^{(r)} ku)\, a_x^{(s)} - (a^{(s)} ku)\, a_x^{(r)}\} \\ &= N\{(a^{(r)} a^{(s)} k)\, u_x - (a^{(r)} a^{(s)} u)\, k_x\}. \end{aligned}$$

Now $N(a^{(r)}a^{(s)}k)$ is a term in

$$(\phi_1, k_x^n)^{p-1, q+1}$$

where ϕ_1 is deduced from ϕ by changing $a_x^{(r)}a_x^{(s)}$ into $(a^{(r)}a^{(s)}u)$ and further

$$N(a^{(r)}a^{(s)}u)\,k_x$$

is a term in $\{\phi_2, k_x^n\}^{p-1, q}$

where ϕ_2 is the same as ϕ_1.

In the second case let u_{a_i} and u_{a_j} be the additional factors altered in the two terms, then

$$(N_1 - N_2) = N'\{u_{a_i}k_{a_j} - u_{a_j}k_{a_i}\}$$

$$= -N'\{a_i a_j \overline{ku}\}$$

and this latter is a term in

$$\{\phi_3, k_x^n\}^{p+1, q-1}$$

where ϕ_3 is deduced from ϕ by changing $u_{a_i}u_{a_j}$ into $(a_i a_j x)$.

Now if we call the sum of the order and class of a function its *grade* it is evident that ϕ_1, ϕ_2, ϕ_3 are each of grade less by unity than that of ϕ.

Further between any two terms of the transvectant we can insert a number of others such that any two of the whole sequence are adjacent in our sense of the word and accordingly we have the important theorem:

"*The difference between any two terms of the same transvectant can be expressed in terms of transvectants of functions of lower grade than ϕ with k_x^n.*"

Thus if we consider our function ϕ of degree $r-1$ in ascending order of grade we need only retain one term out of each transvectant that we consider—or if we please the sum of any number of terms will equally serve our purpose and in particular the transvectant itself might be used. It follows at once that if a transvectant contains a single reducible term it may be neglected entirely.

Again, if there be a linear relation among a number of the forms of degree $r-1$ there will be a linear relation among the transvectants of given index formed from them, so that we need only consider linearly independent forms of degree $r-1$. In particular, zero forms of degree $r-1$ can be entirely put out of account.

A knowledge of the irreducible system up to and including degree $r-1$ therefore gives us immediately all the forms ϕ of which transvectants need be considered, for we have only to include the irreducible forms of degree $r-1$ and such simple products of the others as are of total degree $r-1$.

We have now effected our purpose of making the method at present under discussion of real value, and we proceed to illustrate it by reference to the complete system for a single quadratic.

235. Quadratics. We have here five different sets of indices for transvectants, namely

$$p=1, q=0; \quad p=0, q=1; \quad p=2, q=0; \quad p=1, q=1; \quad p=0, q=2.$$

Consider now how far products need be taken into account, if $p+q=1$ then all products may be neglected because only one factor is modified and hence some terms of the transvectant of a product are certainly reducible. If $p=2, q=0$, then a transvectant of $\phi_1\phi_2$ with k_x^2 will contain reducible terms unless the orders of ϕ_1 and ϕ_2 are both unity, also for $p=0, q=2$, we need only consider in like manner products of two forms whose class is unity. If $p=1$, $q=1$, we need only consider the product of two forms when one is of zero order and the other of zero class. Throughout products of more than two forms need not be taken into account.

Further, in every case pure invariants of degree $r-1$ can give rise to no new forms.

236. Single Quadratic Form. Of the first degree we have

$$a_x^2 = b_x^2 = c_x^2 = \ldots$$

Proceeding to the second degree we have

$$(abu)\, a_x b_x \text{ for } p=1, q=0$$

and
$$(abu)^2 \text{ for } p=2, q=0$$

of which the first is zero.

Third degree. From $(abu)^2$ we get

$$(abc)(abu)\, c_x \, (01)$$
$$(abc)^2 \, (02)$$

and from $a_x^2 b_x^2$ we can only get reducible forms.

Now
$$(abc)\{(abu)\, c_x\} = \tfrac{1}{3}(abc)\{(abu)\, c_x + (bcu)\, a_x + (abc)\, u_x\}$$
$$= \tfrac{1}{3}(abc)^2 u_x^2$$

so that of the third degree we have only the invariant $(abc)^2$.

Fourth degree. From $(abu)^2 c_x^2$ we need only consider

$$(abd)(abu)(cdu)\, c_x$$

arising from $p=1$ and $q=1$.

This is $\tfrac{1}{4}(abu)(cdu)\{(abd)\, c_x + (bdc)\, a_x + (dca)\, b_x + (cab)\, d_x\} = 0$.

For further forms we need only consider transvectants of products of powers of c_x^2 and $(abu)^2$ with $p+q \not> 2$.

These all contain reducible terms and hence there are no new forms so that the complete system consists of a_x^2, $(abu)^2$ and $(abc)^2$ as already indicated.

For the case of three quadratics and incidentally two see Baker, *Camb. Phil. Trans.* Vol. xv.

CHAPTER XIV.

APOLARITY (*continued*).

237. Apolar Conics. Two conics S and S' whose equations in point and line coordinates are respectively

$$S \equiv a_x^2 \equiv ax_1^2 + bx_2^2 + cx_3^2 + 2fx_2x_3 + 2gx_3x_1 + 2hx_1x_2 = 0,$$

and $\Sigma' \equiv u_{\alpha'}^2 \equiv A'u_1^2 + B'u_2^2 + C'u_3^2 + 2F'u_2u_3 + 2G'u_3u_1 + 2H'u_1u_2 = 0$

are said to be apolar when the invariant $a_{\alpha'}^2$, or what is the same thing,

$$aA' + bB' + cC' + 2fF' + 2gG' + 2hH'$$

vanishes.

This relation between the two conics is not a symmetrical one, inasmuch as it arises from the point equation of one and the line equation of the other; it is convenient to have an alternative name shewing the exact relation between the curves. For reasons to be explained later we shall say that S is harmonically inscribed in S', and that S' is harmonically circumscribed to S.

The curves are also apolar when $a'_{\alpha}{}^2 = 0$, but in this case S' is harmonically inscribed in S.

As is well known from the geometry of conics, $a_{\alpha'}^2 = 0$ is the condition that there should exist an infinite number of triangles self-conjugate to S and circumscribed to S', or an infinite number of triangles inscribed in S and self-conjugate to S'—in fact the equations $a_x^2 = 0$ and $u_{\alpha'}^2 = 0$ can be so transformed that the first has no product terms and the second has no square terms, or that the first has no square terms and the second has no product terms.

The relation $a_{\alpha'}^2 = 0$ is linear in the coefficients of the equations

$$a_x^2 = 0$$

and

$$u_{\alpha'}^2 = 0,$$

hence a conic apolar to the conics $S_1, S_2, \ldots S_r$ is apolar to any conic

$$\lambda_1 S_1 + \lambda_2 S_2 + \ldots + \lambda_r S_r = 0.$$

Further if these r conics are linearly independent there are $(6-r)$ linearly independent conics apolar to them. In particular there is a unique conic apolar to (harmonically circumscribed to) five given linearly independent conics.

The same remarks apply to ρ given conics

$$\Sigma_1' = 0, \quad \Sigma_2' = 0, \ldots \Sigma_\rho' = 0,$$

and in particular there is a unique conic apolar to (harmonically inscribed in) five given conics.

238. Particular Cases. (i) If $a_x^2 = 0$ represents two straight lines and $a_{\alpha'}^2 = 0$, the two lines are conjugate with respect to $u_{\alpha'}^2 = 0$. If $a_x^2 = 0$ represents two straight lines coinciding in l then the line l touches the conic $u_{\alpha'}^2 = 0$.

(ii) If $u_{\alpha'}^2 = 0$ represents two points then these points are conjugate with respect to $a_x^2 = 0$. If $u_{\alpha'}^2 = 0$ represents two points coinciding in l then the point l lies on the conic $a_x^2 = 0$.

All these statements can be verified immediately by using the apolar condition expressed in terms of actual coefficients or symbolically, *e.g.* if $a_x^2 = v_x w_x$ the apolar condition is

$$0 = \{a_x^2 u_{\alpha'}^2\}^{0,2} = (v_x w_x, \ u_{\alpha'}^2)^{0,2} = v_{\alpha'} w_{\alpha'},$$

which is the condition that the lines $v_x = 0$, $w_x = 0$ should be conjugate with respect to $u_{\alpha'}^2 = 0$.

239. Ex. (i). *If two pairs of opposite vertices of a complete quadrilateral are conjugate with respect to a given conic so also is the third pair.*

Let the conic be $a_x^2 = 0$, and suppose the two pairs of opposite vertices are given tangentially by

$$u_p = 0, \ u_{p'} = 0; \quad u_q = 0, \ u_{q'} = 0.$$

The general equation of a conic inscribed in the quadrilateral is then

$$\lambda u_p u_{p'} + \mu u_q u_{q'} = 0,$$

and since $u_p u_{p'} = 0$ and $u_q u_{q'} = 0$ are both apolar to $a_x^2 = 0$ it follows that every conic inscribed in the quadrilateral is apolar to $a_x^2 = 0$. But the third pair of opposite vertices is one such conic, hence these remaining vertices are conjugate with respect to the given conic. We shall call such a quadrilateral a quadrilateral harmonically inscribed in the conic $a_x^2 = 0$.

Ex. (ii). *Four conics have in general one common harmonic quadrilateral.*

Let the conics be S_1, S_2, S_3, S_4, then the apolar system is of the type

$$\lambda_1\Sigma_1+\lambda_2\Sigma_2=0,$$

consequently the apolar conics in general all touch four fixed straight lines. The opposite vertices of the quadrilateral formed by these lines taken in pairs constitute conics of the apolar system, and hence pairs of opposite vertices are conjugate with respect to each of our four conics. Hence the quadrilateral is harmonically inscribed in each of the given conics.

Ex. (iii). *A triangle ABC and its polar triangle with respect to a conic are in perspective.* For if the polars of B and C meet the sides CA and AB respectively in Q, R, and the line QR meet BC in P, then the quadrilateral formed by BC, CA, AB and the line PQR has two pairs of opposite vertices, viz. (B, Q), (C, R) conjugate with respect to the conic; therefore (A, P) are conjugate with respect to the conic, or the polar of A meets BC in P. Thus the polars of A, B, C meet the opposite sides in three collinear points, and they therefore form a triangle in perspective with ABC.

Ex. (iv). If $\qquad u_x^{(1)}=0,\ u_x^{(2)}=0,\ u_x^{(3)}=0,\ u_x^{(4)}=0$

be the sides of a quadrilateral harmonic with respect to the conic $S=0$, then we have

$$S=\lambda_1 u^{(1)}{}_x{}^2+\lambda_2 u^{(2)}{}_x{}^2+\lambda_3 u^{(3)}{}_x{}^2+\lambda_4 u^{(4)}{}_x{}^2.$$

For let two pairs of opposite vertices be (pp') and (qq'); then apolar to the conics (pp') and (qq') we have the five conics

$$u^{(1)}{}_x{}^2=0,\ u^{(2)}{}_x{}^2=0,\ u^{(3)}{}_x{}^2=0,\ u^{(4)}{}_x{}^2=0 \text{ and } S=0.$$

But the first four are linearly independent and hence S is a linear combination of them.

240. Some interesting applications can also be made to the metrical geometry of conics.

In fact, suppose that the tangential equation of the circular points at infinity (I, J) is

$$\phi\equiv u_\gamma{}^2=0.$$

Then a conic apolar to ϕ has I, J for conjugate points and is therefore a rectangular hyperbola.

Again, the tangential equation of a circle whose centre is p is of the form

$$u_p{}^2=\lambda\phi$$

where λ varies with the radius.

If a circle C be apolar to a conic

$$\Sigma = u_a^2 = 0$$

then the director circle of the conic cuts the circle C orthogonally.

Use rectangular Cartesian coordinates, and let the equations of the conic and circle respectively be

$$Al^2 + 2Hlm + Bm^2 + 2Gl + 2Fm + C = 0,$$

and
$$x^2 + y^2 + 2gx + 2fy + c = 0,$$

so that we have
$$A + B + 2gG + 2fF + cC = 0.$$

But the equation of the director circle of the conic is

$$C(x^2 + y^2) - 2Gx - 2Fy + A + B = 0,$$

and this cuts the given circle at right angles if

$$-2g\frac{G}{C} - 2f\frac{F}{C} - \frac{A+B}{C} - c = 0,$$

i.e. if
$$2gG + 2fF + cC + A + B = 0$$

which is precisely the condition of apolarity.

The director circle of a conic inscribed in a triangle cuts the self-conjugate circle orthogonally.

For since the self-conjugate circle has the triangle for a self-conjugate triangle and the conics are inscribed in the triangle, each of the conics is apolar to the circle. Hence their director circles cut the self-polar circle at right angles. Or thus,—the system apolar to the inscribed conics is of the form

$$\lambda p_x^2 + \mu q_x^2 + \nu r_x^2 = 0,$$

where $p_x = 0$, $q_x = 0$, $r_x = 0$ represent the sides of the triangle. By suitably choosing λ, μ, ν this equation may be made to represent a circle, and from the form of its equation it is the self-polar circle of the triangle.

The locus of the centre of a circle which has two fixed pairs of conjugate lines is a rectangular hyperbola.

In fact, suppose the lines are

$$\left.\begin{matrix} p_x = 0, & q_x = 0 \\ r_x = 0, & s_x = 0 \end{matrix}\right\}.$$

The system apolar to the tangential system of conics having these two pairs of conjugate lines is

$$\lambda p_x q_x + \mu r_x s_x = 0.$$

There is one value of the ratio $\lambda : \mu$ for which this represents a rectangular hyperbola.

Let $S = 0$ be the rectangular hyperbola in question and let

$$u_p^2 = \lambda\phi$$

be one of the circles.

Then S is apolar to ϕ, because $S = 0$ is a rectangular hyperbola, and as it is apolar to

$$u_p^2 = \lambda\phi$$

it is apolar to u_p^2. Hence the point p must lie on S and therefore S is the centre locus of the circles.

In general when a rectangular hyperbola S is apolar to a circle Σ the centre of the circle lies on the rectangular hyperbola.

241. Apolar Curves in general. The two curves whose equations are

$$\left.\begin{aligned} f &\equiv a_x^m = 0 \\ \text{and} \qquad \phi &\equiv u_\alpha^n = 0 \end{aligned}\right\}$$

are said to be apolar when the form $a_\alpha^n a_x^{m-n}$, which we denote by ψ, is identically zero.

Except for a numerical multiple we have

$$\begin{aligned} \psi &= \left(\frac{\partial^2}{\partial u_1 \partial x_1} + \frac{\partial^2}{\partial u_2 \partial x_2} + \frac{\partial^2}{\partial u_3 \partial x_3}\right)^n a_x^m u_\alpha^n \\ &= (a_x^m,\ u_\alpha^n)^{0,n}. \end{aligned}$$

The following are analogous to theorems on binary forms.

I. *The form ϕ is apolar to any polar of f whose order is not less than n.*

For $$\{a_x^{m-r} a_y^r, u_\alpha^n\}^{0,n} = a_\alpha^n a_x^{m-r-n} a_y^r$$
and this is zero as we see by polarizing the identity

$$a_\alpha^n a_x^{m-n} = 0$$

r times with respect to y.

II. *The form f is apolar to any form which contains ϕ as a factor and whose class does not exceed m.*

For if ϕ' be any form of class n' we have

$$\{f, \phi\phi'\}^{0, n+n'} = \{(f, \phi)^{0,n}, \phi'\}^{0,n'}$$
$$= 0,$$

since $(f, \phi)^{0,n}$ vanishes identically. Hence f is apolar to $\phi\phi'$. We have supposed that $m \geqslant n$ hitherto. Exactly similar remarks apply to the case in which $n > m$.

The search for the forms of given class (n) apolar to a given form f is facilitated by the fact that the necessary and sufficient conditions for ϕ are that it should be apolar to every $(m-n)$th polar of f.

For an $(m-n)$th polar is

$$a_x^n a_y^{m-n}$$

and this is apolar to u_α^n if

$$a_\alpha^n a_y^{m-n} = 0.$$

But if this relation be true for all values of y then the form ϕ is apolar to f.

242. Ex. (i). *A ternary cubic has three linearly independent apolar conics.*

For the first polars of the cubic are linear combinations of

$$\frac{\partial f}{\partial x_1}, \frac{\partial f}{\partial x_2}, \frac{\partial f}{\partial x_3}$$

which are three linearly independent quadratic forms. Hence there are three linearly independent conics apolar to all first polars and therefore apolar to the cubic itself.

Ex. (ii). *A ternary quartic has an apolar conic only when the determinant of the coefficients of its second differential coefficients vanishes.*

For an apolar conic must be apolar to all second polars and they are linear combinations of

$$\frac{\partial^2 f}{\partial x_1^2}, \frac{\partial^2 f}{\partial x_2^2}, \frac{\partial^2 f}{\partial x_3^2}, \frac{\partial^2 f}{\partial x_2 \partial x_3}, \frac{\partial^2 f}{\partial x_3 \partial x_1}, \frac{\partial^2 f}{\partial x_1 \partial x_2}.$$

In general there is no conic apolar to each of these six, but there will be an apolar conic if the six be not linearly independent, *i.e.* if the determinant of six rows and six columns be zero.

243. General Theory of Curves which possess an Apolar Conic. By using suitable coordinates the analysis of ternary forms apolar to a given conic may be reduced to that of binary forms.

Suppose that the fixed conic is

$$x_1x_3 - x_2^2 = 0$$

or in line coordinates $\quad 4u_1u_3 - u_2^2 = 0.$

There is no loss of generality in taking the equations in this form, because by suitably choosing the triangle of reference, the equation of a proper conic can be always reduced to the form

$$x_1x_3 - x_2^2 = 0.$$

Thus we may take for the parametric representation of points on the conic

$$x_1 = \nu_1^2, \quad x_2 = \nu_1\nu_2, \quad x_3 = \nu_2^2,$$

and we shall call this point (x_1, x_2, x_3) the point ν.

If the line $u_x = 0$ meet the conic in the points (λ, μ) the quantities λ, μ are given by

$$u_1\nu_1^2 + u_2\nu_1\nu_2 + u_3\nu_2^2 = 0$$

so that

$$\left.\begin{aligned} u_1 &= \lambda_2\mu_2 \\ u_2 &= -(\lambda_1\mu_2 + \lambda_2\mu_1) \\ u_3 &= \lambda_1\mu_1 \end{aligned}\right\} \quad \text{......................(A)}$$

except for a constant factor.

Hence we may regard the quantities

$$\lambda_2\mu_2, \quad -(\lambda_1\mu_2 + \lambda_2\mu_1), \quad \lambda_1\mu_1,$$

as the coordinates of the line, and a homogeneous relation of order m connecting the u's becomes a homogeneous symmetrical relation of order $2m$ between the λ's and μ's. Thus any symmetrical relation between λ and μ is equivalent to the tangential equation of a certain curve whose class is one-half of the order of the given relation.

The coordinates of the tangent at the point ν are

$$\nu_2^2, \quad -2\nu_1\nu_2, \quad \nu_1^2$$

and hence the points of contact of the tangents from the point x to the curve are given by

$$x_1\nu_2^2 - 2x_2\nu_1\nu_2 + x_3\nu_1^2 = 0,$$

so that we may take

$$\left.\begin{aligned} x_1 &= \lambda_1\mu_1 \\ x_2 &= \tfrac{1}{2}(\lambda_1\mu_2 + \lambda_2\mu_1) \\ x_3 &= \lambda_2\mu_2 \end{aligned}\right\} \quad \text{....................(B).}$$

Consequently a homogeneous symmetrical relation between the λ's and μ's of order $2n$ represents a curve of order n. Being given a symmetrical relation we therefore deduce two curve equations from it, one in line coordinates and the other in point coordinates. The curves represented are reciprocal with respect to the fundamental conic because, taking λ, μ to be fixed, the line u given by (A) is the line joining them, and the point x given by (B) is the intersection of the tangents at λ and μ, *i.e.* the pole of the line u.

This method of representing a point by the parameters of the tangents drawn from it to a fixed conic and a line by the parameters of the points in which it meets the conic was practically used by Hesse and first explicitly used by Darboux.

244. By means of this system of coordinates we can readily find all curves apolar to the given conic.

I. Suppose that $$u_\gamma^n = 0$$
is a class curve apolar to the conic, then we have
$$(\gamma_2^2 - \gamma_1\gamma_3)\, u_\gamma^{n-2} \equiv 0.$$

Thus $(\gamma_2^2 - \gamma_1\gamma_3)$ multiplied by any function of the γ's is zero if it be interpretable, hence this symbolical expression must be zero, and we may write
$$\gamma_1 = a_2^2,$$
$$\gamma_2 = - a_1a_2,$$
$$\gamma_3 = a_1^2,$$
and our equation is
$$(a_2^2u_1 - a_1a_2u_2 + a_1^2u_3)^n = 0.$$

The a's are now the only symbols used, and it is clear that as any expression of degree n in the γ's represents an actual quantity, any expression of degree $2n$ in the a's is an actual coefficient, or in other words the a's are the symbols of a binary form of order $2n$.

On introducing the λ's and μ's our equation becomes
$$\{a_2^2\lambda_2\mu_2 + a_1a_2(\lambda_1\mu_2 + \lambda_2\mu_1) + a_1^2\lambda_1\mu_1\}^n = 0,$$
or
$$\{(a_1\lambda_1 + a_2\lambda_2)(a_1\mu_1 + a_2\mu_2)\}^n = 0,$$
that is finally
$$a_\lambda^n a_\mu^n = 0.$$

The binary $2n$-ic of which the a's are symbols has an important significance, for if we make $\lambda = \mu$, the line u_x is the tangent to the fundamental conic at the point λ; consequently the equation

$$a_\lambda^{2n} = 0$$

gives the parameters of the points of contact of the $2n$ common tangents of the apolar curve and the conic.

Conversely, when the equation

$$a_\lambda^{2n} = 0$$

is given, the equation $\quad a_\lambda^n a_\mu^n = 0$

is uniquely determined by polarizing, and hence we have the theorem that a class curve apolar to a conic is uniquely determined when its common tangents with the conic are given.

By proving this theorem from first principles, and then observing that

$$a_\lambda^n a_\mu^n = 0,$$

or its equivalent

$$(a_2^2 u_1 - a_1 a_2 u_2 + a_1^2 u_3)^n = 0,$$

certainly represents a curve apolar to a conic, we can shew that any apolar curve may be reduced to the form

$$a_\lambda^n a_\mu^n = 0,$$

without using a parametric representation of the symbolical equation

$$\gamma_1 \gamma_3 - \gamma_2^2 = 0 *.$$

II. Suppose that the curve

$$c_x^n = 0$$

of order n is apolar to the given conic

$$4u_1 u_3 - u_2^2 = 0,$$

then we must have

$$(4c_1 c_3 - c_2^2)\, c_x^{n-2} \equiv 0,$$

and reasoning as before, we have

$$4c_1 c_3 - c_2^2 = 0.$$

* See Schlesinger, *Math. Ann.* Band XXII.

We may now use the parametric representation

$$c_1 = a_1^2,$$
$$c_2 = 2a_1a_2,$$
$$c_3 = a_2^2,$$

and the equation of the curve becomes

$$(a_1^2x_1 + 2a_1a_2x_2 + a_2^2x_3)^n = 0,$$

or introducing the λ's and μ's

$$\{a_1^2\lambda_1\mu_1 + a_1a_2(\lambda_1\mu_2 + \lambda_2\mu_1) + a_2^2\lambda_2\mu_2\}^n = 0,$$

i.e. $$a_\lambda^n a_\mu^n = 0,$$

and as before the a's are the symbols of the binary $2n$-ic a_λ^{2n}, which, equated to zero, gives the points of intersection of the apolar curve and the conic.

Example. To find the conic apolar to $x_1x_3 - x_2^2 = 0$ which touches the tangents to this conic at the points given by $\nu_1^4 - \nu_2^4 = 0$.

Here $$a_\lambda^{2n} = 0 \text{ is } \lambda_1^4 - \lambda_2^4 = 0,$$

thence $$a_\lambda^n a_\mu^n = 0$$

is $$\left(\mu\frac{\partial}{\partial\lambda}\right)^2(\lambda_1^4 - \lambda_2^4),$$

or $$\lambda_1^2\mu_1^2 - \lambda_2^2\mu_2^2 = 0.$$

On using the substitutions

$$u_1 = \lambda_2\mu_2, \text{ etc.}$$

the equation becomes $$u_3^2 - u_1^2 = 0,$$

or $$(u_3 + u_1)(u_3 - u_1) = 0,$$

so that the conic consists of the two points (1, 0, 1) (1, 0, −1), and in fact it is easy to see that these points are conjugate with respect to the conic.

245. Theorems on conics apolar to the fundamental conic. The equation of a conic apolar to $x_1x_3 - x_2^2 = 0$, and touching the tangents at the points given by

$$a_\tau^4 = 0,$$

is equivalent to $$a_\lambda^2 a_\mu^2 = 0,$$

and hence to $$u_\gamma^2 = 0,$$

where $$u_\gamma \equiv a_\lambda a_\mu.$$

Now suppose that A, B, C, D are four points $(\lambda, \mu, \nu, \rho)$ on the fundamental conic, and that the lines AB, CD whose equations are

$$v_x = 0 \text{ and } w_x = 0,$$

are conjugate with respect to the apolar conic $u_\gamma^2 = 0$.

Then since the condition of conjugacy is $v_\gamma w_\gamma = 0$ and

$$v_\gamma \equiv a_\lambda a_\mu, \quad w_\gamma \equiv a_\nu a_\rho,$$

we have

$$a_\lambda a_\mu a_\nu a_\rho = 0,$$

that is the quartic giving λ, μ, ν, ρ is apolar to $a_\tau^4 = 0$.

This is one of the simplest geometrical representations of forms apolar to a given form. From the symmetry of the result, we see that each pair of opposite sides of the quadrangle $ABCD$ are conjugate with respect to the apolar conic.

Thus there is an infinite number of quadrangles inscribed in the fundamental conic and harmonic to the apolar conic; the four vertices are given by forms apolar to the form

$$a_\tau^4 = 0.$$

Now if λ, μ, ν be chosen so that ρ is arbitrary, any line through A is conjugate to BC, so that A is the pole of BC, and hence ABC is a self-conjugate triangle of the apolar conic $u_\gamma^2 = 0$. In this case the cubic giving λ, μ, ν is apolar to $a_\tau^4 = 0$, and therefore to every first polar of this form; hence there is an infinite number of triangles inscribed in the fundamental conic and self-conjugate with respect to the apolar conic, and their vertices are given by the singly infinite number of cubic forms apolar to $a_\tau^4 = 0$.

Next suppose that the linear factors of the quartic giving λ, μ, ν, ρ are l_τ, m_τ, n_τ, r_τ, then

$$a_\tau^4 = Ll_\tau^4 + Mm_\tau^4 + Nn_\tau^4 + Rr_\tau^4,$$

where L, M, N, R are independent of τ.

By polarizing we obtain the identity

$$a_\xi^2 a_\eta^2 = L(l_\xi l_\eta)^2 + M(m_\xi m_\eta)^2 + N(n_\xi n_\eta)^2 + R(r_\xi r_\eta)^2,$$

where ξ, η are any two points on the conic.

Now by means of the usual substitutions

$$u_1 = \xi_2\eta_2,$$
$$u_2 = -(\xi_1\eta_2 + \xi_2\eta_1),$$
$$u_3 = \xi_1\eta_1,$$

the left-hand side becomes u_γ^2.

Consider next the term $\quad l_\xi l_\eta$.

Since $l_1 = \lambda_2$ and $l_2 = -\lambda_1$, this becomes

$$l_1^2 u_3 - l_1 l_2 u_2 + l_2^2 u_1$$

or
$$\lambda_1^2 u_1 + \lambda_1\lambda_2 u_2 + \lambda_2^2 u_3,$$

and λ_1^2, $\lambda_1\lambda_2$, λ_2^2, are the coordinates of A, so that

$$l_\xi l_\eta = u_A,$$

where $u_A = 0$ is the tangential equation of A.

Hence we have

$$u_\gamma^2 \equiv L u_A^2 + M u_B^2 + N u_C^2 + R u_D^2,$$

and the conic is represented as the sum of four squares.

In particular, if A, B, C be the vertices of a self-conjugate triangle, we obtain in like manner

$$u_\gamma^2 \equiv L u_A^2 + M u_B^2 + N u_C^2.$$

These results are well known and easily obtained otherwise, but the methods here used may be applied with equal success to more difficult problems as we shall presently shew.

Exactly the same reasoning applies to a conic c_x^2 apolar to

$$4u_1u_3 - u_2^2 = 0,$$

and now the triangles are circumscribed to the fundamental conic and self-conjugate to the apolar conic.

246. Condition of apolarity of two conics apolar to the standard conic.

Suppose the two conics are

$$c_x^2 = 0, \quad u_\gamma^2 = 0,$$

that the first meets the standard conic in the points $a_\lambda^4 = 0$, and the second touches the tangents to the standard conic in the points $b_\lambda^4 = 0$.

Then we have

$$c_x \equiv a_\lambda a_\mu \text{ and } u_\gamma \equiv b_\lambda b_\mu,$$

and in particular

$$\left.\begin{aligned} c_1 &= a_1^2 \\ c_2 &= 2a_1a_2 \\ c_3 &= a_2^2 \end{aligned}\right\}, \quad \left.\begin{aligned} \gamma_1 &= b_2^2 \\ \gamma_2 &= -b_1b_2 \\ \gamma_3 &= b_1^2 \end{aligned}\right\},$$

hence
$$c_\gamma = a_1^2b_2^2 + a_2^2b_1^2 - 2a_1a_2b_1b_2 = (ab)^2,$$

and the conics are apolar when $c_\gamma^2 = 0$, that is when

$$(ab)^4 = 0,$$

or when the two binary quartics a_λ^4 and b_λ^4 are apolar. This gives another simple geometrical representation of apolar quartics.

247. To many of the theorems developed for conics there are analogues for all curves possessing an apolar conic. For brevity we introduce a definition.

If the equation of a curve of order m be

$$f(x_1, x_2, x_3) = 0$$

and f can be written as a linear combination of the forms

$$u_x^{(r)\,m}, \qquad r = 1, 2, \ldots n,$$

then the n lines $\qquad u_x^{(r)} = 0$

are said to form a *conjugate n-line* with respect to the curve.

In like manner if the tangential equation of a curve of class m be

$$\phi(u_1, u_2, u_3) = 0$$

and ϕ can be written as a linear combination of the forms

$$u_{x_{(r)}}^{\;m}, \qquad r = 1, 2, \ldots n,$$

then the n points $\qquad u_{x_{(r)}} = 0$

are said to form a *conjugate n-point* with respect to the curve.

Suppose then that the curve

$$c_x^n = 0$$

is apolar to the fundamental conic

$$x_1x_3 - x_2^2 = 0,$$

the equation may be written

$$a_\lambda^n a_\mu^n = 0$$

where $\qquad a_\lambda^{2n} = 0$

gives the $2n$ points in which the curve meets the conic.

If a_λ^{2n} be apolar to the form

$$b_\lambda^r = 0$$

whose linear factors are

$$p_\lambda^{(1)}, \; p_\lambda^{(2)}, \; \ldots \; p_\lambda^{(r)},$$

then a_λ^{2n} can be expressed as a linear combination of

$$p_{\lambda^{(1)}}{}^{2n},\ p_{\lambda^{(2)}}{}^{2n},\ \ldots\ p_{\lambda^{(r)}}{}^{2n},$$

so that $a_\xi{}^n a_\eta{}^n$ is a linear combination of

$$p_{\xi^{(1)}}{}^n p_{\eta^{(1)}}{}^n, \text{ etc.}$$

and hence just as in the case of the conic it follows that the tangents to the conic at the points given by

$$p_{\lambda^{(1)}} = 0,\ p_{\lambda^{(2)}} = 0,\ \ldots\ p_{\lambda^{(r)}} = 0,$$

i.e. by

$$b_\lambda{}^r = 0,$$

form a conjugate r-line with respect to the curve whose equation can accordingly be expressed as a sum of r nth powers.

The above will suffice to indicate the general principles which we shall now apply to the ternary cubic and quartic.

248. Ternary Cubic. A cubic curve, as we have seen, § 242, always possesses an infinite number of apolar conics. Take the fundamental conic for one of these and let

$$c_x{}^3 = 0$$

be the equation of the cubic.

This may be written

$$a_\lambda{}^3 a_\mu{}^3 = 0$$

and meets the conic in the points given by

$$a_\lambda{}^6 = 0.$$

This binary sextic has three linearly independent second polars, and therefore a singly infinite number of apolar binary quartics but not in general an apolar cubic.

Hence a ternary cubic may be written in an infinite number of ways as the sum of four cubes for each apolar conic it possesses, but not in general as the sum of three cubes for an arbitrary apolar conic.

The condition that the binary sextic

$$a_\lambda{}^6 \equiv a'_\lambda{}^6 \equiv a''_\lambda{}^6 \equiv a'''_\lambda{}^6$$

may have an apolar cubic is that the determinant formed by the coefficients of its third differential coefficients may be zero, *i.e.* that any four third polars may be linearly dependent.

This condition is

$$(aa')^2 (a'a'')^2 (a''a)^2 (aa''')^2 (a'a''')^2 (a''a''')^2 = 0. \qquad (\S\ 187)$$

Now
$$a_1^2 = c_1,$$
$$2a_1 a_2 = c_2,$$
$$a_2^2 = c_3,$$

hence
$$(aa')(a'a'')(a''a)$$

$$= \begin{vmatrix} a_1^2 & a_1 a_2 & a_2^2 \\ a'^2_1 & a_1' a_2' & a'^2_2 \\ a''^2_1 & a_1'' a_2'' & a''^2_2 \end{vmatrix} = \tfrac{1}{2}(cc'c'')$$

and the condition reduces to

$$(cc'c'')(c'c''c''')(c''c'''c)(c'''cc') = 0.$$

This is an invariant of the cubic

$$c_x^3 = c'^3_x = c''^3_x = c'''^3_x$$

and hence however we choose the apolar conic we cannot reduce the cubic to the sum of three cubes unless a certain invariant of degree four vanishes.

249. Ternary Quartic. Here there is no apolar conic unless the six second differential coefficients are linearly dependent, *i.e.* unless a certain invariant called the *Catalecticant* vanishes.

Now if the quartic can be written as the sum of five fourth powers it must have an apolar conic, because a conic can be chosen apolar to any five fourth powers—in fact we have only to describe a conic touching the five straight lines represented by the linear forms.

Hence in general a ternary quartic cannot be expressed as the sum of five fourth powers.

But if the catalecticant be zero there is an apolar conic and, taking it for a fundamental conic, the equation of the quartic may be written

$$a_\lambda^4 a_\mu^4 = 0,$$

where $a_\lambda^8 = 0$ gives the points of intersection with the conic.

Now a singly infinite number of quintics can be found apolar to a binary octavic, hence in this case the quartic curve has a singly infinite number of conjugate five-lines, and all such lines touch the apolar conic.

250. We shall conclude this chapter with a brief account of the class of invariant forms known as combinants, confining ourselves to binary forms.

An invariant or covariant of any number of binary forms

$$f_1, f_2, \dots f_r$$

of the same order is said to be a combinant if it be unaltered, except as regards a factor independent of the forms, when each form f is replaced by a linear combination of the type

$$l_1 f_1 + l_2 f_2 + \dots + l_r f_r,$$

in which the l's are constants.

For example in the case of two binary forms we have

$$(l_1 f_1 + l_2 f_2, \; m_1 f_1 + m_2 f_2)$$
$$= (lm)(f_1 f_2),$$

so that the Jacobian of two binary forms is a combinant.

For the sake of brevity we shall deal with the combinants of three binary forms

$$f_1 = a_0 x_1^n + n a_1 x_1^{n-1} x_2 + \dots + a_n x_2^n$$
$$f_2 = b_0 x_1^n + n b_1 x_1^{n-1} x_2 + \dots + b_n x_2^n$$
$$f_3 = c_0 x_1^n + n c_1 x_1^{n-1} x_2 + \dots + c_n x_2^n.$$

A combinant is not only unaltered by a linear substitution effected on the variables, but also by a linear substitution of the type

$$a_r' = l_1 a_r + m_1 b_r + n_1 c_r$$
$$b_r' = l_2 a_r + m_2 b_r + n_2 c_r$$
$$c_r' = l_3 a_r + m_3 b_r + n_3 c_r$$

effected on the coefficients.

Regarded as a function of the coefficients the combinant is therefore an invariant of the linear forms

$$a_r \xi_1 + b_r \xi_2 + c_r \xi_3, \qquad r = 1, \; 2 \dots\dots n,$$

because if we put

$$\xi_1 = l_1 \xi_1' + l_2 \xi_2' + l_3 \xi_3' \text{ etc.}$$

we find $$a_r' = l_1 a_r + m_1 b_r + n_1 c_r \text{ etc.}$$
which are the substitutions above.

Hence, by the fundamental theorem on the symbolical representation of invariants, a combinant, as far as the coefficients a, b, c are concerned, is a rational integral function of determinants of the type

$$\begin{vmatrix} a_p & a_q & a_r \\ b_p & b_q & b_r \\ c_p & c_q & c_r \end{vmatrix}.$$

A like result applies to any number of binary forms.

Thus for example for two quadratics

$$a_0 x_1^2 + 2a_1 x_1 x_2 + a_2 x_2^2$$
$$b_0 x_1^2 + 2b_1 x_1 x_2 + b_2 x_2^2,$$

a combinant is a function of

$$(a_0 b_1 - a_1 b_0),\ (a_1 b_2 - a_2 b_1),\ (a_0 b_2 - a_2 b_0),$$

as far as the coefficients are concerned.

But the Jacobian is

$$(a_0 b_1 - a_1 b_0) x_1^2 + (a_0 b_2 - a_2 b_0) x_1 x_2 + (a_1 b_2 - a_2 b_1) x_2^2,$$

and hence a combinant is a rational integral function of the coefficients of the Jacobian and the variables. Hence any combinant is an invariant form of the Jacobian, and therefore the complete system of combinants in this case consists of the Jacobian and its discriminant—the latter is equivalent to the resultant of the two original forms.

It is easy to form any number of combinants of two binary forms, for

(i) An invariant or covariant of a combinant is itself a combinant, since it is manifestly an invariant form and further involves the coefficients of the original forms in the manner peculiar to combinants.

(ii) Let f_1 and f_2 be two binary forms, I_0 an invariant form of f_1 and the corresponding form for $\lambda_1 f_1 + \lambda_2 f_2$,

$$\lambda_1^m I_0 + \lambda_1^{m-1} \lambda_2 I_1 + \ldots + \lambda_2^m I_m,$$

then an invariant of this expression considered as a binary form in (λ_1, λ_2) is a combinant of f_1 and f_2.

For such an invariant is unaltered when we effect a linear substitution on the x's because each I is an invariant form; and it is unaltered when we effect a linear substitution on the coefficients because it is an invariant of the form in λ written above.

251. Combinants naturally occur in the discussion of rational curves as we shall now shew.

Suppose such a curve is parametrically represented by

$$\xi_1 = a_x^n \equiv f_1$$
$$\xi_2 = b_x^n \equiv f_2$$
$$\xi_3 = c_x^n \equiv f_3, \qquad \text{(cf. § 196)}$$

then the curve is unaltered by a linear substitution effected on the x's since its equation is found by eliminating the x's.

Now if a set of points on the curve be defined by some projective property the equation giving their parameters is derived from f_1, f_2, f_3 in a definite way, hence if by means of a linear substitution f_1, f_2, f_3 become f_1', f_2', f_3' the transformed equation for the parameters is derived from f_1', f_2', f_3' in the same way as its original form was derived from f_1, f_2, f_3.

Thus if the equation be $C=0$ it follows that C is a covariant of f_1, f_2, f_3.

Next, keeping the parameters fixed, to change the triangle of reference we replace ξ_1, ξ_2, ξ_3 by linear functions of themselves, so that f_1, f_2, f_3 are replaced by linear combinations of the form

$$l_1 f_1 + l_2 f_2 + l_3 f_3.$$

Now the equation giving the parameters of the set of points must be independent of the triangle of reference, for such points depend on the curve itself, and the parameters of every point of the curve are unchanged when we alter the triangle of reference; hence C is not only a covariant but a combinant of the forms f_1, f_2, f_3, and the rational curve is the natural geometrical representation of the system of combinants.

The curve can be equally well defined by the system of forms apolar to f_1, f_2, f_3, because these determine f_1, f_2, f_3, and the projective properties of the curve are also given by the combinants of the apolar system of forms.

We are therefore led to the theorem that the combinants of two apolar systems of forms are identical, and in fact a rigorous algebraic proof of its truth will be found in Meyer's *Apolarität*, § 11.

As an example the reader may verify that in the quartic curve

$$\xi_1 = a_x^4,\ \xi_2 = b_x^4,\ \xi_3 = c_x^4,$$

the points of inflexion are given by

$$(bc)(ca)(ab)\, a_x^2 b_x^2 c_x^2 = 0,$$

and that, if d_x^4 and e_x^4 be two forms belonging to the apolar system, they are also given by

$$(de)\, d_x^3 e_x^3 = 0.$$

The first equation follows from the ordinary methods of the differential calculus—the second from the fact that the conditions of collinearity of four points are easily expressed by means of the apolar system; if λ be a point of inflexion, and μ the point in which the inflexional tangent meets the curve again, we have $d_\lambda^3 d_\mu = 0$, $e_\mu^3 e_\mu = 0$ so that μ may be eliminated.

The full discussion of the theory of combinants would lead us too far from the methods of the present treatise, and accordingly we shall content ourselves with the explanation of a "translation-principle" connecting the combinants of binary forms with the covariants of ternary forms.

252. It will be convenient to change the notation and to suppose a rational curve given by

$$\left.\begin{aligned} \xi_1 &= a_1 x_1^n + n b_1 x_1^{n-1} x_2 + \ldots\ldots + k_1 x_2^n \\ \xi_2 &= a_2 x_1^n + n b_2 x_1^{n-1} x_2 + \ldots\ldots + k_2 x_2^n \\ \xi_3 &= a_3 x_1^n + n b_3 x_1^{n-1} x_2 + \ldots\ldots + k_3 x_2^n \end{aligned}\right\}.$$

Consider the problem of finding the locus of the point of intersection of two straight lines which meet the curve in two sets of n points given by binary forms for which a certain combinant is zero.

Let the two lines be

$$u_\xi = 0,\ v_\xi = 0,$$

and denote by ξ their point of intersection.

The two binary forms are

$$a_u x_1^n + n b_u x_1^{n-1} x_2 + \ldots + k_u x_2^n$$

and

$$a_v x_1^n + n b_v x_1^{n-1} x_2 + \ldots + k_v x_2^n,$$

and any combinant is a function of determinants of the type

$$\begin{vmatrix} a_u & b_u \\ a_v & b_v \end{vmatrix}.$$

Now $$a_u b_v - a_v b_u = (ab\xi),$$

hence the equation of the locus is found by changing (ab) in the expression of the vanishing combinant into $(ab\xi)$.

For example if two lines meet the cubic curve

$$\xi_1 = a_1 x_1^3 + 3b_1 x_1^2 x_2 + 3c_1 x_1 x_2^2 + d_1 x_2^3$$
etc.

in two apolar sets of points we have

$$(a_u d_v - a_v d_u) - 3(b_u c_v - b_v c_u) = 0$$

or $$(ad\xi) - 3(bc\xi) = 0,$$

and hence the locus of their common point ξ is a straight line.

It is evident that if A be a point of inflexion then the tangent at A and any line through A satisfy the conditions of the problem, so that all the points of inflexion of the curve lie on this straight line.

As a second example let us find the equation of the cubic. Here the two straight lines meet on the curve and the vanishing combinant is the λ eliminant of the two binary forms.

Following Bezout's method, the eliminant of

$$px_1^3 + qx_1^2 x_2 + rx_1 x_2^2 + sx_2^3$$

and $$p'x_1^3 + q'x_1^2 x_2 + r'x_1 x_2^2 + s'x_2^3$$

is $$\begin{vmatrix} (pq') & (pr') & (ps') \\ (pr') & (ps') + (qr') & (qs') \\ (ps') & (qs') & (rs') \end{vmatrix} = 0,$$

and hence making $p = a_u$, $q = 3b_u$ etc. the equation required is

$$\begin{vmatrix} 3(ab\xi) & 3(ac\xi) & (ad\xi) \\ 3(ac\xi) & (ad\xi) + 9(bc\xi) & 3(bd\xi) \\ (ad\xi) & 3(bd\xi) & 3(cd\xi) \end{vmatrix} = 0.$$

It is clear that a similar method applies to the curve of the nth degree.

CHAPTER XV.

TYPES.

253. IT was proved in § 35 that the effect of operating with

$$\left(A\frac{\partial}{\partial B}\right) \equiv A_0\frac{\partial}{\partial B_0} + A_1\frac{\partial}{\partial B_1} + \dots + A_n\frac{\partial}{\partial B_n}$$

on a covariant Φ of a simultaneous system of binary forms, which includes a_x^n and b_x^n where

$$a_x^n \equiv (A_0, A_1, \dots A_n \between x_1, x_2)^n$$
$$b_x^n \equiv (B_0, B_1, \dots B_n \between x_1, x_2)^n,$$

is itself a covariant of the system.

All covariants thus obtained from Φ are said to be of *the same type* as Φ. In other words two covariants are said to be of the same type if one of them is obtainable from the other by means of operators of this kind. For example the invariants

$$(f, f)^2, \ (f, \phi)^2, \ (\phi, \phi)^2$$

of two quadratics f, ϕ are all of the same type.

It should be noticed that this connection between two covariants is not necessarily reciprocal; two covariants Φ_1, Φ_2, where Φ_2 is obtainable from Φ_1 by operators of the required kind, are of the same type, even if Φ_1 is not so obtainable from Φ_2. Thus if $F(a, a', \dots)$ is a simultaneous covariant of a system of quantics which includes $f = a_x^n = a'^n_x$, $\phi = b_x^n$, and if F is of the second degree in the coefficients of f but does not contain those of ϕ, the covariant

$$F(a, b, \dots) + F(b, a, \dots) = \left(\phi\frac{\partial}{\partial f}\right) F(a, a', \dots)$$

is of the same type as $F(a, a', \ldots)$; but here

$$F(a, a', \ldots) = \left(f \frac{\partial}{\partial \phi}\right) F(a, b, \ldots)$$

and we see that $F(a, a', \ldots)$ and $F(a, b, \ldots)$ are of the same type.

It will be seen in this way that two covariants Φ_1, Φ_2 may each be of the same type as a third covariant Φ, although neither Φ_1 nor Φ_2 is obtainable from the other by an operator of the kind considered. In view of this the further statement is necessary that covariants which are each of the same type as a third covariant are (by definition) of the same type as each other.

254. Every covariant of degree m, of one or more quantics, is of the same type as a covariant which is linear in the coefficients of each of m quantics—the number of quantics in the system being, of course, increased if necessary. Any such representative covariant is, for convenience, called a *type*; a type is then a covariant which is linear in the coefficients of each of the quantics concerned, it being understood that these are not special quantics of the system and that the word type is used in a purely formal sense.

Thus for three quadratics (§ 139 A)

$$(ab)(bc)(ca)$$

is an irreducible type, and furnishes only one invariant of the system ; $(ab)^2$ is also an irreducible type and furnishes six invariants.

It should be noticed that if f_1, f_2, f_3 are the quadratics, and

$$J_{1,2} = (f_1, f_2),$$

the invariant

$$(J_{1,2}, f_3)^2$$

is not of the same type as $(ab)^2$, because $J_{1,2}$ is not one of the fundamental quantics of the system.

Consider the covariants of a simultaneous system of binary forms of the same order. When the number of binary forms is indefinitely increased, the number of irreducible covariants will also be increased without limit; in fact the number of irreducible covariants belonging to any one type will be indefinitely increased. The question arises—does the number of irreducible types increase indefinitely too? This question has been answered in the negative by Peano*. Peano's theorem is the following: *Every type of a system of binary n-ics which does not furnish irreducible covariants for a system of n n-ics is reducible, with the single possible exception of the invariant type*

$$\begin{vmatrix} A_0 & A_1 & \dots & A_n \\ B_0 & B_1 & \dots & B_n \\ \dots & \dots & \dots & \dots \\ K_0 & K_1 & \dots & K_n \end{vmatrix},$$

where

$$(A_0, A_1, \dots A_n \between x_1, x_2)^n$$
$$(B_0, B_1, \dots B_n \between x_1, x_2)^n$$
$$\dots\dots\dots\dots\dots$$
$$(K_0, K_1, \dots K_n \between x_1, x_2)^n$$

are $n+1$ n-ics. But if this invariant is reducible, all types are reducible which do not furnish irreducible covariants for $n-1$ n-ics.

A proof of this theorem is given in the next chapter.

255. As was pointed out in § 21 there are two principles by means of which the reduction of a covariant has to be attempted, viz.:

(i) The fundamental identities

$$(bc)(ad) + (ca)(bd) + (ab)(cd) = 0$$
$$(bc)\, a_x + (ca)\, b_x + (ab)\, c_x = 0.$$

(ii) The fact that the interchange of two symbols which refer to the same quantic does not alter the actual value of a symbolical product.

To effect the reduction of a type the first of these two principles must alone be employed.

256. The quadratic types. The quadratics will be denoted, as usual, by

$$a_x^2,\ b_x^2,\ c_x^2,\ \dots.$$

* *Atti di Torino*, t. XVII. p. 580 (1881). See also Jordan (*Liouville*, 1876, 2 Sér. III.), who proved that the number of irreducible types belonging to any simultaneous system of forms, the order of each of which is less than some fixed number n, is finite.

For invariants the only symbolical products to be considered are

$$(ab)^2$$

$$(ab)(bc)(ca)$$

$$(ab)(bc)(cd)(da)$$

$$\ldots\ldots\ldots\ldots\ldots\ldots$$

The first two of these are irreducible, for the fundamental identities give us no relations by which we may reduce them.

The other invariant types are all reducible. For

$$2(ab)(bc)(cd)(da) = (ab)^2(cd)^2 + (bc)^2(da)^2 - (ac)^2(bd)^2;$$

operate on this identity with

$$e_1 \frac{\partial}{\partial a_1} + e_2 \frac{\partial}{\partial a_2},$$

then

$$(ab)(bc)(cd)(de) + (eb)(bc)(cd)(da)$$
$$= -(cd)^2(ab)(be) - (bc)^2(ad)(de) + (bd)^2(ac)(ce).$$

But

$$(ab)(bc)(cd)(de) - (eb)(bc)(cd)(da) = (bc)(cd)(db)(ae).$$

Hence

$$2(ab)(bc)(cd)(de)$$
$$= (bc)(cd)(db)(ae) - (cd)^2(ab)(be) - (bc)^2(ad)(de) + (bd)^2(ac)(ce).$$

By means of these two identities all the invariant types of degree greater than three are at once reduced. Now any covariant of a system of quantics of even order must be itself of even order (§ 20); hence any covariant type of the quadratic may always be obtained by replacing one or more letters in the symbolical expression for some invariant type by the variable. For example, from $(ab)^2$ we obtain a_x^2 on replacing b_1 by $-x_2$ and b_2 by x_1. Hence the irreducible covariant types are

$$a_x^2,\ (ab)\,a_x b_x.$$

The quadratic has then only four irreducible types (compare § 139 A),

$$(ab)^2,\ (ab)(bc)(ca),$$

$$a_x^2,\ (ab)\,a_x b_x.$$

The second is in fact the determinant type referred to above, for as has already been pointed out,

$$(ab)(bc)(ca) = -\begin{vmatrix} a_1^2 & a_1a_2 & a_2^2 \\ b_1^2 & b_1b_2 & b_2^2 \\ c_1^2 & c_1c_2 & c_2^2 \end{vmatrix}.$$

257. The cubic types. It is possible to obtain the complete system of types for binary forms of a given order by a method almost identical with that of Chapter VI. for covariants of a single binary form. The reductions in this method are generally very difficult to obtain. The cubic types, however, can be thus obtained simply. It is thought unnecessary to go through the general argument, the alterations in Chapter VI. to meet the case of types being mainly verbal. It should be noticed that the finiteness of the complete irreducible system of types could thus be demonstrated.

Let a_x^3, b_x^3, c_x^3, ... be the cubics. The symbol F will be used to denote any one of them indifferently. The types of degree two are

$$(ab)\,a_x^2b_x^2 = J,\ (ab)^2\,a_xb_x = H,\ (ab)^3.$$

Consider first the types of grade unity. These all contain a factor (ab), and hence are terms of transvectants of J with types of grade not greater than unity.

In fact any such type is a term of a transvectant of the form

$$(\overline{J_1J_2 \dots J_r},\ F_1F_2 \dots F_s)^\lambda,$$

where the bar over the left-hand member indicates any type obtained by convolution from the product there written down. It follows at once that every type of unit grade can be expressed as a sum of numerical multiples of such transvectants. Now by §§ 74, 75 any type obtained by convolution from $J_1, J_2, \dots J_r$ is of grade two at least. Hence the only irreducible types of unit grade are expressible as transvectants of the form

$$(J_1J_2 \dots J_r,\ F_1F_2 \dots F_s)^\lambda.$$

If $\lambda = 1$ this is clearly reducible—for J is a Jacobian.

If $\lambda > 1$, this contains a term of grade two.

Therefore the only irreducible type of unit grade is J.

Now the irreducible types of the quadratic H are

$$H, \ (H_1, H_2) = K, \ (H_1 H_2)^2, \ (H_1 H_2)(H_2 H_3)(H_3 H_1).$$

Hence the types of grade two are expressible as transvectants of the form

$$(H_1 H_2 \ldots H_\alpha K_1 K_2 \ldots K_\beta, \ F_1 F_2 \ldots F_\gamma J_1 J_2 \ldots J_\delta)^\lambda$$

or of these multiplied by invariant types.

In the first place we notice that K and J are Jacobians and hence we may suppose that neither β nor δ exceeds unity.

We have the following types to consider:

$$(H, F), \ (H, F)^2, \ (H_1 H_2, F)^3, \ (H_1 H_2 H_3, F_1 F_2)^6$$
$$(H, J)^2, \ (H_1 H_2, J)^3, \ (H_1 H_2, J)^4$$
$$(K, F)^2, \ (HK, F)^3, \ (H_1 H_2 K, F_1 F_2)^6, \ (K, J)^2, \ (HK, J)^3, \ (HK, J)^4.$$

Of these $(H_1 H_2 H_3, F_1 F_2)^6$ contains the term

$$(a_1 b_1)^2 (a_2 b_2)^2 (a_3 b_3)^2 (a_1 c_1)(b_1 c_1)(a_2 c_2)(b_2 c_2)(a_3 c_1)(b_3 c_2)$$
$$= (a_1 b_1)(b_1 c_1)(c_1 a_1) . (a_2 b_2)(b_2 c_2)(c_2 a_2) . (a_3 b_3)^2 (a_1 b_1)(a_2 b_2)(a_3 c_1)(b_3 c_2)$$
$$= \tfrac{1}{2} \begin{vmatrix} (a_1 a_2)^2 & (a_1 b_2)^2 & (a_1 c_2)^2 \\ (b_1 a_2)^2 & (b_1 b_2)^2 & (b_1 c_2)^2 \\ (c_1 a_2)^2 & (c_1 b_2)^2 & (c_1 c_2)^2 \end{vmatrix} (a_3 b_3)^2 (a_1 b_1)(a_2 b_2)(a_3 c_1)(b_3 c_2), \ldots \ (\S\ 77).$$

This is a sum of terms obtained by convolution from products of four types H, and hence is reducible. In exactly the same way the type $(H_1 H_2 K, F_1 F_2)^6$ may be reduced.

The type $(H, J)^2$ contains a term

$$(ab)^2 (bc)(ad)(cd) c_x d_x$$
$$= -\tfrac{1}{2} (ab) \{(ab)^2 (cd)^2 + (bc)^2 (ad)^2 - (ac)^2 (bd)^2\} c_x d_x,$$

which may be expressed in terms of the type $(H_1 H_2)$ and reducible forms.

The type $(K, J)^2$ contains the term

$$(H_1 H_2)(H_2 c)(H_1 d)(cd) c_x d_x$$
$$= -\tfrac{1}{2} \{(H_1 H_2)^2 (cd)^2 + (H_2 c)^2 (H_1 d)^2 - (H_2 d)^2 (H_1 c)^2\} c_x d_x,$$

and hence is reducible.

The type $(H_1 H_2, J)^3$ contains the term

$$(H_1, (H_2, J)^2) = \Sigma\, (H_1, (H_2, H_3)) + \text{reducible terms},$$

and hence is reducible. In the same way the types $(H_1 H_2, J)^4$, $(HK, J)^3$, $(HK, J)^4$ may be reduced.

The type

$$(K, F)^2 = (H_1H_2)(H_2F)(H_1F)F_x$$
$$= (H_1F)^2(H_2F)H_{2x} - (H_2F)^2(H_1F)H_{1x};$$

and $$2(H_1H_2, F)^3 = (H_1F)^2(H_2F)F_x + (H_2F)^2(H_1F)F_x.$$

Hence both $(K, F)^2$ and $(H_1H_2, F)^2$ can be expressed in terms of the type

$$(H_1F)^2(H_2F)F_x = L_x.$$

Lastly, $(HK, F)^3$ contains the term

$$(a_1b_1)^2(a_2b_2)^2(a_3b_3)^2(b_1a_2)(a_1c)(a_3c)(b_3c)b_{2x}$$
$$= ((a_1b_1)^2(a_3b_3)^2(b_1a_2)(a_1c)(a_3c)(b_3c)a_{2x}^2, b_{2x}^3)^2.$$

Now the left-hand member of this transvectant is a type of degree six and order two; looking back at the possible types of this order we see that it must be of the form

$$\lambda L_1L_2 + \mu(ab)^3K + \nu(H_1H_2)^2H_3.$$

The second and third of these terms contain invariant factors, and can therefore only lead to reducible terms in the above transvectant. Also

$$2L_1L_2 = 2(ab)^2(ac)(bc)c_x \,.\, (de)^2(df)(ef)f_x$$
$$= (ab)(de)c_xf_x \begin{vmatrix} (ad)^2 & (ae)^2 & (af)^2 \\ (bd)^2 & (be)^2 & (bf)^2 \\ (cd)^2 & (ce)^2 & (cf)^2 \end{vmatrix}$$
$$= \Sigma(H_1H_2)^2H_3 + \Sigma(H_1H_2)(H_2H_3)H_{1x}H_{3x}$$
$$= \Sigma(H_1H_2)^2H_3.$$

The type $(HK, F)^3$ is thus reducible.

The complete system of irreducible cubic types is then

$$(ab)^3,\ (ab)^2(bc)(cd)^2(da),\ (ab)^2(bc)(cd)^2(de)(ef)^2(fa),$$
$$(ab)^2(ac)(bc)c_x,\ (ab)^2(cd)^2(ae)(be)(ce)d_x,$$
$$(ab)^2a_xb_x,\ (ab)^2(bc)(cd)^2a_xd_x,$$
$$a_x^3,\ (ab)^2(bc)a_xc_x^2,$$
$$(ab)a_x^2b_x^2,$$

there being ten types in all.

The system of irreducible concomitants for two cubics may be obtained from the system of types or else directly, they will be found in the works of Clebsch and Gordan. The syzygies between them have been obtained by von Gall (*Math. Ann.* Bd. XXXI.).

258. Perpetuants. The irreducible seminvariants (§ 32) of the binary form of infinite order are called *perpetuants*. The complete system of perpetuants for one binary form of infinite order has been obtained by Macmahon* and Stroh†. The system is, of course, infinite in extent, but the individual members of it have all been identified.

The complete system of perpetuants for any simultaneous system of binary quantics of infinite order was obtained by Macmahon‡; and in particular the perpetuant types may be at once obtained from this paper.

The method by which these results were obtained, does not fall within the scope of this book. The results have been obtained more recently by means of the symbolical notation which has been here developed; and this investigation§ we shall follow.

259. A covariant is completely defined when the determinant factors in its symbolical expression are known; it will be convenient to use this part of the symbolical expression only. In dealing with forms of infinite order, it must be remembered that the complete expression for a covariant contains each of the factors a_x, b_x, ... raised to an indefinitely high power.

The identity

$$(bc)\, a_x + (ca)\, b_x + (ab)\, c_x = 0$$

may now be written

$$(bc) + (ca) + (ab) = 0.$$

By means of this identity any factor (bc), in a covariant, which does not contain a may be replaced by

$$(ac) - (ab),$$

i.e. by factors which do contain a. Thus all covariant types may be expressed in terms of those which are of the form

$$(ab)^\lambda (ac)^\mu (ad)^\nu \ldots$$

where a is any one of the letters chosen at will.

* *Proc. Lond. Math. Soc.*, vol. XXVI. See also *Am. Journal*, vols. VII. VIII.

† *Math. Ann.*, Bd. XXXVI.

‡ *Camb. Phil. Soc. Trans.*, vol. XIX. pp. 234—248.

§ Grace, *Proc. Lond. Math. Soc.*, vol. XXXV.

The types of this form are all linearly independent, for no linear algebraical identity can connect their symbolical expressions.

Hence if all reducible types were expressed in terms of types of this form, we should be able to write down the perpetuant types.

It must be remembered that a is a perfectly definite quantic of the system. Further that the remaining quantics concerned in any particular covariant will be considered in a particular order *determined beforehand.*

260. Consider the types of degree three; if w be the weight, we know that

$$\begin{aligned}(bc)^w &= \{(ac)-(ab)\}^w \\ &= (ac)^w - w\,(ab)\,(ac)^{w-1} + \ldots\ldots + (-1)^w\,(ab)^w.\end{aligned}$$

Hence the covariant $(ab)\,(ac)^{w-1}$ is expressible in terms of reducible covariants and of covariants in which the index of (ab) is greater than unity. Hence all perpetuant types of degree three are expressible in terms of the types

$$(ab)^\lambda\,(ac)^\mu,\ \ \lambda \nless 2,\ \ \mu \nless 1.$$

It should be noticed that of the three quantics concerned any one may be chosen to correspond to a, b or c respectively.

Further, the only reducible covariants of degree three and weight w are represented by

$$(bc)^w,\ (ab)^w,\ (ac)^w,$$

and hence the seminvariants $(ab)^\lambda\,(ac)^\mu\ (\lambda \nless 2,\ \mu \nless 1)$ are both independent and irreducible.

261. Types of degree four may all be expressed in terms of the independent forms

$$(ab)^\lambda\,(ac)^\mu\,(ad)^\nu.$$

If λ or μ be less than 2, then as in the previous paragraph the index of (ab) or (ac), as the case may be, can be increased at the expense of the index of (ad).

Thus, since the reducible covariant

$$(ab)^\lambda (cd)^{w-\lambda} = (ab)^\lambda \{(ad) - (ac)\}^{w-\lambda}$$
$$= (ab)^\lambda (ad)^{w-\lambda} - \binom{w-\lambda}{1} (ab)^\lambda (ac) (ad)^{w-\lambda-1}$$
$$+ \binom{w-\lambda}{2} (ab)^\lambda (ac)^2 (ad)^{w-\lambda-2} - \dots,$$

the covariant $(ab)^\lambda (ac) (ad)^{w-\lambda-1}$ can be expressed in terms of reducible forms and of covariants

$$(ab)^\lambda (ac)^\mu (ad)^{w-\mu-\lambda} \qquad (\mu > 1).$$

When both λ and μ are greater than unity, say $\lambda + \mu = M$, we may express, by means of Stroh's series § 64, the products

$$(ab)^\lambda (ac)^\mu$$

in terms of the following three sets:

(i) $(ab)^M, (ab)^{M-1} (ac), \dots (ab)^4 (ac)^{M-4}$;

(ii) $(ac)^M, (ac)^{M-1} (ab)$;

(iii) $(bc)^M, (bc)^{M-1} (ab)$.

The products contained in (ii) and (iii) need not be considered, for the corresponding covariants can be expressed linearly in terms of reducible covariants and of covariants in which the number of factors involving a, b, c only is greater than $\lambda + \mu$. These latter forms can be dealt with in the same way.

Thus we see that ultimately we can express all the covariants of degree four in terms of reducible covariants and of such as have the factor $(ab)^4$. Further we have seen that we may suppose the coefficient of (ac) to be greater than unity: hence all covariants of degree four can be expressed in terms of reducible covariants and of covariants of the form

$$(ab)^\lambda (ac)^\mu (ad)^\nu$$

where $\lambda \nless 4$, $\mu \nless 2$, $\nu \nless 1$, and the arrangement of the letters a, b, c, d has been fixed beforehand.

262. The theorem can now be proved in general by induction. We shall assume that all covariant types, of a system of binary forms of infinite order, which are of degree $n+1$ or less, can be expressed linearly in terms of reducible covariants and of covariants of the form

$$(aa_1)^{\lambda_1}(aa_2)^{\lambda_2}\dots(aa_n)^{\lambda_n},$$

where $\lambda_1 \nless 2^{n-1},\quad \lambda_2 \nless 2^{n-2},\ \dots\ \lambda_n \nless 1,$

and the arrangement of the letters $a, a_1, a_2, \dots a_n$ is fixed.

For degree $n+2$ we have only to consider the covariants of the form

$$(aa_1)^{\lambda_1}(aa_2)^{\lambda_2}\dots(aa_{n+1})^{\lambda_{n+1}} = (aa_1)^{\lambda_1}(aa_2)^{\lambda_2}R.$$

Now $(aa_2)^{\lambda_2}R$ is of the same symbolical form as a covariant of degree $n+1$ of the system; hence, using the result for that degree, we may, if $\lambda_2 < 2^{n-1}$, express it in terms of covariants of the same form but for which the index of (aa_2) is not less than 2^{n-1}, and of reducible covariants. In the same way, if $\lambda_1 < 2^{n-1}$, the index of (aa_1) in the product $(aa_1)^{\lambda_1}R$ can be increased.

Let $\lambda_1+\lambda_2 = M \nless 2\,.\,2^{n-1},\quad N \equiv 2^{n-1};$

then, as before, by means of Stroh's series all products $(aa_1)^{\lambda}(aa_2)^{\mu}$ can be expressed in terms of the $M+1$ following products:

$$\text{(i)}\quad (aa_1)^M,\ (aa_1)^{M-1}(aa_2),\ \dots\ (aa_1)^{2N}(aa_2)^{M-2N},$$

$$\text{(ii)}\quad (aa_2)^M,\ (aa_2)^{M-1}(aa_1),\ \dots\ (aa_2)^{M-N+1}(aa_1)^{N-1},$$

$$\text{(iii)}\quad (a_1a_2)^M,\ (a_1a_2)^{M-1}(aa_1),\ \dots\ (a_1a_2)^{M-N+1}(aa_1)^{N-1}.$$

The products contained in (ii) and (iii) need not be considered, for the corresponding covariants have factors of the form $(aa_1)^{\rho}R$ where $\rho < 2^{n-1}$; hence these covariants can be expressed in terms of reducible forms and of products which contain a greater number of factors involving a, a_1, a_2 only.

The products contained in (i) all contain the factor $(aa_1)^{2N}$. Hence all covariants of degree $n+2$ are expressible in terms of reducible forms and of covariants which have the symbolical factor $(aa_1)^{2N}$. But these can, by an application of the assumed result for degree $n+1$, be expressed in terms of reducible forms and of the covariants

$$(aa_1)^{\lambda_1}(aa_2)^{\lambda_2}\dots(aa_{n+1})^{\lambda_{n+1}},$$

where $\lambda_1 \nless 2^n,\quad \lambda_2 \nless 2^{n-1},\ \dots\ \lambda_{n+1} \nless 1.$

The theorem is then true for degree $n+2$ if it is true for degree $n+1$; it has been proved for degrees three and four and is therefore true for all degrees.

263. It should be noticed that it has not been proved that the covariants retained are irreducible. It is practically certain that this is so, but no rigorous proof has yet been given. The number of covariants retained which are of degree $n+1$ and weight w may be found as follows. If

$$w < 2^{n-1} + 2^{n-2} + \ldots + 1,$$

i.e. if $w < 2^n - 1$, all the covariants are reducible. If $w \not< 2^n - 1$, the covariants retained are of the form

$$(aa_1)^{2^{n-1}} (aa_2)^{2^{n-2}} \ldots (aa_n) . R,$$

where R is any product

$$(aa_1)^{\lambda_1} (aa_2)^{\lambda_2} \ldots (aa_n)^{\lambda_n},$$

and $$\lambda_1 + \lambda_2 + \ldots + \lambda_n = w - 2^n + 1.$$

Hence the number required is the coefficient of x^{w-2^n+1} in the expansion of $(1-x)^{-n}$—this being the number of homogeneous products of dimensions $w - 2^n + 1$ of n letters. This is equal to the coefficient of x^w in the expansion of

$$\frac{x^{2^n-1}}{(1-x)^n}.$$

This *generating function* for perpetuant types is the same as that obtained by Macmahon's methods.

264. The results thus obtained for perpetuant types are of great use in obtaining either the types or the ordinary covariants of a binary form of finite order. All that was required in the course of the argument was that the weight of the covariant under consideration should not exceed the order of the quantic—or quantics. Thus any covariant of weight w and of degree δ of the binary n-ic which is such that

$$2^{\delta-1} - 1 > w \not> n,$$

is reducible.

In § 114 the system of forms A_3 for a single binary form of order $\geqslant 12$ was discussed. The above considerations of weight, alone shew that the following forms which were there retained are reducible

$$(ab)^6(bc)^2(cd)^2(de), \quad (ab)^6(bc)^3(cd)^2(de), \quad (ab)^6(bc)(cd)^4(de)^2;$$

the argument applies to the last of these three covariants only if the order of the binary quantic a_x^n is greater than 12, it will be seen later that this is indeed reducible for the 12-ic.

The remaining forms of the system would not be reducible as perpetuants and hence we cannot hope to reduce them for forms of finite order; the two forms

$$(ab)^6(bc)^3(cd)^2, \quad (ab)^6(bc)(cd)^4,$$

however, are congruent mod. $(ab)^6$, as will be presently proved.

265. The theorem for covariants of forms of finite order corresponding to that which has been proved for perpetuant types is the following*.

All covariants which are of the first degree in the coefficients of each of the quantics

$$a_{1_x}^{n_1}, \quad a_{2_x}^{n_2}, \ \ldots \ a_{\delta_x}^{n_\delta}$$

can be expressed linearly in terms of

(i) *covariants of the form*

$$(a_1a_2)^{\lambda_1}(a_2a_3)^{\lambda_2}\ldots(a_{\delta-1}a_\delta)^{\lambda_\delta},$$

where $\quad \lambda_1 \nless 2^{\delta-2}, \quad \lambda_2 \nless 2^{\delta-3}, \ \ldots \ \lambda_\delta \nless 1,$

and the arrangement of the letters $a_1, a_2, \ldots a_\delta$ is fixed;

(ii) *covariants which have a symbolical factor*

$$(a_ha_k)^\lambda(a_ka_l)^{n_k-\lambda};$$

(iii) *products of covariants of lower total degree.*

The proof of this theorem follows that for perpetuants very closely. We first assume that it is true when the total degree of the covariant considered is not greater than $\delta-1$, and prove it when this total degree is δ.

Now the covariants to be considered can be expressed in terms of transvectants

$$(a_{1_x}^{n_1}, C_{\delta-1})^\mu \ \ldots\ldots\ldots\ldots\ldots\ldots\ldots\ldots\text{(I)},$$

* Young, *Proc. Lond. Math. Soc.* 1903.

where $C_{\delta-1}$ is a covariant of the first degree in the coefficients of each of the quantics $a_{1_x}^{n_1}, a_{2_x}^{n_2}, \ldots a_{\delta_x}^{n_\delta}$.

On the assumption made $C_{\delta-1}$ can be expressed linearly in terms of covariants of the second class, of covariants of the form

$$(a_2 a_3)^{\lambda_2} (a_3 a_4)^{\lambda_3} \ldots (a_{\delta-1} a_\delta)^{\lambda_\delta},$$

and of products of covariants of lower total degree.

If $C_{\delta-1}$ is of the second class the transvectant (I), and each of its terms, must be of the second class.

If $C_{\delta-1}$ is a product of two covariants P, Q, then the total degree of P being less than δ we may express it in terms of covariants of the form

$$(ab)^{\mu_1} (bc)^{\mu_2} \ldots (fg)^{\mu_i},$$

and of covariants of the second and third classes.

If $\mu + \mu_1$ be greater than the order of the form a the transvectant contains a term of the second class; if it be not greater than this order the transvectant contains a reducible term.

If P is of the second class the transvectant itself belongs to the second class; and if P is of the third class, we may take one of its factors and proceed as before.

If $C_{\delta-1}$ belongs to the first class, we may take

$$C_{\delta-1} = (a_2 a_3)^{\lambda_2} (a_3 a_4)^{\lambda_3} \ldots (a_{\delta-1} a_\delta)^{\lambda_\delta},$$

then when $\lambda_2 + \mu \not< n_2$, the transvectant contains a term belonging to the second class; but when $\lambda_2 + \mu < n_2$ it contains the term

$$(a_1 a_2)^{\mu} (a_2 a_3)^{\lambda_2} (a_3 a_4)^{\lambda_3} \ldots (a_{\delta-1} a_\delta)^{\lambda_\delta},$$

and hence covariants of this form alone need be considered.

266. Let us now, for the sake of shortness, write

$$a_{4_x}^{n_4-\lambda_4} q_y^{\rho}$$

$$\equiv (a_4 a_5)^{\lambda_4} (a_5 a_6)^{\lambda_5} \ldots (a_{\delta-1} a_\delta)^{\lambda_{\delta-1}} a_{4_x}^{n_4-\lambda_4} a_{5_y}^{n_5-\lambda_4-\lambda_5} \ldots a_{\delta_y}^{n_\delta-\lambda_{\delta-1}}.$$

Then we shall proceed to shew that, if $\kappa < 2^{\delta-3}$, the transvectant

$$((a_1 a_2)^{\mu} (a_2 a_3)^{\kappa}, a_{4_x}^{n_4-\lambda_4} q_y^{\rho})^r{}_{y=x}$$

can be linearly expressed in terms of covariants of the second and third classes, and of covariants which contain a greater number of factors involving a_1, a_2, a_3 only.

The two sets of covariants $(a_1a_2)^\mu (a_2a_3)^\kappa$, and $((a_1a_2)^\mu, a_{3_x}^{n_3})^\kappa$, where $\mu+\kappa$ has a fixed value and κ takes all possible values less than $2^{\delta-3}$, are equivalent.

Hence if
$$(a_1a_2)^\mu \equiv \alpha_x^{n_1+n_2-2\mu},$$
the above transvectants may be linearly expressed in terms of the following,
$$((\alpha a_3)^\kappa, a_{4_x}^{n_4-\lambda_4} q_y^\rho)^r.$$

Any one of these transvectants is a covariant of unit degree in the coefficients of each of the $\delta-1$ quantics
$$\alpha_x^{n_1+n_2-2\mu},\quad a_{3_x}^{n_3},\quad a_{4_x}^{n_4}, \ldots\ a_{\delta_x}^{n_\delta};$$
it can therefore, by hypothesis, be expressed in terms of covariants of the form
$$(\alpha a_3)^{\mu_1} (a_3a_4)^{\mu_2} \ldots (a_{\delta-1}a_\delta)^{\mu_{\delta-2}},$$
where
$$\mu_1 \not< 2^{\delta-3},\quad \mu_2 \not< 2^{\delta-4},\ \ldots\ \mu_{\delta-2} \not< 1,$$
and of covariants belonging to the second and third classes. Thus the number of factors involving a_1, a_2, a_3 only, can be increased when $\kappa < 2^{\delta-3}$.

It should be noticed that covariants of the second class here include those which contain the factor
$$(a_h\alpha)^\nu (\alpha a_k)^{n_1+n_2-2\mu-\nu}.$$

It is easy to see that such a covariant belongs to the second class in the enunciation; for, we may suppose
$$\nu \geqslant n_1+n_2-2\mu-\nu,\quad n_1 \leqslant n_2,$$
and therefore
$$\nu \geqslant n_1-\mu,$$
the covariant considered then contains the factor
$$(a_ha_1)^{n_1-\mu} (a_1a_2)^\mu.$$

267. The covariant
$$(a_1a_2)^\mu (a_2a_3)^{\lambda_2} (a_3a_4)^{\lambda_3} \ldots (a_{\delta-1}a_\delta)^{\lambda_{\delta-1}}$$
is a term of the transvectant
$$((a_1a_2)^\mu (a_2a_3)^{\lambda_2}, a_{4_x}^{n_4-\lambda_4} q_y^\rho)^{\lambda_3}{}_{y=x},$$
and hence differs from the whole transvectant or from any one of

its terms by covariants in which the number of factors involving a_1, a_2, a_3 only is greater than $\lambda_2 + \mu$.

By § 266, we see that we may suppose that neither λ_2 nor μ is less than $2^{\delta-3}$; and hence that $\lambda_2 + \mu \nless 2^{\delta-2}$. The covariant

$$(a_1 a_2)^{\mu} (a_2 a_3)^{\lambda_2}$$

can be linearly expressed in terms of the covariants

$$\text{(i)}\quad (a_1 a_2)^{\lambda_2+\mu}, (a_1 a_2)^{\lambda_2+\mu-1} (a_2 a_3), \ldots (a_1 a_2)^{2^{\delta-2}} (a_2 a_3)^{\lambda_2+\mu-2^{\delta-2}},$$

$$\text{(ii)}\quad (a_1 a_3)^{\lambda_2+\mu}, (a_1 a_3)^{\lambda_2+\mu-1} (a_3 a_2), \ldots (a_1 a_3)^{\lambda_2+\mu-2^{\delta-3}+1} (a_3 a_2)^{2^{\delta-3}-1},$$

$$\text{(iii)}\quad (a_2 a_3)^{\lambda_2+\mu}, (a_2 a_3)^{\lambda_2+\mu-1} (a_3 a_1), \ldots (a_2 a_3)^{\lambda_2+\mu-2^{\delta-3}+1} (a_3 a_1)^{2^{\delta-3}-1}.$$

Transvectants of a covariant from one of the last two rows with $a_{4_x}^{n_4-\lambda_4} q_y^{\rho}$ can be expressed in terms of covariants which contain a greater number of factors involving a_1, a_2, a_3 only. Hence we may ultimately express all covariants in question linearly in terms of covariants having a factor $(a_1 a_2)^{\lambda_1}$, where $\lambda_1 \nless 2^{\delta-2}$, and of covariants belonging to the second and third classes. Proceeding, as in § 266, with the covariants which have a factor $(a_1 a_2)^{\lambda_1}$ where $\lambda_1 \nless 2^{\delta-2}$, we see that all covariants may be expressed linearly in terms of covariants of the form

$$(a_1 a_2)^{\lambda_1} (a_2 a_3)^{\lambda_2} \ldots (a_{\delta-1} a_\delta)^{\lambda_\delta},$$

where $$\lambda_1 \nless 2^{\delta-2},\ \lambda_2 \nless 2^{\delta-3},\ \ldots\ \lambda_\delta \nless 1,$$

and of covariants of the second and third classes.

Thus the theorem is true when the total degree of the covariant is δ, provided that it is true when this total degree is less than δ.

268. It remains to shew that this theorem is true when the total degree is three.

The covariants to be considered are

$$(a_1 a_2)^{\lambda_1} (a_2 a_3)^{\lambda_2} (a_3 a_1)^{\lambda_3}.$$

Unless $\lambda_1 + \lambda_2 + \lambda_3$ is less than each of the numbers n_1, n_2, n_3, this covariant belongs to the second class. For let

$$\lambda_1 + \lambda_2 + \lambda_3 \geqslant n_2,$$

then by means of the identity

$$(a_3 a_1) = -(a_1 a_2) - (a_2 a_3)$$

the above covariant can be expressed in terms of the forms

$$(a_1 a_2)^r (a_2 a_3)^{n_2 - r} (a_3 a_1)^{\lambda_1 + \lambda_2 + \lambda_3 - n_2},$$

and belongs to the second class.

If $\lambda_1 + \lambda_2 + \lambda_3$ is less than each of n_1, n_2, n_3 the argument used for perpetuants may be repeated here word for word. The theorem is then true for total degree three, and therefore for any total degree.

269. If all the quantics are of the same order n we obtain a theorem concerning covariant types of a simultaneous system of binary n-ics.

In this case the covariants of the second class contain a factor of the form $(ab)^\lambda (bc)^{n-\lambda}$, and hence a factor of the form

$$(ab)^\mu (bc)^{n-\mu} (ca)^\rho,$$

where $$\mu \nless \frac{n}{2}, \quad \rho \nless 2n - 3\mu,$$

(see § 68).

Hence *all covariant types of a system of binary n-ics can be expressed linearly in terms of*

(i) *Covariants of the form*

$$(a_1 a_2)^{\lambda_1} (a_2 a_3)^{\lambda_2} \dots (a_{\delta-1} a_\delta)^{\lambda_\delta},$$

where $$\lambda_1 \nless 2^{\delta-2}, \; \lambda_2 \nless 2^{\delta-3}, \; \dots \; \lambda_\delta \nless 1,$$

and the order of the letters is fixed beforehand.

(ii) *Covariants which have a factor of the form*

$$(ab)^\lambda (bc)^{n-\lambda} (ca)^\rho,$$

where $$\lambda \nless \frac{n}{2}, \quad \rho \nless 2n - 3\lambda.$$

(iii) *Products of covariants of lower total degree.*

270. The theorem just proved expresses all irreducible covariants, of grade $< \frac{n}{2}$, in terms of a certain number of forms, which, there is good reason to believe, are irreducible when n is infinite. If this be so, these forms are certainly irreducible for finite values of n.

However, it does not follow that we cannot express them in terms of covariants of higher grade or else of covariants of the second class.

In this connection we shall prove the following: *all covariant types of the binary n-ic can be expressed linearly in terms of covariants of the form*

$$(a_1 a_2)^{\lambda_1} (a_2 a_3)^{\lambda_2} \dots (a_{\delta-1} a_\delta)^{\lambda_{\delta-1}},$$

where $\lambda_1 \geqslant 2\lambda_2,\ \lambda_2 > \lambda_3,\ \lambda_3 > \lambda_4,\ \dots\ \lambda_{\delta-2} > \lambda_{\delta-1},$

and of covariants belonging to the second and third classes.

Using the previous theorem we see that covariants of the form

$$C = (a_1 a_2)^{\mu_1} (a_2 a_3)^{\mu_2} \dots (a_{\delta-1} a_\delta)^{\mu_{\delta-1}}$$

alone need be considered.

Let $\mu_1 \not> 2\mu_2$, then C is a term of

$$((a_1 a_2)^{\mu_1} (a_2 a_3)^{\mu_2}, (a_4 a_5)^{\mu_4} \dots (a_{\delta-1} a_\delta)^{\mu_{\delta-1}} a_{4_x}^{n-\delta_4} a_{5_y}^{n-\delta_4-\delta_5} \dots)^{\mu_3}{}_{y=x},$$

and differs from any other term by covariants which involve a greater number of factors containing a_1, a_2, a_3 only.

Then by Stroh's theorem, we may express $(a_1 a_2)^{\mu_1} (a_2 a_3)^{\mu_2}$ in terms of covariants of the form

$$(ab)^{\lambda_1} (bc)^{\lambda_2},$$

where $\lambda_1 \not< 2\lambda_2,\quad \lambda_1 + \lambda_2 = \mu_1 + \mu_2,$

and a, b, c are the letters a_1, a_2, a_3 in some order.

Let $(ab)^{\lambda_1} \equiv \alpha_x^{2n-2\lambda_1},$

we have then to consider covariants of the form

$$(\alpha a_3)^{\lambda_2} (a_3 a_4)^{\mu_3} \dots (a_{\delta-1} a_\delta)^{\mu_{\delta-1}}.$$

If $\lambda_2 < \mu_3$ we may consider the transvectant

$$((\alpha a_3)^{\lambda_2} (a_3 a_4)^{\mu_3}, (a_5 a_6)^{\mu_5} \dots (a_{\delta-1} a_\delta)^{\mu_{\delta-1}} a_{5_x}^{n-\mu_5} a_6^{n-\mu_5-\mu_6})^{\mu_4}{}_{y=x}.$$

Then $(\alpha a_3)^{\lambda_2} (a_3 a_4)^{\mu_3}$ can be expressed linearly in terms of $(a_3 a_4)^{\lambda_2+\mu_3}$, and of members of the sets

$$(\alpha a_3)^{\kappa} (a_3 a_4)^{\lambda_2+\mu_3-\kappa},\ (\alpha a_4)^{\kappa} (a_4 a_3)^{\lambda_2+\mu_3-\kappa},$$

where $$\kappa > \frac{\lambda_2 + \mu_3}{2}.$$

We may proceed in exactly the same way at every step, and so prove the theorem.

It will be noticed here that the order of the letters in the covariants

$$(a_1 a_2)^{\lambda_1} (a_2 a_3)^{\lambda_2} \dots (a_{\delta-1} a_\delta)^{\lambda_{\delta-1}}$$

is not fixed.

271. The Maximum Order of a covariant. Let us consider the covariant types of a system of quantics of which none of the orders exceeds n. By § 265, these types can all be expressed in terms of covariants of three kinds. Consider the covariants of the second kind. These contain a factor of the form

$$(a_1 a_2)^{\lambda} (a_2 a_3)^{n_2 - \lambda}.$$

We may suppose that $\lambda \not< n_2 - \lambda$; the order of the covariant

$$(a_1 a_2)^{\lambda} a_{1_x}^{n_1 - \lambda} a_{2_x}^{n_2 - \lambda}$$

is then $$n_1 + n_2 - 2\lambda \not> n_1 \leqslant n.$$

Now let us introduce a new symbol, for each covariant whose order does not exceed n. Covariants of the second kind are thus at once reduced in degree. Covariants thus reduced may themselves be expressed in terms of covariants of the three different kinds. The covariants of the second kind may again be reduced, and so on. Hence finally we have expressed the system of covariants in terms of covariants of the form

$$(a_1 a_2)^{\lambda_1} (a_2 a_3)^{\lambda_2} \dots (a_{\delta-1} a_\delta)^{\lambda_{\delta-1}},$$

where $$\lambda_1 \not< 2^{\delta-2},\ \lambda_2 \not< 2^{\delta-3},\ \dots\ \lambda_{\delta-1} \not< 1,$$

—$a_{1_x}^{n_1}$, $a_{2_x}^{n_2}$, ... $a_{\delta_x}^{n_\delta}$ being either members of the original system of quantics or covariants of that system whose order does not exceed n—; and of products of covariants of lower degree.

The covariant of maximum order must then be of the form

$$(a_1 a_2)^{\lambda_1} (a_2 a_3)^{\lambda_2} \dots (a_{\delta-1} a_\delta)^{\lambda_{\delta-1}} a_{1_x}^{n-\lambda_1} a_{2_x}^{n_2 - \lambda_1 - \lambda_2} \dots a_{\delta_x}^{n_\delta - \lambda_{\delta-1}},$$

where $$\lambda_1 \not< 2^{\delta-2},\ \lambda_2 \not< 2^{\delta-3},\ \dots\ \lambda_{\delta-1} \not< 1,$$

and $n_1, n_2, \dots n_\delta$ are all equal to or less than n.

The order of this for a given value of δ is a maximum when

$$\lambda_1 = 2^{\delta-2},\ \lambda_2 = 2^{\delta-3},\ \dots\ \lambda_{\delta-1} = 1,$$

$$n_1 = n_2 = \dots = n_\delta = n.$$

In this case the order is

$$n\delta - 2(1 + 2 + \ldots + 2^{\delta-2}) = n\delta - 2^\delta + 2.$$

The maximum order is then the greatest of the numbers

$$n, \; 2n-2, \; 3n-6, \; \ldots \; n\delta - 2^\delta + 2, \; \ldots.$$

It is easy to see that if $n = 2^\lambda + n_1$, where $n_1 < 2^\lambda$, then this maximum order is

$$(\lambda+1)(2^\lambda + n_1) - 2^{\lambda+1} + 2 = (\lambda - 1)\, 2^\lambda + n_1(\lambda+1) + 2.$$

Comparison with perpetuants shews at once, that if the results for these are absolutely accurate, then the maximum order just obtained is always reached—even for a single quantic of order n except for the case $n = 3$.

Ex. (i). The covariant $(ab)^6 (bc) (cd)^4 (de)^2$ of the twelvic referred to in § 264 is of weight 13 and hence must be reducible to covariants of the second and third classes; it is evidently reducible in the usual sense of the word.

Ex. (ii). Shew that the following covariants of the ten-ic can be expressed in terms of reducible covariants and of covariants of higher grade:

$$(ab)^4 (bc)^3 (cd),$$
$$(ab)^6 (bc)^3 (cd)^3,$$
$$(ab)^6 (bc)^3 (cd)^2 (de).$$

References to papers by Jordan and Sylvester on the problem of § 271 and allied problems regarding weight and degree will be found in Meyer. The limits hitherto given are much too high for large values of n.

CHAPTER XVI.

GENERAL THEOREMS ON QUANTICS.

272. In this chapter certain results are obtained by an application of the theory, or rather the notation of the theory, of finite substitution groups. So little knowledge of this subject is required, that, for the sake of readers unacquainted with it, we shall start from the commencement, and prove the few well-known theorems required.

In the first place a function of n variables

$$f(x_1, x_2, x_3, \dots x_n)$$

is under consideration; the function

$$f(x_2, x_1, x_3, \dots x_n)$$

is derived from this by the interchange of the two variables x_1 and x_2. The operation by which the latter function is obtained from the former is called a substitution, it is usually denoted by the symbol (x_1x_2). Thus we may write

$$f(x_2, x_1, x_3, \dots x_n) = (x_1x_2) f(x_1, x_2, x_3, \dots x_n).$$

A more general example of a substitution is the operation by which the arrangement of variables

$$x_1, x_2, \dots x_n$$

is changed to $\qquad y_1, y_2, \dots y_n,$

the y's being the variables $x_1, x_2, \dots x_n$ arranged in some order. This substitution is often written

$$\begin{pmatrix} x_1 & x_2 \dots & x_n \\ y_1 & y_2 \dots & y_n \end{pmatrix};$$

thus $\qquad \begin{pmatrix} x_1 & x_2 \dots & x_n \\ y_1 & y_2 \dots & y_n \end{pmatrix} f(x_1, x_2, \dots x_n) = f(y_1, y_2, \dots y_n).$

Substitutions which represent merely the interchange of two variables are called transpositions; thus the substitution (x_1x_2) introduced above is a transposition.

The product of two substitutions. Let s_1, s_2 be any two substitutions of the letters $x_1, x_2, \ldots x_n$, the meaning here attached to the product s_1s_2 is that it is an operation which when applied to a function of $x_1, x_2, \ldots x_n$ is equivalent to the operation first of s_2 on this function and then of s_1 on the resulting function. (The usual convention is that the substitution on the left is the first to operate, but the above is more convenient for our present purpose.) Thus

$$s_1s_2f(x_1, x_2, \ldots x_n) = s_1[s_2f(x_1, x_2, \ldots x_n)].$$

The effect of s_2 is merely to produce a rearrangement of the variables, the effect of s_1 on the resulting function is to produce a fresh rearrangement; thus the product of two substitutions is a substitution.

It will be seen at once that substitutions obey the distributive law, for

$$\begin{aligned} s_1[s_2s_3]f(x_1, x_2, \ldots x_n) &= s_1[s_2s_3f(x_1, x_2, \ldots x_n)] \\ &= s_1[s_2\{s_3f(x_1, x_2, \ldots x_n)\}] \\ &= [s_1s_2]s_3f(x_1, x_2, \ldots x_n). \end{aligned}$$

On the other hand substitutions are not in general commutative, for example:

$$\begin{aligned} (x_1x_2)(x_1x_3)f(x_1, x_2, x_3) &= (x_1x_2)f(x_3, x_2, x_1) \\ &= f(x_3, x_1, x_2), \end{aligned}$$

but

$$\begin{aligned} (x_1x_3)(x_1x_2)f(x_1, x_2, x_3) &= (x_1x_3)f(x_2, x_1, x_3) \\ &= f(x_2, x_3, x_1). \end{aligned}$$

Any substitution can be represented as a product of transpositions.

For any rearrangement of the letters $x_1, x_2, \ldots x_n$ can be produced, first by an interchange of x_1 and one other letter by which x_1 takes its new position, next by an interchange of x_2 and another letter by which x_2 is brought to its new position, and so on.

It will be found that a substitution can be represented as a product of transpositions in a great number of ways, *e.g.*

$$(x_1x_2) = (x_2x_3)(x_1x_3)(x_2x_3);$$

but *the number of transpositions in a product which represents a given substitution is always even or always odd.* For consider the function

$$\Delta = \begin{vmatrix} x_1^{n-1} & x_1^{n-2} \dots 1 \\ x_2^{n-1} & x_2^{n-2} \dots 1 \\ \dots\dots & \dots\dots \\ x_n^{n-1} & x_n^{n-2} \dots 1 \end{vmatrix} = \prod_{r,s} (x_r - x_s);$$

the effect of any transposition operating on Δ is merely to change its sign. Hence

$$s\Delta = \pm\Delta$$

according as s is a product of an even or odd number of substitutions. Substitutions will be called even or odd according as the number of transpositions of which they are composed is even or odd.

Consider any rearrangement of the letters $x_1, x_2, \dots x_n$; let x_r be the letter which takes the place of x_1; x_s that which takes the place of x_r; x_t that which takes the place of x_s, and so on; we must sooner or later arrive at a stage when x_1 is the letter which takes the place of x_u the last of the series. The substitution which replaces x_1 by x_r, x_r by x_s, x_s by x_t and so on, and finally x_u by x_1 is usually written $(x_1 x_r x_s x_t \dots x_u)$ and is called a *cycle.* It is evident from the definition that

$$(x_1 x_r x_s x_t \dots x_u) = (x_r x_s x_t \dots x_u x_1).$$

The rearrangement considered may be produced so far as the letters in the cycle are concerned by operating with $(x_1 x_r x_s x_t \dots x_u)$. Let x_a be one of the letters not contained in this cycle, then we may suppose that x_β in the new arrangement takes the place of x_a and proceed as before. Thus we see that the rearrangement may be produced by operating with a number of independent cycles, *i.e.* cycles such that no two contain a common letter. *Hence any substitution is equal to a product of a number of independent cycles.*

That operation which leaves any function operated on unaltered is called the identical substitution and is written 1. The product

$$(x_1 x_2) . (x_1 x_2) = (x_1 x_2)^2$$

leaves every function unaltered, hence

$$(x_1 x_2)^2 = 1.$$

Consider the rearrangement of the letters $x_1, x_2, \ldots x_n$ produced by any substitution s; there is a perfectly definite substitution which will change this new arrangement back to the old arrangement. This is called the *inverse* substitution of s and is written s^{-1}. In virtue of the definition of s^{-1} we see that

$$s^{-1}sf(x_1, x_2, \ldots x_n) = f(x_1, x_2, \ldots x_n),$$

and hence $$s^{-1}s = 1.$$

Again it is to be observed that the result of operating on the new arrangement of the letters with ss^{-1} is to leave it unaltered, hence

$$ss^{-1} = 1.$$

Consider the powers of any substitution,

$$s, s^2, s^3, \ldots;$$

they are all substitutions, and since the number of different substitutions of n letters is finite—in fact $n!$, the number of possible arrangements of those letters—these powers cannot be all different. Hence for some values of h, k,

$$s^h = s^k.$$

In virtue of the associative law, we can write

$$s^l s^m = s^{l+m}.$$

Hence $$s^{h+l} = s^{k+l}.$$

Further, whatever substitution σ may be,

$$\sigma . s^h = \sigma s^k;$$

hence if $\sigma = (s^{-1})^h$, we see that

$$(s^{-1})^h . s^h = (s^{-1})^{h-1} s^{-1} s s^{h-1} = (s^{-1})^{h-1} s^{h-1} = \ldots = 1;$$

and hence $$1 = s^{k-h}.$$

We may suppose that $k > h$, hence among the positive powers of s we must find the identical substitution. Let p be the smallest positive index for which $s^p = 1$, then p is called the *order* of the substitution.

The substitution $$\sigma s \sigma^{-1}$$

is said to be conjugate to s.

In σ let the letter which replaces x_r be denoted by x_r'; let

$$s = \begin{pmatrix} x_1 & x_2 \dots x_n \\ y_1 & y_2 \dots y_n \end{pmatrix},$$

where $y_1, y_2, \dots y_n$ are the letters $x_1, x_2, \dots x_n$ arranged in some order.

Now $$\sigma = \begin{pmatrix} x_1 & x_2 \dots x_n \\ x_1' & x_2' \dots x_n' \end{pmatrix} = \begin{pmatrix} y_1 & y_2 \dots y_n \\ y_1' & y_2' \dots y_n' \end{pmatrix},$$

and $$\sigma^{-1} = \begin{pmatrix} x_1' & x_2' \dots x_n' \\ x_1 & x_2 \dots x_n \end{pmatrix}.$$

Hence

$$\sigma s \sigma^{-1} = \begin{pmatrix} y_1 & y_2 \dots y_n \\ y_1' & y_2' \dots y_n' \end{pmatrix} \begin{pmatrix} x_1 & x_2 \dots x_n \\ y_1 & y_2 \dots y_n \end{pmatrix} \begin{pmatrix} x_1' & x_2' \dots x_n' \\ x_1 & x_2 \dots x_n \end{pmatrix}$$

$$= \begin{pmatrix} y_1 & y_2 \dots y_n \\ y_1' & y_2' \dots y_n' \end{pmatrix} \begin{pmatrix} x_1' & x_2' \dots x_n' \\ y_1 & y_2 \dots y_n \end{pmatrix} = \begin{pmatrix} x_1' & x_2' \dots x_n' \\ y_1' & y_2' \dots y_n' \end{pmatrix}.$$

Therefore $\sigma s \sigma^{-1}$ is the substitution which would be obtained by operating on the expression for s with the substitution σ. It must then be a product of cycles each having the same number of letters as the cycles of s; and is obtained from s by permuting the letters. Such substitutions are called *similar*. Every substitution similar to s is obtained from s by a suitable permutation of the letters, and is therefore of the form $\sigma s \sigma^{-1}$.

Now if s be any cycle $(x_1 x_2 \dots x_k)$ then

$$s^{-1} = (x_k \dots x_2 x_1)$$

as may easily be verified; and hence s^{-1} is similar to s. But every substitution is a product of a number of independent cycles, the inverse substitution is then the product of the inverse cycles; hence any substitution is similar to its inverse.

273. If m substitutions $s_1, s_2, \dots s_m$ are such that the product of any two of them is itself one of the m substitutions, these m substitutions are said to form a group.

Thus as may be at once verified

$$1, (x_1 x_2);\ 1, (x_1 x_2 x_3), (x_1 x_3 x_2)$$

are groups.

The number of substitutions included in a group is called the *order*, the number of letters affected is called the *degree* of the group.

Thus the two groups written down are of order 2 degree 2, and of order 3 degree 3 respectively.

Now the n letters $x_1, x_2, \ldots x_n$ can be arranged in $n!$ ways, hence the total number of substitutions affecting n letters is $n!$. These substitutions obviously form a group, it is of degree n and of order $n!$. This group is called the symmetric group for the n letters $x_1, x_2, \ldots x_n$.

It is useful to have a symbol by which to denote this group, the symmetric group for the letters $x_1, x_2, \ldots x_n$ will be written

$$\{x_1 x_2 \ldots x_n\}.$$

More particularly this symbol will be used to denote the sum of all the substitutions of the symmetric group.

Again the product of two even substitutions is obviously an even substitution, hence the even substitutions which affect n letters form a group. This group is called the alternating group. Let

$$s_1, s_2, \ldots s_m$$

be the members of the alternating group, and

$$\sigma_1, \sigma_2, \ldots \sigma_{m'}$$

the remaining substitutions which affect the letters $x_1, x_2, \ldots x_n$.

Then these latter substitutions are all odd; and hence the product of any two of them is an even substitution.

Now if t_1, t_2, t_3 be any three substitutions and

$$t_1 t_2 = t_1 t_3,$$

then
$$t_1^{-1} t_1 t_2 = t_1^{-1} t_1 t_3,$$

and hence
$$t_2 = t_3.$$

By hypothesis the substitutions $s_1, s_2, \ldots s_m, \sigma_1, \sigma_2, \ldots \sigma_{m'}$ are all different, hence the substitutions

$$\sigma_1 s_1, \sigma_1 s_2, \ldots \sigma_1 s_m, \sigma_1^2, \sigma_1 \sigma_2, \ldots \sigma_1 \sigma_{m'}$$

are all different. But the former set include all the substitutions of the n letters, hence the latter must do so too. Hence the even substitutions

$$\sigma_1^2, \sigma_1 \sigma_2, \ldots \sigma_1 \sigma_{m'}$$

form the alternating group. And therefore

$$m = m' = \tfrac{1}{2} n!.$$

The symbol $$\{x_1x_2 \dots x_n\}'$$
will be used to denote the sum of the even substitutions minus the sum of the odd substitutions of the letters

$$x_1, x_2, \dots x_n.$$

This will be called the negative symmetric group, and on the other hand

$$\{x_1x_2 \dots x_n\}$$

will be called the positive symmetric group of the n letters.

For example

$$\{x_1x_2x_3\} = 1 + (x_1x_2x_3) + (x_1x_3x_2) + (x_1x_2) + (x_2x_3) + (x_3x_1),$$

$$\{x_1x_2x_3\}' = 1 + (x_1x_2x_3) + (x_1x_3x_2) - (x_1x_2) - (x_2x_3) - (x_3x_1).$$

274. As an illustration of the notation just introduced we remark that the determinant

$$\begin{vmatrix} a_1 & a_2 & \dots & a_n \\ b_1 & b_2 & \dots & b_n \\ \dots & \dots & \dots & \dots \\ \dots & \dots & \dots & \dots \\ k_1 & k_2 & \dots & k_n \end{vmatrix}$$

may be written

$$\{ab \dots k\}' a_1b_2 \dots k_n,$$

the substitutions being supposed to affect the letters and not the suffixes. Or adopting a double suffix notation we may write

$$\begin{vmatrix} a_{1,1} & a_{1,2} & \dots & a_{1,n} \\ a_{2,1} & a_{2,2} & \dots & a_{2,n} \\ \dots & \dots & \dots & \dots \\ \dots & \dots & \dots & \dots \\ a_{n,1} & a_{n,2} & \dots & a_{n,n} \end{vmatrix} = \{a_1a_2 \dots a_n\}' a_{1,1}a_{2,2} \dots a_{n,n},$$

where the first suffix only appears in the substitutions and is alone affected by them.

Or again

$$\begin{vmatrix} a^p & b^p & c^p \\ a^q & b^q & c^q \\ a^r & b^r & c^r \end{vmatrix} = \{abc\}' a^p b^q c^r.$$

As an example of the use of the positive symmetric group, referring to § 44, we observe that the rth polar of the form

$$a_x{}^n = a_{1_x} a_{2_x} \dots a_{n_x}$$

may be conveniently written

$$a_x^{n-r} a_y^r = \frac{1}{n!} \{\alpha_1 \alpha_2 \ldots \alpha_n\} \alpha_{1_y} \alpha_{2_y} \ldots \alpha_{r_y} \alpha_{r+1_x} \ldots \alpha_{n_x}.$$

275. Consider any function F of the coefficients of certain linear binary forms α_x, β_x, γ_x, which is homogeneous and linear in the coefficients of each form separately. We may write

$$F = \alpha_1 \beta_1 \phi_1 + \alpha_1 \beta_2 \phi_2 + \alpha_2 \beta_1 \phi_3 + \alpha_2 \beta_2 \phi_4,$$

where the ϕ's are functions of the coefficients of the linear forms γ_x, δ_x, ... of the same character as F.

Then $$\{\alpha\beta\}' F = (\alpha\beta)[\phi_2 - \phi_3];$$

i.e. $\{\alpha\beta\}' F$ is the product of $(\alpha\beta)$ and a function which does not contain the coefficients of α_x or β_x.

Again we may write

$$F = \Sigma \alpha_r \beta_s \gamma_t \phi_{r, s, t}, \quad (r, s, t = 1, 2),$$

but here the suffixes r, s, t can never be all different.

Now

$$\{\alpha\beta\gamma\}' \alpha_r \beta_s \gamma_t = \begin{vmatrix} \alpha_r & \alpha_s & \alpha_t \\ \beta_r & \beta_s & \beta_t \\ \gamma_r & \gamma_s & \gamma_t \end{vmatrix},$$

hence $$\{\alpha\beta\gamma\}' F = 0.$$

In the same way if α_{1_x}, α_{2_x}, ... be any p-ary linear forms, where

$$\alpha_{r_x} = \alpha_{r,1} x_1 + \alpha_{r,2} x_2 + \ldots + \alpha_{r,p} x_p,$$

and F be any function homogeneous and linear in the coefficients of each,

$$\{\alpha_1 \alpha_2 \ldots \alpha_{p+1}\}' F = 0,$$

$$\{\alpha_1 \alpha_2 \ldots \alpha_p\}' F = \begin{vmatrix} \alpha_{1,1} & \alpha_{1,2} & \ldots & \alpha_{1,p} \\ \alpha_{2,1} & \alpha_{2,2} & \ldots & \alpha_{2,p} \\ \ldots & \ldots & \ldots & \ldots \\ \ldots & \ldots & \ldots & \ldots \\ \alpha_{p,1} & \alpha_{p,2} & \ldots & \alpha_{p,p} \end{vmatrix} . \psi,$$

where the substitutions affect the first suffixes only, and ψ is a function of the coefficients of

$$\alpha_{p+1_x}, \alpha_{p+2_x}, \ldots.$$

The expression $\mid \alpha_1 \alpha_2 \ldots \alpha_p \mid$ will be used as an abbreviation for the determinant just written down.

Again, if $r > p$,

$$\{\alpha_1 \alpha_2 \ldots \alpha_r\}' F = 0;$$

and if $r < p$, it is easy to see that

$$\{\alpha_1 \alpha_2 \ldots \alpha_r\}' F = \sum_\lambda \Delta_\lambda \psi_\lambda,$$

where Δ_λ is one of the determinants of the matrix

$$\left| \begin{array}{cccc} a_{1,1} & a_{1,2} & \ldots & a_{1,p} \\ a_{2,1} & a_{2,2} & \ldots & a_{2,p} \\ \multicolumn{4}{c}{\ldots\ldots\ldots\ldots\ldots\ldots\ldots} \\ a_{r,1} & a_{r,2} & \ldots & a_{r,p} \end{array} \right|.$$

It is unnecessary to suppose that $a_{r,1}, a_{r,2}, \ldots a_{r,p}$ are the coefficients of a linear p-ary form a_{r_x}. The facts just established are true if F is homogeneous and linear in each of any m sets of quantities

$$\begin{array}{cccc} a_{1,1} & a_{1,2} & \ldots & a_{1,p} \\ a_{2,1} & a_{2,2} & \ldots & a_{2,p} \\ \multicolumn{4}{c}{\ldots\ldots\ldots\ldots\ldots\ldots\ldots} \\ a_{m,1} & a_{m,2} & \ldots & a_{m,p} \end{array}$$

there being p quantities in each set; and a one-to-one correspondence between the members of any two sets.

Thus in particular: *If F is a function homogeneous and linear in the coefficients of each of m binary n-ics*

$$(a_{1,0}, a_{1,1}, \ldots a_{1,n} \between x_1, x_2)^n, \ldots (a_{m,0}, a_{m,1}, \ldots a_{m,n} \between x_1, x_2)^n,$$

m being greater than $n+1$, then

$$\{\alpha_1 \alpha_2 \ldots \alpha_{n+2}\}' F = 0 \quad \ldots\ldots\ldots\ldots\ldots\ldots\ldots\ldots\text{(i)},$$

$$\{\alpha_1 \alpha_2 \ldots \alpha_{n+1}\}' F = \mid \alpha_1 \alpha_2 \ldots \alpha_{n+1} \mid . F_1 \quad \ldots\ldots\text{(ii)},$$

$$\{\alpha_1 \alpha_2 \ldots \alpha_n\}' F = \mid \alpha_1 \alpha_2 \ldots \alpha_n \beta \mid \quad \ldots\ldots\ldots\ldots\text{(iii)},$$

where the substitutions affect the first suffixes only of the coefficients $a_{p,q}$; $\mid \alpha_1 \alpha_2 \ldots \alpha_{n+1} \mid$ is the determinant of $n+1$ rows and columns formed by the coefficients of the $n+1$ quantics concerned;

$$\mid \alpha_1 \alpha_2 \ldots \alpha_n \beta \mid$$

is the same determinant with quantities $\beta_0, \beta_1, \ldots \beta_n$ replacing

$$a_{n+1,0}, a_{n+1,1}, \ldots a_{n+1,n},$$

these quantities being homogeneous and linear in the coefficients of the quantics represented by $\alpha_{n+1}, \alpha_{n+2}, \ldots \alpha_m$; *and where* F_1 *is homogeneous and linear in the coefficients of the quantics represented by* $\alpha_{n+2}, \ldots \alpha_m$.

The first two of the above results are sufficiently clear from what has already been said. As regards the last we observe that

$$\{\alpha_1\alpha_2 \ldots \alpha_n\}' F = \sum_\lambda \Delta_\lambda \beta_\lambda,$$

where Δ_λ is one of the determinants of the matrix

$$\begin{vmatrix} \alpha_{1,0} & \alpha_{1,1} & \ldots & \alpha_{1,n} \\ \alpha_{2,0} & \alpha_{2,1} & \ldots & \alpha_{2,n} \\ \cdots & \cdots & \cdots & \cdots \\ \alpha_{n,0} & \alpha_{n,1} & \ldots & \alpha_{n,n} \end{vmatrix}.$$

That is Δ_λ may be taken to be the minor of $\alpha_{n+1,\lambda}$ in the determinant

$$| \alpha_1\alpha_2 \ldots \alpha_{n+1} |,$$

and hence $$\{\alpha_1\alpha_2 \ldots \alpha_n\}' F = | \alpha_1\alpha_2 \ldots \alpha_n\beta |.$$

276. Let the quantics of the last paragraph be represented symbolically thus

$$(\alpha_{r,0}, \alpha_{r,1}, \ldots \alpha_{r,n} \between x_1, x_2)^n = [a_x^{(r)}]^n, \quad (r = 1, 2, \ldots m).$$

Then the determinant $| \alpha_1\alpha_2 \ldots \alpha_{n+1} |$ is an invariant, for it may be written symbolically

$$\prod_{r,s} (a^{(r)}a^{(s)}) \qquad (r, s = 1, 2, \ldots m;\ r \neq s).$$

Hence if F is an invariant, then F_1 is also an invariant.

Again we can shew that if F is an invariant, then

$$(\beta_0, \beta_1, \ldots \beta_n \between x_1, x_2)^n$$

is a covariant. For let A_r be the minor of $\alpha_{n+1,r}$ in the determinant

$$| \alpha_1\alpha_2 \ldots \alpha_{n+1} |.$$

Then since F is an invariant

$$\Sigma\beta_r A_r$$

is also an invariant.

Now we know that $\Sigma a_{n+1,r} A_r$ and $\Sigma a_{n+1,r} \binom{n}{r} x_1^{n-r} x_2^r$ are both invariantive, and hence that the quantities

$$A_0, A_1, \dots A_n$$

and
$$x_1^n, \binom{n}{1} x_1^{n-1} x_2, \dots \binom{n}{n} x_2^n$$

form two cogredient sets. Therefore, since $\Sigma \beta_r A_r$ is an invariant, $\Sigma \beta_r \binom{n}{r} x_1^{n-r} x_2^r$ must also be invariantive. In other words

$$(\beta_0, \beta_1, \dots \beta_n \between x_1, x_2)^n$$

is a covariant.

If F were a covariant, and contained the variables x_1, x_2, then $\beta_0, \beta_1, \dots \beta_n$ would also contain these quantities; in this case the form

$$(\beta_0, \beta_1, \dots \beta_n \between y_1, y_2)^n$$

is a covariant in two sets of variables.

277. As examples of the use of the results just established we may instance the fundamental identity for binary forms

$$\{abc\}' (ab)(cd) = 0;$$

that for ternary forms

$$\{abcd\}' (abc)(def) = 0,$$

or
$$\{abcd\}' (abe)(cdf) = 0;$$

those for quaternary forms

$$\{abcde\}' (abcd)(efgh) = 0,$$

$$\{abcde\}' (abfg)(cdhi)(ejkl) = 0,$$

and so on.

Let I be an invariant linear in the coefficients of each of $n+1$ binary n-ics; then with the notation of § 276 we have

$$\{\alpha_1 \alpha_2 \dots \alpha_{n+1}\}' I = \lambda \,|\, \alpha_1 \alpha_2 \dots \alpha_{n+1} \,|$$

where λ is some constant, possibly zero.

If n is odd we may take

$$I = (a^{(1)} a^{(2)})^n (a^{(3)} a^{(4)})^n \dots (a^{(n)} a^{(n+1)})^n.$$

In this case, provided the $n+1$ quantics are all linearly independent,

$$\{\alpha_1 \alpha_2 \dots \alpha_{n+1}\}' I$$

is different from zero, for all terms of I in which two coefficients have the same suffixes are destroyed by the operator, and the rest are obtained from

$$[\alpha_{1,0}\,\alpha_{2,n}]\left[-\binom{n}{1}\alpha_{3,1}\,\alpha_{4,n-1}\right]\left[\binom{n}{2}\alpha_{5,2}\,\alpha_{6,n-2}\right]\dots\dots$$

$$\left[(-1)^{\frac{n-1}{2}}\binom{n}{\frac{n-1}{2}}\alpha_{n,\frac{n-1}{2}}\,\alpha_{n+1,\frac{n+1}{2}}\right]$$

by interchanging both the first suffixes of pairs of brackets in all possible ways keeping the order of the three first suffixes unaltered—so that the first suffixes of any one bracket are always of the form $(2r-1)$, $(2r)$ and in this order—and by then interchanging the first suffixes inside individual brackets, each such interchange being accompanied by a change of sign. But both these operations are effected by

$$\{\alpha_1\alpha_2\dots\alpha_{n+1}\}'$$

—since the first only requires even substitutions. Hence

$$\{\alpha_1\alpha_2\dots\alpha_{n+1}\}'\,I=\lambda\,\{\alpha_1\alpha_2\dots\alpha_{n+1}\}'\,\alpha_{1,0}\,\alpha_{2,n}\,\alpha_{3,1}\,\alpha_{4,n-1}\dots\alpha_{n,\frac{n-1}{2}}\,\alpha_{n+1,\frac{n+1}{2}}$$

$$=\pm\lambda\,|\,\alpha_1\alpha_2\dots\alpha_{n+1}\,|$$

where λ is not zero. Hence when n is odd

$$|\,\alpha_1\alpha_2\dots\alpha_{n+1}\,|$$

is reducible.

Again, if $n=4$, it is easy to see that this invariant is irreducible. Let us suppose that it be reducible, then

$$|\,\alpha_1\alpha_2\alpha_3\alpha_4\alpha_5\,|=\Sigma I_2 I_3$$

where I_r is an invariant of degree r.

Now I_2 must be of the form $(a^{(r)}a^{(s)})^4$, where r, s are two of the numbers 1, 2, 3, 4, 5; hence

$$\{\alpha_1\alpha_2\alpha_3\alpha_4\alpha_5\}'\,I_2I_3=0,$$

and therefore

$$\{\alpha_1\alpha_2\alpha_3\alpha_4\alpha_5\}'\,|\,\alpha_1\alpha_2\alpha_3\alpha_4\alpha_5\,|$$
$$=5\,!\,|\,\alpha_1\alpha_2\alpha_3\alpha_4\alpha_5\,|=0.$$

This we know to be untrue in general, hence the hypothesis, that the invariant in question is reducible, is false.

278. Let s be any substitution of the symmetric group

$$\{a_1 a_2 \dots a_n\},$$

then
$$\{a_1 a_2 \dots a_n\} s = \{a_1 a_2 \dots a_n\}.$$

For $\{a_1 a_2 \dots a_n\} s$ contains $n!$ terms, which are all different (if σ and σ' are different substitutions, then σs and $\sigma' s$ are also different), and are all members of the positive symmetric group

$$\{a_1 a_2 \dots a_n\}.$$

Similarly
$$s \{a_1 a_2 \dots a_n\} = \{a_1 a_2 \dots a_n\}.$$

Now any purely formal relation between substitutions will still hold good if the sign of every transposition be changed; hence

$$s \{a_1 a_2 \dots a_n\}' = \pm \{a_1 a_2 \dots a_n\}'$$

according as s is an even or odd substitution, in particular

$$(a_1 a_2) \{a_1 a_2 \dots a_n\}' = - \{a_1 a_2 \dots a_n\}'.$$

Again if $\{a_1 a_2 \dots a_n\}$ be any positive symmetric group and

$$\{a_1 a_2 b_3 \dots b_m\}'$$

be a negative symmetric group, the two groups having a pair of common letters, then

$$\begin{aligned}\{a_1 a_2 \dots a_n\} &\{a_1 a_2 b_3 \dots b_m\}' \\ &= \{a_1 a_2 \dots a_n\} (a_1 a_2) [-(a_1 a_2) \{a_1 a_2 b_3 \dots b_m\}'] \\ &= - \{a_1 a_2 \dots a_n\} \{a_1 a_2 b_3 \dots b_m\}' = 0.\end{aligned}$$

Similarly
$$\{a_1 a_2 b_3 \dots b_m\}' \{a_1 a_2 \dots a_n\} = 0.$$

Thus if two symmetric groups, one positive the other negative, have a pair of common letters, their product is always zero.

279. The following purely formal theorem enables us to establish various results relating to invariants.

Let the letters $a_1, a_2, \dots a_n$ be arranged in any manner in horizontal rows, so that each row has its first letter in the same vertical column, its second letter in a second vertical column, and so on; and so that no row contains more letters than any row above it.

Thus for four letters $a_1 a_2 a_3 a_4$ the five possible kinds of arrangement of the tableau would be

$$\begin{array}{ccccc} a_1 a_2 a_3 a_4; & a_1 a_2 a_3; & a_1 a_2; & a_2 a_1; & a_1 \\ & a_4 & a_3 a_4 & a_3 & a_2 \\ & & & a_4 & a_3 \\ & & & & a_4 \end{array}$$

Then form the substitutional expression

$$S = \Sigma G_1 G_2 \ldots G_h \Gamma_1' \Gamma_2' \ldots \Gamma_k'$$

such that G_1 is the positive symmetric group of the letters of the first row, G_2 that of the letters of the second row, and so on, G_h being that of the letters of the last row; and that Γ_1' is the negative symmetric group of the letters of the first column, Γ_2' that of the letters of the second column, and so on, Γ_k' being that of the letters of the last column (in case a row or column contains only one letter, it is understood that the positive or negative symmetric group of a single letter is unity).

Let us suppose that in the tableau considered there are α_1 letters in the first row, α_2 in the second, and so on; where owing to the conditions laid down

$$\left.\begin{array}{c} \alpha_1 + \alpha_2 + \ldots + \alpha_h = n \\ \alpha_1 \not< \alpha_2 \not< \alpha_3 \ldots \not< \alpha_h, \ \alpha_1 = k \end{array}\right\} \ldots\ldots\ldots\ldots\ldots (\text{I});$$

let $T_{\alpha_1, \alpha_2, \ldots \alpha_h}$ be the sum of the $n!$ expressions S obtained by permuting the letters in the tableau in all possible ways, the numbers $\alpha_1, \alpha_2, \ldots \alpha_h$ of letters in the various rows remaining fixed.

Then $$\Sigma A_{\alpha_1, \alpha_2, \ldots \alpha_h} T_{\alpha_1, \alpha_2, \ldots \alpha_h} = 1$$

where the summation extends to all possible values of the numbers $\alpha_1, \alpha_2, \ldots \alpha_h$ which satisfy the conditions (I); and $A_{\alpha_1, \alpha_2, \ldots \alpha_h}$ is a numerical coefficient which can be uniquely determined.

For two letters we have

$$\begin{aligned} 1 &= \tfrac{1}{4} T_2 + \tfrac{1}{4} T_{1,1} \\ &= \tfrac{1}{2} \{a_1 a_2\} + \tfrac{1}{2} \{a_1 a_2\}'. \end{aligned}$$

For three letters we have

$$\begin{aligned} 1 &= \tfrac{1}{36} T_3 + \tfrac{1}{9} T_{2,1} + \tfrac{1}{36} T_{1,1,1} \\ &= \tfrac{1}{6} \{a_1 a_2 a_3\} + \tfrac{1}{9} \Sigma \{a_1 a_2\}' \{a_1 a_3\} + \tfrac{1}{6} \{a_1 a_2 a_3\}', \end{aligned}$$

as can easily be verified.

280. Let the T's be arranged in the order defined by the convention that $T_{\alpha_1, \alpha_2, \dots \alpha_h}$ comes before $T_{\beta_1, \beta_2, \dots \beta_{h'}}$ when the first of the differences

$$\alpha_1 - \beta_1, \; \alpha_2 - \beta_2, \; \dots$$

which does not vanish is positive.

Now if S be one of the $n!$ expressions of which $T_{\alpha_1, \alpha_2, \dots \alpha_h}$ is the sum, then $T_{\alpha_1, \alpha_2, \dots \alpha_h}$ is obtained from S by permuting the letters in all possible ways and taking the sum. Hence if when S is expanded as a sum of substitutions, any particular substitution s occurs in it, then in $T_{\alpha_1, \alpha_2, \dots \alpha_h}$ the sum of all the substitutions of the symmetric group of the letters $a_1, a_2, \dots a_n$ similar to s must occur. Hence defining $t_{\beta_1, \beta_2, \dots \beta_k}$ to be the sum of all those substitutions which are formed of k cycles of orders $\beta_1, \beta_2, \dots \beta_k$ respectively, it follows that

$$T_{\alpha_1, \alpha_2, \dots \alpha_h} = \Sigma \lambda_{\beta_1, \beta_2, \dots \beta_k} t_{\beta_1, \beta_2, \dots \beta_k} \quad \dots\dots\dots\dots(\text{II}),$$

where $\lambda_{\beta_1, \beta_2, \dots \beta_k}$ is a numerical coefficient.

If cycles of order unity, which are equivalent to the identical substitution, be introduced, we may suppose that the suffixes of $t_{\beta_1, \beta_2, \dots \beta_k}$ satisfy the conditions

$$\beta_1 + \beta_2 + \dots + \beta_k = n$$

$$\beta_1 \not< \beta_2 \not< \beta_3 \dots \not< \beta_k.$$

The t's are now defined by numbers which obey exactly the same conditions as those which define the T's. The number of t's must then be equal to the number of T's, let us say equal to M. Now the equations (II) may be regarded as a system of linear equations expressing the t's in terms of the T's. Hence if these equations are all independent

$$t_{\beta_1, \beta_2, \dots \beta_k} = \Sigma \mu_{\alpha_1, \alpha_2, \dots \alpha_h} T_{\alpha_1, \alpha_2, \dots \alpha_h}$$

and in particular

$$1 = t_{1, 1, \dots 1} = \Sigma A_{\alpha_1, \alpha_2, \dots \alpha_h} T_{\alpha_1, \alpha_2, \dots \alpha_h}.$$

If these equations are not all linearly independent, there must be a relation of the form

$$\Sigma BT = 0.$$

In order to prove the impossibility of such a relation, it will be shewn that

(i) $$T_{\alpha_1, \alpha_2, \dots \alpha_h} T_{\beta_1, \beta_2, \dots \beta_{h'}} = 0$$

when $T_{\alpha_1, \alpha_2, \dots \alpha_h}$ comes after $T_{\beta_1, \beta_2, \dots \beta_{h'}}$, the T's being arranged in the order defined above; and that

(ii) $$T^2_{\alpha_1, \alpha_2, \dots \alpha_h} \neq 0.$$

Let $S = PN$ be one of the $n!$ expressions of which $T_{\alpha_1, \alpha_2, \dots \alpha_h}$ is the sum, where P denotes the product of the positive symmetric groups, and N that of the negative symmetric groups. Similarly let $S' = P'N'$ be one of the expressions of which $T_{\beta_1, \beta_2, \dots \beta_{h'}}$ is the sum. If one of the groups of P' contains a pair of letters contained in any one group of N then, § 278,

$$NP' = 0.$$

Consider the tableau by means of which S is formed, α_1 is the number of letters in the top row, it is also the number of columns, and consequently it is the number of the positive symmetric groups in P. Again α_2 is the number of letters in the second row, and hence $\alpha_1 - \alpha_2$ is the number of columns which contain one letter only. Similarly $\alpha_2 - \alpha_3$ is the number of columns containing exactly two letters, and in general $\alpha_i - \alpha_{i+1}$ is the number of columns containing exactly i letters.

If $\beta_1 > \alpha_1$ there are more letters in the first row of the tableau for S' than there are columns in the tableau for S. Hence one group at least of the product P' contains a pair of letters belonging to the same group of N; and therefore

$$NP' = 0.$$

In order that NP' may be other than zero, we must then have $\beta_1 \not> \alpha_1$. If this condition be satisfied but

$$\beta_1 + \beta_2 > \alpha_1 + \alpha_2,$$

we see that there are more letters in the first two rows of the tableau for S' than can be arranged in the tableau for S with the condition that no three occur in the same column. In this case some group of N must contain three of the letters of the first two groups of P', and therefore two of the letters belonging to one of these groups. Hence if

$$\beta_1 + \beta_2 > \alpha_1 + \alpha_2,$$

then

$$NP' = 0.$$

Again if $$\beta_1 + \beta_2 + \beta_3 > \alpha_1 + \alpha_2 + \alpha_3,$$

some one group of N must contain four of the letters belonging to the first three groups of P', and again NP' vanishes.

Proceeding thus we see that NP' is always zero unless

$$\alpha_1 \not< \beta_1, \ \alpha_1 + \alpha_2 \not< \beta_1 + \beta_2, \ \ldots \ \alpha_1 + \alpha_2 + \ldots + \alpha_i \not< \beta_1 + \beta_2 + \ldots + \beta_i.$$

We deduce that if the first of the differences

$$\alpha_1 - \beta_1, \ \alpha_2 - \beta_2, \ \ldots$$

which is other than zero is negative, then NP' is zero, *i.e.* if $T_{\alpha_1, \alpha_2, \ldots \alpha_h}$ comes after $T_{\beta_1, \beta_2, \ldots \beta_{h'}}$ then

$$NP' = 0;$$

and hence

$$T_{\alpha_1, \alpha_2, \ldots \alpha_h} T_{\beta_1, \beta_2, \ldots \beta_{h'}} = (\Sigma PN)(\Sigma P'N') = 0.$$

Next we must prove that

$$T_{\alpha_1, \alpha_2, \ldots \alpha_h} T_{\alpha_1, \alpha_2, \ldots \alpha_h} = T^2_{\alpha_1, \alpha_2, \ldots \alpha_h} \neq 0.$$

For this purpose it is only necessary to shew that the coefficient of the identical substitution is other than zero. Now

$$T_{\alpha_1, \alpha_2, \ldots \alpha_h} = \Sigma \lambda_{\beta_1, \beta_2, \ldots \beta_k} t_{\beta_1, \beta_2, \ldots \beta_k}$$

but every substitution is similar to its reciprocal substitution; hence if s be any substitution contained in $t_{\beta_1, \beta_2, \ldots \beta_k}$, s^{-1} must also be contained in this expression. It follows that in $T_{\alpha_1, \alpha_2, \ldots \alpha_h}$ both s and s^{-1} have the same numerical coefficient. In $T^2_{\alpha_1, \alpha_2, \ldots \alpha_h}$ the only products which produce the identical substitution are those obtained when a substitution s is taken in the first T, and s^{-1} in the second. Hence the required coefficient of the identical substitution is of the form $\Sigma\lambda^2$, which, being the sum of a number of positive terms, cannot be zero.

Let us now suppose that there is a relation of the form

$$\Sigma BT = 0,$$

and let $\lambda T_{a_1, a_2, \dots a_h}$ be the first term of this relation, the terms being arranged in the order defined above. Then

$$[\Sigma BT]\, T_{a_1, a_2, \dots a_h} = 0.$$

And hence by the relation (i)

$$\lambda T^2_{a_1, a_2, \dots a_h} = 0;$$

therefore by (ii) $\lambda = 0.$

Thus no T can be the first term in such a relation; in other words the equations (II) are linearly independent. Hence

$$1 = \Sigma A_{a_1, a_2, \dots a_h} T_{a_1, a_2, \dots a_h} \quad \dots\dots\dots\dots\text{(III)}.$$

Q.E.D.

281. The coefficients in this series have been calculated* but as their values are not of importance for our present purpose we merely quote the formula

$$A_{a_1, a_2, \dots a_h} = \left(\frac{\prod\limits_{r, s} (a_r - a_s - r + s)}{\prod\limits_{r} (a_r + h - r)!} \right)^2.$$

The substitutions $s_1 s_2$ and $s_2 s_1$ are similar, for

$$s_1 s_2 = s_2^{-1} (s_2 s_1)\, s_2,$$

their coefficients in the expansion of $T_{a_1, a_2, \dots a_h}$ are therefore the same. Hence since

$$T_{a_1, a_2, \dots a_h} = \Sigma PN,$$

it follows that also

$$T_{a_1, a_2, \dots a_h} = \Sigma NP.$$

If $T_{\beta_1, \beta_2 \dots \beta_{h'}} = \Sigma P'N'$ comes before $T_{a_1, a_2 \dots a_h}$ it has been shewn that

$$NP' = 0.$$

Hence also $$P'N = 0,$$

and $$N'P'NP = 0,$$

and $$[\Sigma N'P']\,[\Sigma NP] = 0,$$

and therefore $$T_{\beta_1, \beta_2, \dots \beta_{h'}} T_{a_1, a_2, \dots a_h} = 0.$$

That is, the product of any two different T's is zero.

* Young, *Proc. Lond. Math. Soc.* vol. XXXIV. p. 361.

Now multiply the relation (III) by $T_{a_1, a_2, \dots a_h}$, we obtain at once

$$T_{a_1, a_2, \dots a_h} = A_{a_1, a_2, \dots a_h} T^2_{a_1, a_2, \dots a_h}.$$

282. By polarizing the form

$$a_x^{(1)} a_x^{(2)} \dots a_x^{(m)} = F$$

once with respect to each of the sets of variables

$$\begin{array}{cccc} x_1^{(1)} & x_2^{(1)} & \dots & x_n^{(1)} \\ x_1^{(2)} & x_2^{(2)} & \dots & x_n^{(2)} \\ \dots & \dots & \dots & \dots \\ x_1^{(m)} & x_2^{(m)} & \dots & x_n^{(m)} \end{array}$$

we obtain an expression which may be written

$$F_1 = \frac{1}{n!} \{x^{(1)} x^{(2)} \dots x^{(m)}\} a^{(1)}{}_{x^{(1)}} a^{(2)}{}_{x^{(2)}} \dots a^{(m)}{}_{x^{(m)}}$$

$$= \frac{1}{n!} \{a^{(1)} a^{(2)} \dots a^{(m)}\} a^{(1)}{}_{x^{(1)}} a^{(2)}{}_{x^{(2)}} \dots a^{(m)}{}_{x^{(m)}}.$$

If in F_1 each of the sets of variables is replaced by the original set $x_1, x_2, \dots x_n$, we obtain the form F from which we started. Neither the passage from F to F_1 nor that from F_1 to F affects the invariant properties of F, so that we may regard F and F_1 as equivalent.

Similarly if $f(a^{(1)}, a^{(2)}, \dots a^{(m)})$ be an invariant linear in the coefficients of each of m quantics of the same order, whose coefficients are

$$a_0^{(r)}, \; a_1^{(r)}, \; \dots \qquad (r = 1, 2, \dots m),$$

then
$$\{a^{(1)} a^{(2)} \dots a^{(m)}\} f(a^{(1)}, a^{(2)}, \dots a^{(m)})$$

is an invariant which may be obtained by means of Aronhold operators from an invariant of a single quantic.

Again if P be the product of h positive symmetric groups, which between them contain all the letters $a^{(1)}, a^{(2)}, \dots a^{(m)}$, but no two contain the same letter, then

$$Pf(a^{(1)}, a^{(2)}, \dots a^{(m)})$$

is an invariant which may be obtained by means of Aronhold operators from an invariant of h quantics.

283. Peano's Theorem. Let $F(a^{(1)}, a^{(2)}, \ldots a^{(m)})$ be a covariant linear in the coefficients of each of m binary n-ics

$$(a_0^{(r)}, a_1^{(r)}, \ldots a_n^{(r)} \between x_1, x_2)^n; \quad r = 1, 2, \ldots m.$$

Operate on F with the two sides of the identity § 280 (III.)

$$1 = \Sigma A_{a_1, a_2, \ldots a_h} T_{a_1, a_2, \ldots a_h},$$

then
$$F = \Sigma A_{a_1, a_2, \ldots a_h} T_{a_1, a_2, \ldots a_h} F.$$

Now
$$T_{a_1, a_2, \ldots a_h} = \Sigma PN,$$

where N has a factor of the form

$$\{a^{(1)} a^{(2)} \ldots a^{(h)}\}'.$$

If $h > n+1$, then by § 275,

$$\{a^{(1)} a^{(2)} \ldots a^{(h)}\}' F = 0,$$

if $h = n+1$,

$$\{a^{(1)} a^{(2)} \ldots a^{(h)}\}' F = | a^{(1)} a^{(2)} \ldots a^{(h)} | . R,$$

if $h = n$,

$$\{a^{(1)} a^{(2)} \ldots a^{(h)}\}' F = | a^{(1)} a^{(2)} \ldots a^{(h)} Q |,$$

where $(Q_0, Q_1, \ldots Q_n \between x_1, x_2)^n$ is a covariant of the forms considered.

Hence when $h > n+1$,

$$T_{a_1, a_2, \ldots a_h} F = 0;$$

when $h = n+1$, $T_{a_1, a_2, \ldots a_h} F$ is a sum of terms each of which contains a factor of the form $| a^{(1)} a^{(2)} \ldots a^{(n+1)} |$; when $h = n$, $T_{a_1, a_2, \ldots a_h} F$ is a sum of terms of the form $| a^{(1)} a^{(2)} \ldots a^{(n)} Q |$.

Now the number of positive symmetric groups in P is h; hence by § 282, $T_{a_1, a_2, \ldots a_h} F$ is a sum of terms each of which is obtainable by means of Aronhold operators from a covariant of only h different n-ics.

Let us suppose that F is a covariant type which does not give any irreducible covariant, unless we are considering a system of more than n n-ics. Then

$$F = \Sigma A_{a_1, a_2, \ldots a_h} T_{a_1, a_2, \ldots a_h} F,$$

if $h < n+1$, $T_{a_1, a_2, \ldots a_h} F$ is reducible for it is obtained by Aronhold operators from reducible forms: if $h > n+1$

$$T_{a_1, a_2, \ldots a_h} F = 0;$$

and if $h = n + 1$, $T_{a_1, a_2, \ldots a_h} F$ is a sum of terms each of which contains a factor of the form $| a^{(1)} a^{(2)} \ldots a^{(n+1)} |$, and is thus reducible unless $m = n + 1$.

Hence *every type of the binary n-ic which does not furnish an irreducible covariant for a system of n n-ics is reducible, with the possible exception of the invariant type*

$$| a^{(1)} a^{(2)} \ldots a^{(n+1)} |.$$

Further it has been shewn that, if $h = n$, $T_{a_1, a_2, \ldots a_h} F$ is equal to a sum of terms of the form $| a^{(1)} a^{(2)} \ldots a^{(n)} Q |$. But if the invariant $| a^{(1)} a^{(2)} \ldots a^{(n+1)} |$ is reducible, the invariant

$$| a^{(1)} a^{(2)} \ldots a^{(n)} Q |$$

is reducible in the same way. Hence when the above invariant is reducible every irreducible type furnishes an irreducible covariant for a system of $n - 1$ n-ics.

It has been shewn § 277 that this invariant is reducible when n is odd; thus the covariants of any number of cubics can be obtained by means of Aronhold operators from those for two cubics.

Similar results may be obtained in exactly the same way for ternary forms, or for forms involving any number of variables.

Thus all covariants of any number of ternary n-ics may be obtained by means of Aronhold operators from the system for $\binom{n+2}{2} - 1$ ternary n-ics with the possible exception of the invariant determinant of $\binom{n+2}{2}$ rows and columns, each row being formed by the coefficients of one n-ic.

Ex. Shew that the determinant formed by the coefficients of six conics is an irreducible invariant of the system.

284. Peano's theorem was first proved by means of an expansion due to Capelli*, which is virtually the same as the expansion used here, but is expressed in terms of polar operators.

* "Sur les Opérations dans la Théorie des formes Algébriques," *Math. Ann.* Bd. 37. See also "Lezioni sulla Teoria delle Forme Algebriche," Ch. I. § XXIII. by the same writer.

It is sufficient, in order to make the comparison clear, to point out that if f be a function linear in each of n sets of q-ary variables, $a_1, a_2, \dots a_n$; then

$$\{a_1 a_2 \dots a_h\}' f = \Sigma \Delta_\lambda f_\lambda$$

where Δ_λ is one of the determinants of the matrix

$$\begin{vmatrix} a_{1,1} & a_{1,2} \cdots\cdots & a_{1,q} \\ a_{2,1} & a_{2,2} \cdots\cdots & a_{2,q} \\ \cdots & \cdots & \cdots \\ a_{h,1} & a_{h,2} \cdots\cdots & a_{h,q} \end{vmatrix},$$

and in fact is equal to $H_{a_1, a_2, \dots a_h} f$ where

$$H_{a_1, a_2, \dots a_h} = \Sigma \Delta_\lambda D_\lambda,$$

D_λ being the determinant of the matrix

$$\begin{vmatrix} \frac{\partial}{\partial a_{1,1}} & \frac{\partial}{\partial a_{1,2}} \cdots\cdots & \frac{\partial}{\partial a_{1,q}} \\ \frac{\partial}{\partial a_{2,1}} & \frac{\partial}{\partial a_{2,2}} \cdots\cdots & \frac{\partial}{\partial a_{2,q}} \\ \cdots & \cdots & \cdots \\ \frac{\partial}{\partial a_{h,1}} & \frac{\partial}{\partial a_{h,2}} \cdots\cdots & \frac{\partial}{\partial a_{h,q}} \end{vmatrix}$$

obtained by writing $\frac{\partial}{\partial a_{r,s}}$ for $a_{r,s}$ in each element of Δ_λ.

In the case of binary forms, if $a_x^m b_y^n$ be polarized so as to obtain a function of $m+n$ sets of binary variables, and then the identity § 280 (III.) be applied, we obtain a proof of Gordan's series, § 52, which is a particular case of the series of Capelli. The coefficients in Gordan's series are not apparent, but it is possible to obtain them by these methods*.

Thus if $\qquad f = a_{x^{(1)}} a_{x^{(2)}} \dots a_{x^{(m)}} b_{y^{(1)}} b_{y^{(2)}} \dots b_{y^{(n)}},$

then $\qquad T_{a_1, a_2, \dots a_h} f = 0,$

when $h > 2$.

If $h = 2$, we need only consider those expressions PN which are of the form

$$PN = \{x^{(1)} x^{(2)} \dots x^{(m)} y^{(1)} \dots y^{(i)}\} \{y^{(i+1)} y^{(i+2)} \dots y^{(n)}\} \{x^{(1)} y^{(i+1)}\}' \dots \{x^{(n-i)} y^{(n)}\}';$$

* See Young, *Proc. Lond. Math. Soc.* vol. XXXIII.

and in this case

$$PN.f = P.(x^{(1)}y^{(i+1)})\ldots(x^{(n-i)}y^{(n)})(ab)^{n-i}a_{x^{(n-i+1)}}\ldots a_{x^{(m)}}b_{y^{(1)}}\ldots b_{y^{(i)}},$$

which may be obtained by polarization from

$$(xy)^{n-i}(ab)^{n-i}a_x^{m+i-n}b_x^{i}.$$

In just the same way we see that, if x, y, z be ternary variables, $a_x^m b_y^n c_z^p$ can be expanded in a series each term of which may be obtained by polarization from a term of the form

$$\lambda(xyz)^i(abc)^i(ab\widehat{xy})^{j_1}(bc\widehat{xy})^{j_2}(ca\widehat{xy})^{j_3}a_x^{k_1}b_x^{k_2}c_x^{k_3},$$

where

$$(ab\widehat{xy}) = \begin{vmatrix} a_1 & b_1 & x_2y_3 - x_3y_2 \\ a_2 & b_2 & x_3y_1 - x_1y_3 \\ a_3 & b_3 & x_1y_2 - x_2y_1 \end{vmatrix}.$$

Let

$$u_1 = x_2y_3 - x_3y_2$$
$$u_2 = x_3y_1 - x_1y_3$$
$$u_3 = x_1y_2 - x_2y_1,$$

then

$$u_z = 0$$

is the condition that the point z lies on the straight line joining the points x and y. Hence u_1, u_2, u_3 are really the coordinates of a straight line.

We see then that all covariants of ternary forms which contain any number of variables can be expressed in terms of polars of covariants which contain the variables x and u only. The u's were introduced for geometrical reasons § 207, but it is now apparent that they are necessary to make the analytical theory complete.

285. Let $x^{(1)}, x^{(2)}, \ldots x^{(n)}$ represent n sets of q-ary variables

$$x_1^{(r)}, x_2^{(r)}, \ldots x_q^{(r)};\ r = 1, 2, \ldots n.$$

Then if F be a function linear in each of these sets, we see, in the same way as before, that PNF is a function obtainable by polarization from a function, not necessarily linear, of

1st the single set $\quad x_1^{(1)}, x_2^{(1)}, \ldots x_q^{(1)}$,

2nd the determinants of the matrix

$$\begin{vmatrix} x_1^{(1)} & x_2^{(1)} & \ldots\ldots & x_q^{(1)} \\ x_1^{(2)} & x_2^{(2)} & \ldots\ldots & x_q^{(2)} \end{vmatrix},$$

3rd the determinants of the matrix

$$\begin{vmatrix} x_1^{(1)} & x_2^{(1)} & \ldots\ldots & x_q^{(1)} \\ x_1^{(2)} & x_2^{(2)} & \ldots\ldots & x_q^{(2)} \\ x_1^{(3)} & x_2^{(3)} & \ldots\ldots & x_q^{(3)} \end{vmatrix},$$

and so on, and lastly of the determinant

$$\begin{vmatrix} x_1^{(1)} & x_2^{(1)} & \ldots\ldots & x_q^{(1)} \\ x_1^{(2)} & x_2^{(2)} & \ldots\ldots & x_q^{(2)} \\ \ldots & \ldots & \ldots & \ldots \\ x_1^{(q)} & x_2^{(q)} & \ldots\ldots & x_q^{(q)} \end{vmatrix}.$$

These may be all regarded as auxiliary variables. It will be useful to denote the variables of the 2nd set, viz. the determinants of the matrix

$$\begin{vmatrix} x_1^{(1)} & x_2^{(1)} & \ldots\ldots & x_q^{(1)} \\ x_1^{(2)} & x_2^{(2)} & \ldots\ldots & x_q^{(2)} \end{vmatrix}$$

by the letter ${}_2x$, with appropriate suffixes. Similarly the third set will be denoted by ${}_3x$ and so on.

Then in taking the complete system of concomitants of q-ary forms we have $q-1$ different kinds of variables which may appear.

The geometric meaning of these variables is easy to obtain. A space of $q-1$ dimensions being under consideration, the variables x or ${}_1x$ represent point coordinates: the variables ${}_2x$ are line coordinates; the variables ${}_3x$ are plane coordinates, and so on.

The linear substitutions by which these auxiliary variables are transformed, when any linear transformation of the point coordinates is made, are easy to find. The variables ${}_ix$ and ${}_{q-i}x$ are contragredient, and in fact

$$\Sigma\, {}_ix\, {}_{q-i}x$$

is an absolute concomitant. As a particular case, when q is even*, this leads to a relation between the variables ${}_{\frac{q}{2}}x$. This remark has already been illustrated for quaternary forms § 220.

* The case of binary forms is of course an exception.

286. Let $a_x^{(1)}, a_x^{(2)}, \ldots a_x^{(n)}$ be any n linear q-ary forms; and F a function linear in the coefficients of each. Then apply the formula

$$F = \Sigma A_{\alpha_1, \alpha_2, \ldots \alpha_h} T_{\alpha_1, \alpha_2, \ldots \alpha_h} F,$$

where substitutions interchange the sets of coefficients

$$a^{(1)}, a^{(2)}, \ldots a^{(n)}.$$

Consider the form PNF, it is a sum of terms each of which is a product of determinants of matrices of the form

$$\left| \begin{array}{llll} a_1^{(1)} & a_2^{(1)} & \ldots\ldots & a_q^{(1)} \\ a_1^{(2)} & a_2^{(2)} & \ldots\ldots & a_q^{(2)} \\ \ldots & \ldots & \ldots & \ldots \\ a_1^{(i)} & a_2^{(i)} & \ldots\ldots & a_q^{(i)} \end{array} \right|.$$

Moreover if N has two negative symmetric groups of degree i and $A_{i,\lambda}$, $B_{i,\mu}$ represent determinants from each of the corresponding matrices (the particular determinant being defined by the second suffix), then every term of NF has a factor of the form

$$A_{i,\lambda} B_{i,\mu}.$$

Hence every term of PNF has a factor of the form

$$A_{i,\lambda} B_{i,\mu} + A_{i,\mu} B_{i,\lambda}$$

as is evident when the tableau from which PN is constructed is considered. But this is a coefficient of the expression

$$[\Sigma A_{i\ i}x]\,[\Sigma B_{i\ i}x]$$

which is a covariant since the a's are contragredient to the x's.

Now if PN is a term of $T_{\alpha_1, \alpha_2, \ldots \alpha_h}$, N contains α_h groups of degree h, $\alpha_{h-1} - \alpha_h$ groups of degree $h-1$, and so on; hence PNF is a linear function of the coefficients of concomitants of the form

$$_hA^{(1)}{}_{_hx}\, {}_hA^{(2)}{}_{_hx} \ldots {}_hA^{(\alpha_h)}{}_{_hx}\, {}_{h-1}A^{(1)}{}_{_{h-1}x} \ldots {}_1A^{(\alpha_1-\alpha_2)}{}_{_1x}.$$

Thus every function of the coefficients of certain linear forms can be expressed in terms of coefficients of concomitants of those forms.

It is unnecessary to assume that the function is linear in the coefficients of each of the forms, in order that the above theorem may be true. For the function can be made linear by means of

Aronhold operators, and after the above process the original coefficients can be restored without affecting the invariantive properties in question.

Further the forms considered may be symbolical, and we at once deduce that *every integral function homogeneous in the coefficients of each of certain q-ary forms can be expressed as a linear function of the coefficients of the concomitants of those forms.*

APPENDIX I.

NOTE ON THE SYMBOLICAL NOTATION.

As we have said in § 82 the notation used in this work is really equivalent to Cayley's hyperdeterminants. The great advance made by the German school lies in the possibility of transforming symbolical expressions, and, of course, in the proof that every invariant form can be represented as a combination of hyperdeterminants. The reader may feel the need of justifying directly the results obtained by manipulating umbral expressions and accordingly we shall indicate how the whole theory can be made to rest on differential operators.

There are different ways of doing this. Salmon has remarked that, f being a binary form of order n, since

$$f = \frac{1}{n!}\left(x_1 \frac{\partial}{\partial y_1} + x_2 \frac{\partial}{\partial y_2}\right)^n f_y,$$

we may regard f as being equal to

$$(\alpha_1 x_1 + \alpha_2 x_2)^n \cdot \frac{f_y}{n!},$$

where
$$\alpha_1{}^p \alpha_2{}^q = \left(\frac{\partial}{\partial y_1}\right)^p \left(\frac{\partial}{\partial y_2}\right)^q.$$

Hence we may suppose that

$$\alpha_1 = \frac{\partial}{\partial y_1}, \quad \alpha_2 = \frac{\partial}{\partial y_2},$$

and that the final operation is on

$$\frac{f_y}{n!}.$$

Any symbolical expression can be thus at once transformed into one exactly like it but involving only differential operators, *e.g.* if

$$f = \alpha_x^m, \quad \phi = \beta_x^n,$$

then

$$(\alpha\beta)^r \alpha_x^{m-r} \beta_x^{n-r}$$

is
$$\left(\frac{\partial^2}{\partial y_1 \partial z_2} - \frac{\partial^2}{\partial z_1 \partial y_2}\right)^r \left(x_1 \frac{\partial}{\partial y_1} + x_2 \frac{\partial}{\partial y_2}\right)^{m-r} \left(x_1 \frac{\partial}{\partial z_1} + x_2 \frac{\partial}{\partial z_2}\right)^{n-r} \frac{f_y}{m\,!} \frac{\phi_z}{n\,!}$$

(see § 82).

In any calculation we may omit the operand while we transform the operator.

After this the reader who wishes to do so will have no difficulty in developing the theory of the symbols when they are regarded as differential operators.

For another method see Kempe, *Proc. L.M.S.* vol. XXIV. p. 102 and Elliott, *Proc. L.M.S.* vol. XXXIII. p. 231.

Some interesting general remarks on the underlying principles of the symbolical notation will be found in Study, *Methoden Ternäre Formen*; and some very curious remarks in Lie-Scheffers, *Vorlesungen über Continuierlichen Grüppen*, p. 720.

This is the most convenient place to give a brief explanation of the so-called Chemico-Algebraic theory—an idea originally due to Sylvester which has perhaps attracted more attention than its intrinsic merits deserve.

In this theory an atom in chemistry corresponds to a binary form in algebra, and the valency of the atom to the order of the form. To each unit in the valency of an atom, in the chemical theory, a bond is supposed to correspond, and each such bond can connect the atom in question with an atom of valency one such as Hydrogen. Thus Oxygen is of valency two, and there exists a compound OH_2 which is written graphically

H—O—H,

there being two bonds proceeding from O and one from each H.

Since each unity in the order of a form gives rise to one possibility of transvection with another form, the analogy is evident

—if we have a binary quadratic o_x^2 and two linear forms h_x, h'_x the formula OH_2 corresponds to the algebraic expression

$$(oh)(oh')$$

an invariant of the three forms.

Then Carbon being of valency four we have the compound (Marsh Gas)

```
              H
              |
CH4 or   H—C—H
              |
              H
```

and this corresponds to the invariant

$$(ch)(ch')(ch'')(ch''')$$

of a binary quartic and four linear forms.

The four hydrogen atoms in CH_4 are supposed on chemical grounds to occupy similar positions* in the structure of the compound and hence CH_4 is more naturally like

$$(ch)^4$$

an invariant of a quartic and a single linear form h_x.

Then the compounds CH_3Cl, CH_2Cl_2 etc. may be supposed like the invariants

$$(ch)^3(ck), \quad (ch)^2(ck)^2$$

where k is written for Cl and we see in the chemistry an analogue to polarizing in algebra.

Guided by the above the reader will have no difficulty in writing down an invariant corresponding to any graphical formula however complicated—in fact the algebraic form of the invariant is only a different (perhaps a more concise) way of writing down the chemical formula.

Difficulties arise when we recollect that some atoms have apparently different valencies illustrated by S in SO_2 and H_2S, for of course a binary form can have only one order. Gordan and Alexeleff suppose that the corresponding algebraic form is then polarized. Thus S in SO_2 would correspond to S_x^4 and S in H_2S to $S_x^2 S_y^2$ and now the degree available for transvection is two.

* For an explanation of this and the other chemical facts we have referred to see Scott, *Chemical Theory*, chap. VI.

The idea has been developed by various writers in the direction of making the algebraic methods graphical (references in Meyer) and lately Gordan and Alexeleff * have written several papers in which the algebra is applied to chemistry; the papers were criticised by Study † and from the objections and the replies the reader may be able to form his own opinion. We venture only to remark that the wonderful feature of the algebra is the capacity for reduction, and that, unless there is something corresponding in chemistry, the whole theory seems to be no more than a superficial analogy. It is of course certain that a reducible invariant often corresponds to a stable compound—moreover the *general* features which lead an algebraist to suppose a form reducible and a chemist to suppose a compound unstable are as nearly opposite in character as they can be.

It has been stated in Chapter I. § 21 that there are two ways of obtaining relations between concomitants symbolically expressed, viz.:

(i) By means of the fundamental identities

$$(bc)\,a_x + (ca)\,b_x + (ab)\,c_x = 0,$$

$$(bc)(ad) + (ca)(bd) + (ab)(cd) = 0;$$

(ii) By means of the fact that a concomitant is left unaltered when a pair of letters which refer to the same quantic is interchanged.

We give here a demonstration of the fact that all relations may be thus obtained ‡.

Let $F(C_i) = 0$ be any identical relation (supposed rational integral and homogeneous) between concomitants C_i of any system of binary forms.

Let $$F(C_i) = \Sigma P_j$$

each term P_j being itself a concomitant. Also let

$$(A_0, A_1, \dots A_n \between x_1, x_2)^n = a_x^n = b_x^n = \dots$$

be one of the quantics of the system. The introduction of the symbolical notation may be effected by operators like

$$a_1^n \frac{\partial}{\partial A_0} + a_1^{n-1} a_2 \frac{\partial}{\partial A_1} + \dots + a_2^n \frac{\partial}{\partial A_n}$$

* *Wiedemann's Annalen der Physik*, 1899, 1900. † In *Wiedemann*.

‡ Cf. Gordan, *Invariantentheorie*, Bd. II. § 117.

operating on $F(C_i)$. These operators do not destroy the property that $F(C_i)$ considered as a function of the coefficients of the quantics is identically zero.

Hence if $F(C_i) = \Sigma P_j$ becomes $\Sigma \Pi_k$, where each term Π_k is a symbolical product of factors of the forms (ab), a_x,

$$\Sigma \Pi_k \equiv 0$$

considered as a function of the symbolical letters.

Owing to the manner in which the symbolical letters were introduced no distinction was possible between two letters which refer to the same quantic, hence $\Sigma \Pi_k$ is unaltered by any interchange of two such letters, and therefore the second method of obtaining relations between forms will have no effect here.

Let $a_1, a_2, a_3, \ldots a_r$ be the symbolical letters, and $n_1, n_2, \ldots n_r$ the orders of the quantics to which they refer—if the forms Π_k are covariants we shall suppose that $a_{r,2} = x_1$, $a_{r,1} = -x_2$ and that n_r is the order of Π_k. Then applying the theorem of § 47, x being replaced by a_1 and y by a_2, we obtain

$$\Sigma \Pi_k = \sum_{j=0}^{n_2} \lambda_j (a_1 a_2)^j \left[a_2 \frac{\partial}{\partial a_1} \right]^{n_2 - j} [\Sigma \Pi_k^{(j)}],$$

where $\Sigma \Pi_k^{(j)}$ does not contain a_2.

Now $\Sigma \Pi_k \equiv 0$, hence every term of this series is identically zero. For if the jth term be the first which does not vanish, we may divide by $(a_1 a_2)^j$, and then put $a_2 = a_1$, whence

$$\lambda_j [\Sigma \Pi_k^{(j)}] = 0.$$

But this series is merely obtained by repeated use of the fundamental identities (see § 46), hence the reductions used so far belong entirely to the two classes mentioned.

It may happen that $\Sigma \Pi_k^{(j)}$ is not identically zero considered as a function of symbolical factors. In this case we may apply the same process again.

Each time we do this the number of letters in the function under consideration is reduced. Hence in $(r-2)$ steps at most the expressions are reduced to expressions which are identically zero when considered as functions of the symbolical factors—for when only two letters are left there is only one possible symbolical factor.

Thus the identity $F(C_i) = 0$ is made to depend entirely on the two fundamental methods of reduction.

APPENDIX II.

ON WRONSKI'S THEOREM AND THE APPLICATION OF TRANSVECTANTS TO DIFFERENTIAL EQUATIONS.

THE form of Wronski's theorem used in § 189 is not the usual one, but the determinant, which there vanishes, can be transformed in the following manner.

If
$$f = a_x^n,$$
then
$$\frac{\partial^r f}{\partial x_1^\lambda \partial x_2^\mu} = \frac{n!}{(n-\lambda-\mu)!} a_x^{n-\lambda-\mu} a_1^\lambda a_2^\mu,$$
and
$$x_2^\mu \frac{\partial^r f}{\partial x_1^\lambda \partial x_2^\mu} = \frac{n!}{(n-\lambda-\mu)!} a_x^{n-\lambda-\mu} a_1^\lambda (a_x - a_1 x_1)^\mu$$
$$= \frac{n!}{(n-\lambda-\mu)!} a_x^{n-\lambda-\mu} a_1^\lambda \{a_x^\mu - \mu a_x^{\mu-1} a_1 x_1 + \ldots\}.$$

It follows that
$$x_2^\mu \frac{\partial^r f}{\partial x_1^\lambda \partial x_2^\mu} = A_0^{(\mu)} \frac{\partial^\lambda f}{\partial x_1^\lambda} + A_1^{(\mu)} x_1 \frac{\partial^{\lambda+1} f}{\partial x_1^{\lambda+1}} + \ldots + A_\mu^{(\mu)} x_1^\mu \frac{\partial^{\lambda+\mu} f}{\partial x_1^{\lambda+\mu}},$$
where the A's are numbers depending on n, r, λ and μ.

Now in § 189 the determinant whose typical row is
$$\frac{\partial^r f'}{\partial x_1^r}, \frac{\partial^r f'}{\partial x_1^{r-1} \partial x_2}, \ldots \frac{\partial^r f'}{\partial x_2^r},$$
vanishes, hence so also does that whose typical row is
$$\frac{\partial^r f'}{\partial x_1^r}, \quad x_2 \frac{\partial^r f'}{\partial x_1^{r-1} \partial x_2}, \ldots x_2^r \frac{\partial^r f'}{\partial x_2^r}.$$

On using the value found above for

$$x_2^{\mu}\frac{\partial^r f'}{\partial x_1^{\lambda}\partial x_2^{\mu}}$$

the typical row of the determinant becomes

$$\frac{\partial^r f'}{\partial x_1^r},\quad A_0^{(1)}\frac{\partial^{r-1}f'}{\partial x_1^{r-1}}+A_1^{(1)}x_1\frac{\partial^r f'}{\partial x_1^r},\ \ldots$$

$$A_0^{(r)}f'+A_1^{(r)}x_1\frac{\partial f'}{\partial x_1}+\ldots+A_r^{(r)}x_1^r\frac{\partial^r f'}{\partial x_1^r},$$

and since this determinant vanishes we infer, on modification of the columns, that the determinant whose typical row is

$$\frac{\partial^r f'}{\partial x_1^r},\ \frac{\partial^{r-1}f'}{\partial x_1^{r-1}},\ldots f'$$

vanishes.

If we replace x_2 by unity in the f's, this is the usual form of Wronski's determinant.

For the sake of completeness, we shall give an easy proof of Wronski's theorem.

If $u_1, u_2, \ldots u_{n+1}$ be $n+1$ functions of a single variable x, and the determinant whose rth row is

$$u_r,\ \frac{du_r}{dx},\ \frac{d^2u_r}{dx^2},\ \ldots\frac{d^nu_r}{dx^n}$$

vanish, then there is an identical relation of the form

$$\lambda_1 u_1+\lambda_2 u_2+\ldots+\lambda_{n+1}u_{n+1}=0,$$

where the λ's are constants.

In fact the vanishing of the determinant is the condition that the u's should be solutions of the same linear differential equation of order n, say

$$p_0\frac{d^ny}{dx^n}+p_1\frac{d^{n-1}y}{dx^{n-1}}+\ldots+p_ny=0.$$

We have therefore to prove that such an equation cannot have $(n+1)$ linearly independent integrals.

The theorem is easy to establish when n is unity, so we assume it true for $n-1$ and proceed inductively.

Now u_1 being a solution of the equation of order n, write

$$y=u_1\int w\,dx.$$

It is quickly seen that w is given by an equation of order $n-1$, say

$$q_0 \frac{d^{n-1} w}{dx^{n-1}} + q_1 \frac{d^{n-2} w}{dx^{n-2}} + \dots + q_n w = 0,$$

and this equation is satisfied by

$$\frac{d}{dx}\left(\frac{u_2}{u_1}\right), \quad \frac{d}{dx}\left(\frac{u_3}{u_1}\right), \dots \frac{d}{dx}\left(\frac{u_{n+1}}{u_1}\right),$$

hence by hypothesis there is a relation of the type

$$\lambda_2 \frac{d}{dx}\left(\frac{u_2}{u_1}\right) + \lambda_3 \frac{d}{dx}\left(\frac{u_3}{u_1}\right) + \dots + \lambda_{n+1} \frac{d}{dx}\left(\frac{u_{n+1}}{u_1}\right) = 0.$$

Integrating and multiplying up by u_1 we obtain a relation of the form

$$\lambda_1 u_1 + \lambda_2 u_2 + \dots + \lambda_{n+1} u_{n+1} = 0$$

between the $(n+1)$ functions u, so Wronski's theorem is completely established.

The application of the result to the $(r+1)$ functions f shews that if x_2 be replaced by unity, there is a relation of the type

$$\lambda_1 f_1 + \lambda_2 f_2 + \dots + \lambda_{r+1} f_{r+1} = 0,$$

and then making each f homogeneous again by the introduction of x_2, the result quoted in § 189 follows at once.

The device of changing from differential coefficients with respect to two variables to differentiation with respect to a single variable is often useful.

For example, the rth transvectant of

$$f = a_x^m, \quad \phi = b_x^n$$

is

$$\psi = (ab)^r a_x^{m-r} b_x^{n-r}.$$

Thus

$$x_2^r \psi = (a_1 b_2 x_2 - a_2 b_1 x_2)^r a_x^{m-r} b_x^{n-r}$$

$$= (a_1 b_x - a_x b_1)^r a_x^{m-r} b_x^{n-r}$$

$$= a_x^{m-r} a_1^r b_x^n - r a_x^{m-r+1} a_1^{r-1} b_x^{n-1} b_1 + \dots + (-1)^r a_x^m b_x^{n-r} b_1^r$$

$$= \frac{(n-r)!}{n!} \frac{\partial^r f}{\partial x_1^r} \phi - r \frac{(n-r+1)!}{n!} \frac{(m-1)!}{m!} \frac{\partial^{r-1} f}{\partial x_1^{r-1}} \frac{\partial \phi}{\partial x_1} + \dots$$

$$+ (-1)^r \frac{(m-r)!}{m!} \frac{\partial^r \phi}{\partial x_1^r} f.$$

Applying this to the nth transvectant of two forms f and ϕ, each of order n, we can reduce the problem of finding a form f apolar to ϕ to the solution of a linear differential equation of order n; it follows at once that there are not more than n linearly independent forms, and moreover from the algebraic theory we infer that all integrals of the equation are polynomials in x_1.

Conversely a differential equation whose coefficients are polynomials can be reduced to a relation between transvectants. To give a simple example, consider the equation

$$P_0 \frac{d^2y}{dx^2} + P_1 \frac{dy}{dx} + P_2 y = 0,$$

the P's being polynomials in x.

If ϕ_1, ϕ_2, ϕ_3 be three forms of orders r_1, r_2, r_3 respectively, and f be a form of order n,

$$(f\phi_1)^0 = f\phi_1,$$

$$x_2 (f\phi_2)^1 = \frac{1}{n} \phi_2 \frac{\partial f}{\partial x_1} - \frac{1}{r_2} f \frac{\partial \phi_2}{\partial x_1},$$

$$x_2^2 (f\phi_3)^2 = \frac{1}{n(n-1)} \phi_3 \frac{\partial^2 f}{\partial x_1^2} - \frac{2}{nr_3} \frac{\partial \phi_3}{\partial x_1} \frac{\partial f}{\partial x_1} + \frac{1}{r_3(r_3-1)} f \frac{\partial^2 \phi_3}{\partial x_1^2}.$$

Now replacing x_2 by unity as usual, we can choose ϕ_1, ϕ_2, ϕ_3, so that

$$x_2^2 (f\phi_3)^2 + x_2 (f\phi_2)^1 + (f\phi_1)^0 = 0$$

is the same as

$$P_0 \frac{d^2 f}{dx_1^2} + P_1 \frac{df}{dx_1} + P_2 f = 0,$$

for the transvectant relation is

$$\frac{1}{n(n-1)} \phi_3 \frac{d^2 f}{dx_1^2} - \frac{2}{nr_3} \frac{df}{dx_1} \frac{d\phi_3}{dx_1} + \frac{1}{r_3(r_3-1)} f \frac{d^2 \phi_3}{dx_1^2}$$
$$+ \frac{1}{n} \phi_2 \frac{df}{dx_1} - \frac{1}{r_2} f \frac{d\phi_2}{dx_1} + f\phi_1 = 0,$$

and we must have

$$\frac{1}{n(n-1)} \phi_3 = P_0,$$

$$\frac{1}{n} \phi_2 - \frac{2}{nr_3} \frac{d\phi_3}{dx_1} = P_1,$$

$$\phi_1 - \frac{1}{r_2} \frac{d\phi_2}{dx_1} + \frac{1}{r_3(r_3-1)} \frac{d^2 \phi_3}{dx_1^2} = P_2.$$

The first equation gives ϕ_3 and its order is that of P_0, the next equation gives ϕ_2 and then the last gives ϕ_1; hence the transformation is always possible, but of course the coefficients depend on the order of the form f that is chosen to represent y.

References to further developments in connection with Differential Equations will be found in Meyer's *Berichte* and in Klein's lithographed lectures on Linear Differential Equations of the Second Order.

APPENDIX III.

JORDAN'S LEMMA.

IN Chapter IV. great use was made of this theorem:—*If* $x+y+z=0$, *then any product of powers of* x, y, z *of order* n *can be expressed linearly in terms of such products as contain one exponent equal to or greater than* $\frac{2n}{3}$.

In § 64 a proof due to Stroh is given. We shall now give a simpler proof in which a much more general theorem is incidentally established.

The general theorem may be stated as follows:—

If
$$a_x, b_x, c_x, \ldots$$
be a system of r distinct linear forms and
$$\alpha, \beta, \gamma, \ldots$$
be r positive integers satisfying the relation
$$\alpha+\beta+\gamma+\ldots=n-r+1,$$
then it is impossible to find binary forms
$$A, B, C, \ldots$$
of orders $\alpha, \beta, \gamma, \ldots$
respectively such that
$$a_x^{n-\alpha}A+b_x^{n-\beta}B+c_x^{n-\gamma}C+\ldots=0 \quad \ldots\ldots\ldots\ldots (\text{I}).$$

In fact, suppose that such an identical relation exists and operate $\alpha+1$ times with
$$a_2\frac{\partial}{\partial x_1}-a_1\frac{\partial}{\partial x_2}=D.$$

It is easily seen, by making a_2 zero, that $D^{\alpha+1}$ annihilates a binary form of order n only when that form contains the factor $a_x^{n-\alpha}$; hence since $a_x, b_x, c_x, \ldots$ are all different, we have

$$b_x^{n-\beta-\alpha-1} B' + c_x^{n-\gamma-\alpha-1} C' + \ldots = 0 \ldots\ldots\ldots\ldots \text{(II)},$$

where $B', C', \ldots$

are of orders $\beta, \gamma, \ldots$

respectively and do not vanish identically unless

$$B, C, \ldots$$

do also.

The relation (II) is of the same form as (I) except that r is changed into $r-1$ and n is changed into $n-\alpha-1$ for

$$\beta+\gamma+\ldots = n-r-\alpha+1 = (n-\alpha-1)-(r-1)+1.$$

Now a relation of the type (I) is impossible when $r=1$ for any value of n unless the form A (the only one occurring) vanishes identically, hence by induction it is impossible for all values of n and the theorem is established.

The forms $A, B, C, \ldots$

involve $\alpha+1, \beta+1, \gamma+1, \ldots$

arbitrary coefficients respectively, or in all

$$\alpha+\beta+\gamma+\ldots+r = (n+1).$$

Any binary form can be expressed linearly in terms of $(n+1)$ linearly independent forms of the same order, and since there is no identical relation of the form

$$a_x^{n-\alpha} A + b_x^{n-\beta} B + c_x^{n-\gamma} C + \ldots = 0,$$

it follows that any binary form of order n can be expressed uniquely in the form

$$a_x^{n-\alpha} A + b_x^{n-\beta} B + c_x^{n-\gamma} C + \ldots,$$

where $a_x, b_x, c_x, \ldots$ are all different, $A, B, C, \ldots$ are of orders $\alpha, \beta, \gamma, \ldots$ respectively, and

$$\alpha+\beta+\gamma+\ldots = n-r+1.$$

Consider now three linear forms

$$x, y, z,$$

where $z = -(x+y)$ and x, y are the variables.

It follows from the above that if

$$\alpha+\beta+\gamma=n-2$$

then any homogeneous expression of x, y, z of order n can be expressed in the form

$$x^{n-\alpha} P + y^{n-\beta} Q + z^{n-\gamma} R,$$

where P, Q, R are of order α, β, γ respectively.

In other words, changing α into $n-\lambda$, β into $n-\mu$ and γ into $n-\nu$, any homogeneous product of order n of x, y, z can be expressed in the form

$$x^{\lambda} P + y^{\mu} Q + z^{\nu} R,$$

where
$$\lambda+\mu+\nu=2n+2,$$

and the expression is unique.

This is Stroh's generalized form of Jordan's lemma given in § 64, and the lemma itself follows at once since we can always choose integers λ, μ, ν satisfying the relation

$$\lambda+\mu+\nu=2n+2$$

and
$$\lambda \geqslant \frac{2n}{3}, \quad \mu \geqslant \frac{2n}{3}, \quad \nu \geqslant \frac{2n}{3}.$$

On expressing P in terms of x and y, Q in terms of y and z, and R in terms of z and x, it follows that any homogeneous product of order n of x, y, z $(x+y+z=0)$ can be expressed linearly in terms of

$$\begin{array}{llll} x^n, & x^{n-1}y, & x^{n-2}y^2, & \ldots\; x^{\lambda}y^{n-\lambda} \\ y^n, & y^{n-1}z, & y^{n-2}z^2, & \ldots\; y^{\mu}z^{n-\mu} \\ z^n, & z^{n-1}x, & z^{n-2}x^2, & \ldots\; z^{\nu}x^{n-\nu} \end{array}$$

and the same would still be true if z were changed into x in the second row.

The reader will have no difficulty in modifying the above so as to obtain another proof of the fact that the general system of apolar forms constructed in § 178 contains n linearly independent forms.

APPENDIX IV.

FURTHER RESULTS ON COVARIANT TYPES.

THE expression given in § 262 for perpetuant types, may be used to determine the perpetuants when the forms are supposed not all different.

The result (using the notation of the paragraph referred to) for one quantic is

$$\begin{aligned}
\lambda_{\delta-1} &= 1 + \xi_{\delta-1} \\
\lambda_{\delta-2} &= 2 + \xi_{\delta-1} + \xi_{\delta-2} \\
&\ldots\ldots\ldots\ldots\ldots\ldots \\
&\ldots\ldots\ldots\ldots\ldots\ldots \\
\lambda_2 &= 2^{\delta-3} + \xi_{\delta-1} + \xi_{\delta-2} + \ldots + \xi_2 \\
\lambda_1 &= 2^{\delta-2} + 2(\xi_{\delta-1} + \xi_{\delta-2} + \ldots + \xi_2 + \xi_1),
\end{aligned}$$

where all the ξ's are zero or positive integers.

For two quantics the covariant can be written in the form

$$(a_1 a_2)^{2\alpha_1} (a_2 a_3)^{\alpha_2} \ldots (a_{i-1} a_i)^{\alpha_{i-1}} (a_i b_1)^{\alpha_i} (b_1 b_2)^{\beta_1} (b_2 b_3)^{\beta_2} \ldots (b_{j-1} b_j)^{\beta_{j-1}},$$

where the a's refer to one quantic and the b's to the other; the indices satisfy the conditions

$$\begin{aligned}
\beta_{j-1} &= 1 + \zeta_j \\
\beta_{j-2} &= 2 + \zeta_j + \zeta_{j-1} \\
&\ldots\ldots\ldots\ldots\ldots\ldots \\
\beta_1 &= 2^{j-2} + \zeta_j + \zeta_{j-1} + \ldots + \zeta_2 \\
\alpha_i &= 2^{j-1} + \zeta_j + \zeta_{j-1} + \ldots + \zeta_1 \\
\alpha_{i-1} &= 2^j + \xi_{i-1} \\
\alpha_{i-2} &= 2^{j+1} + \xi_{i-1} + \xi_{i-2} \\
&\ldots\ldots\ldots\ldots\ldots\ldots \\
\alpha_2 &= 2^{j+i-3} + \xi_{i-1} + \xi_{i-2} + \ldots + \xi_2 \\
2\alpha_1 &= 2^{j+i-2} + 2(\xi_{i-1} + \xi_{i-2} + \ldots + \xi_1).
\end{aligned}$$

For the proof of these results see Grace, "On Perpetuants," *Proc. London Math. Soc.*, vol. XXXV., p. 219.

The reasoning of this paper for the case of a single quantic may be applied to types, with the result that all perpetuant types of degree δ may be expressed in terms of products of types and of types of the form

$$(a_1 a_2)^{\lambda_1} (a_2 a_3)^{\lambda_2} \dots (a_{\delta-1} a_\delta)^{\lambda_{\delta-1}},$$

where the λ's satisfy the conditions laid down above for types for a single quantic, except that

$$\lambda_1 = 2^{\delta-2} + 2(\xi_{\delta-1} + \xi_{\delta-2} + \dots + \xi_2) + \xi_1,$$

and where the order of the letters $a_1, a_2, \dots$ is no longer fixed.

The method of §§ 262, 265 can be extended to the case of any concomitant which is linear in the coefficients of each of certain binary forms

$$a_{1_x}{}^{n_1},\ a_{2_x}{}^{n_2},\ \dots\ a_{\delta_x}{}^{n_\delta}.$$

Thus any such concomitant can be expressed by repeated transvection in terms of

(i) forms of the type

$$((\dots(((a_1 a_2)^{\lambda_1} a_3)^{\lambda_2} a_4)^{\lambda_3} \dots a_{\delta-1})^{\lambda_{\delta-2}} a_\delta)^{\lambda_{\delta-1}},$$

(ii) reducible forms.

The relations satisfied by the indices are however somewhat complicated. They are given as follows:—

A reduction (in the sense of § 265) is always possible when $\lambda_1 < 2^{\delta-2}$ unless one of the following conditions is satisfied:

$$\begin{aligned}
\lambda_i + \Sigma(2\lambda_j + 2^{\delta-j} - n_{j+1}) &> n_1 \quad - (2^{\delta-i} - 1),\\
\lambda_i + \Sigma(2\lambda_j + 2^{\delta-j} - n_{j+1}) &> n_2 \quad - (2^{\delta-i} - 1),\\
\lambda_i \qquad\qquad\qquad\qquad &> n_{i+1} - (2^{\delta-i} - 1),
\end{aligned}$$

where $\delta > i > j > 1$, but i and j are otherwise unrestricted, as also is the number of terms under the sign of summation—in particular there may be none.

In general, the condition that $\lambda_1 = 2^{\delta-2} - \alpha$ may not mean a reduction is that certain positive integers,

$f_1(0), f_2(0), f_3(0)$;

$f_1(\xi_1), f_2(\xi_1), f_3(\xi_1), \quad \xi_1 = 1, 2$;

$f_1(\xi_1, \xi_2), f_2(\xi_1, \xi_2), f_3(\xi_1, \xi_2), \quad \xi_1 = 1, 2 : \xi_2 = 1, 2$;

....

$f_1(\xi_1, \xi_2, \dots \xi_r), f_2(\xi_1, \xi_2, \dots \xi_r), f_3(\xi_1, \xi_2, \dots \xi_r),$
$\xi_1 = 1, 2 : \xi_2 = 1, 2 : \dots \xi_r = 1, 2,$

can be found such that

$$\alpha = f_1(0) + f_2(0) + f_3(0) + f(1) + f(2)$$
$$f(\xi) = f_1(\xi) + f_2(\xi) + f_3(\xi) + f(\xi, 1) + f(\xi, 2)$$

....

$$f(\xi_1, \xi_2, \dots \xi_r) = f_1(\xi_1, \xi_2, \dots \xi_r) + f_2(\xi_1, \xi_2, \dots \xi_r)$$
$$+ f_3(\xi_1, \xi_2, \dots \xi_r) + f(\xi_1, \xi_2, \dots \xi_r, 1) + f(\xi_1, \xi_2, \dots \xi_r, 2);$$

(where the quantities $f(\xi_1, \xi_2, \dots \xi_r)$ are positive integers which satisfy

$$f(\xi_1, \xi_2, \dots \xi_r, \xi_{r+1}) \not> 2^{\delta-r-3} - f_1(\xi_1, \xi_2, \dots \xi_r) - f_3(\xi_1, \xi_2, \dots \xi_r)$$

when $\xi_1 + \xi_2 + \dots + \xi_r + \xi_{r+1} + r$ is odd, and

$$f(\xi_1, \xi_2, \dots \xi_r, \xi_{r+1}) \not> 2^{\delta-r-3} - f_2(\xi_1, \xi_2, \dots \xi_r) - f_3(\xi_1, \xi_2, \dots \xi_r)$$

when $\xi_1 + \xi_2 + \dots + \xi_r + \xi_{r+1} + r$ is even);

$$\lambda_{r+2} + 2^{\delta-r-2} + \phi_1(\xi_1, \xi_2, \dots \xi_r) > n_1 + f_1(\xi_1, \xi_2, \dots \xi_r) - 1^*,$$
$$\lambda_{r+2} + 2^{\delta-r-2} + \phi_2(\xi_1, \xi_2, \dots \xi_r) > n_2 + f_2(\xi_1, \xi_2, \dots \xi_r) - 1,$$
$$\lambda_{r+2} + 2^{\delta-r-2} - f(\xi_1, \xi_2, \dots \xi_r) > n_{r+3} + f_3(\xi_1, \xi_2, \dots \xi_r) - 1;$$

where, finally, the ϕ's are defined by the laws

$$\phi_\eta(\xi_1, \xi_2, \dots \xi_r) + f(\xi_1, \xi_2, \dots \xi_r) - \phi_\eta(\xi_1, \xi_2, \dots \xi_{r-1})$$
$$- f(\xi_1, \xi_2, \dots \xi_{r-1})$$

is zero if $\eta + \xi_1 + \xi_2 + \dots + \xi_r$ is even, but if this sum is odd, the expression considered

$$= (2\lambda_{r+1} + 2^{\delta-r-1} - n_{r+2}) - 2\{f(\xi_1, \xi_2, \dots \xi_{r-1}) - f(\xi_1, \xi_2, \dots \xi_r)\},$$

and

$$\phi_\eta(0) = -\alpha = -f(0),$$
$$\phi_1(1) + f(1) = \mu_2 - 2\{f(0) - f(1)\},$$
$$\phi_2(1) + f(1) = 0,$$
$$\phi_1(2) + f(2) = 0,$$
$$\phi_2(2) + f(2) = \mu_2 - 2\{f(0) - f(2)\}.$$

* In case $f_1(\xi_1, \xi_2, \dots \xi_r)$ is zero, this inequality need not be satisfied; this remark applies also to the other two inequalities.

INDEX.

CAMBRIDGE: PRINTED BY J. & C. F. CLAY, AT THE UNIVERSITY PRESS.

Printed in the United States
By Bookmasters